AN INTRODUCTION TO

ARTHROPODA

AN INTRODUCTION TO ARTHROPODA

By
H.S. Bhamrah
Kavita Juneja

Edited by
Balvinder K.

ANMOL PUBLICATIONS PVT. LTD.
NEW DELHI - 110 002 (INDIA)

ANMOL PUBLICATIONS PVT. LTD.
4374/4B, Ansari Road, Daryaganj
New Delhi - 110 002

An Introduction to Arthropoda

First Edition, 1991
Reprint, 1992, 1999
Second Revised Edition, 2001

ISBN 81-261-0673-5

PRINTED IN INDIA

Published by J.L. Kumar for Anmol Publications Pvt. Ltd., New Delhi - 110 002 and Printed at Mehra Offset Press, Delhi.

Contents

1. Arthropoda : Characters & Classification **1-74**

Branchipus, Streptocephalus, Apus, Daphnia Cyclops, Cypris, Argulus, Lucifer, Mysis, Gammarus, Caprella, Lepas, Balanus, Sacculina, Oniscus, Bopyrus, Squlla, Astacus, Eupagurus, Hippa, Carcinus, Julus, Scolopendra, Lepisma, Isotoma, Schistocerca, Gryllus, Gryllotalpa, Mantis, Phyllum, Carausius, Forficula, Pediculus, Ephemera, Dragonfly, Nepa, Belostoma, Dysdercus, Cicada, Aphid, Danus, Moth, *Coccinella, Vespa, Camponotus, Xenopsylla, Aranea, Sarcoptes, Ixodes, Chelifer.*

2. *Palaemon* **75-126**

Systematic Position, Habits & Habitat, External Morphology, Appendages, Body wall, Body cavity, Musculature, Endoskeleton, Digestive system, Respiratory system, Circulatory system, Excretory system, Nervous system, Endocrine system, Receptors, Reproductive system, Power of regeneration.

3. Prawn Fishery **127-134**

Types of prawns, Prawn culture or farming, Preservation and Processing, Prawn manure, Pollution.

4. Larval forms in Crustaceans **135-143**

Nauplius larva, Metanauplius larva, Cypris larva, Protozoea, Zoea larva, Mysis or Schizopode, Alima larva, Metazoea, Calyptosis megalopa, Phyllosoma larva, Significance of larval forms.

5. Prasitism in Crustacea **144-151**

Effect of parasitism upon parasite, Parasitic Brachiura, Parasitic Cirripedia, Parasitic Isopoda, Parasitic Amphipoda.

6. *Periplaneta americana* **152-186**

Systematic Position, Habits & Habitat, External Characters, Body wall, Endoskeleton, Body cavity, Digestive system, Food & feeding, Circulatory system, Respiratory System, Mechanism of respiration, Excretory system, Nervous system, Sense organs, Endocrine glands, Reproductive system, Ootheca formation, Development, Hatching.

7. Grasshopper or Locust **187-210**

Habits and Habitat, External features, Exoskeleton, Muscular system, Digestive system, Circulatory system, Respiratory system, Excretory system, Nervous system, Sense organs, Reproductive system, Life cycle of grasshopper, Control of grasshopper.

8. *Musca* **211-226**

Systematic position, Habits and Habitat, Morphology, Body cavity, Digestive system, Feeding and Digestion, Respiratory system, Circulatory system, Excretory system, Nervous system, Reproductive sytem, Life History, Copulation, Metamorphosis, Economic importance, Control of house flies, Destruction of adult flies.

9. Mosquitoes **227-239**

Systematic position, Habits and habitat, Morphology, Life history, Larva, Pupa, Imagos, Economic Importance, Control of mosquitoes.

10. *Apis* (Honey bee) **240-259**

Systematic position, Castes of honey bee, Structure of Worker bee, Structure of Queen, Structure of Drone, Digestive system, Circulatory system, Respiratory system, Nervous system, Reproductive system, Life history, Nuptial, flight, Eggs, Larvae, Pupa, Communication, Structure of a hive, Economic Importance.

11. *Cimex* (Bedbug) **260-264**

Systematic positon, Habits & Habitat, Morphology, Life history, Economic importance, Control, Enemis of bedbugs.

12. Termites **265-271**

Systematic position, Habits & Habitat, External features, Polymorphism, Nests, Life history, Economic importance, Control.

13. *Bombyx mori* **272-277**

Systematic position, Types of silkworm, External characters, Life history, Eggs, Larva, Pupa, Imagos, Economic importance, Sericulture, Reling of raw silk from cocoon.

14. Mouth Parts in Insects **278-286**

Biting & chewing type, Chewing and lappiong type, Rasping and sucking type, Piercing and sucking type, Sponging type, Siphoing type.

15. Respiration in Insects **287-293**

Tracheal respiration in terrestrial insects, Close type of respiration in aquatic insects, Cutaneous respiration, Gill respiration, Tracheal gills, Spiracular gills, Blood or anal gills, Open type of respiration in aquatic insects, Air stores, Gaseous plastron, Air tubes, Plant perforating spiracles.

16. Metamorphosis in Insects **294-300**

Ametabolous, Hemimetabolam, Paurometabolous, Holometabolous, Propod larva, Polypod larva, Oligopod larva, Campodeiform larva, Scarabaeiform larva, Apodus larva, Decticous pupae, Adecticous pupae, Hypermetabolous, Internal metamorphosis, Hormones and their role in mettamorphosis.

17. Trilobites **301-306**

Habits and Habitat, External features, Division of body, Head on cephalon, Trunk or Thorax, Pygidium, Development.

18. *Limulus* **307-316**

Systematic position, Distribution, External features, Digestive system, Circulatory system, Excretory system, Nervous system, Sense organs, Reproductive system, Affinities of *Limulus*, Affinities with Arachnids, Affinites with Trilobites, Affinities with Eurypterida.

19. Scorpions **317-342**

Systematic position, Habits & Habitat, External feature, Appendages, Body wall, Endoskeleton, Digestive system, Circulatory system, Respiratory system, Excretory system, Nervous system, Sense organs, Reproductive system, Copulation & Fertilization, Development.

20. *Peripatus* **343-355**

Historical account, Geographical Distribution, Habits and Habitat, External features, Body wall, Digestive system, Circulatory system, Respiratory system, Excretory system, Nervous system, Eye, Reproductive system, Affinities with arthopodes, Its own characters Taxonomic positon.

Index **356-360**

PREFACE

An Introduction to Arthropoda is the revised and enlarged second edition of the book. It has been designed to approach the morphology, anatomy, physiology and development of selected type in a simple and lucid style. The approach to the discussion of all the types is very simple so as to impact to the students a clear and vivid understanding. According to the scheme of treatment the important animal types of the phylum have been dealt with first, and efforts have been made to present their elaborate and uptodate account. General characters and classification and brief description of other important types of the phylum have also been dealt with complete, authentic and uptodate account. Further separate chapters on topics of significance and general interest pertaining to the phylum have also been added to make the treatment more elaborate.

It has been the constant endeavour of the authors to furnish maximum substance, keeping in view the limitations of size of the volume. Efforts have been made to condense the matter as far as practicable. The book features both a text and a laboratory guide. It is hoped that this book will not only meet the requirement of Indian students but will also be useful as a guideline to the teachers in their teaching.

There can be no claim to originality except in the manner of treatment and much of the information has been obtained from the books and scientific journals available in different libraries.

The authors express their thanks to their friends and colleagues whose constant inspiration have initiated them to bring out this book.

In this edition all the errors of first edition have been rectified. New information is incorporated in the light of recent developments and latest research. Teachers and scholars can send their criticism and suggestions to further improve the contents.

Authors

1

Arthropoda : Characters & Classification

Phylum *Arthropoda* (Gr.*arthros* = joint; *pods* = foot) contains the great majority of the known animals, about one million species, and many of them are enormously abundant as individuals. It includes such common and well-known forms as the crabs, shrimps, insects, spiders, scorpions, ticks, centipedes, as well as a host of other less familiar forms. *Cuvier* (1800) placed all these animals together with Annelida under the group Articulata. *Van Siebold* (1845) separated Annelida from Articulata. Annelides were placed under class Vermes while the Crustacea, Arachnida and Insecta were kept under an independent group, the *Arthropoda.*

General characters

1. Arthoropoda are triploblastic, bilaterally symmetrical, metamerically segmented animals.
2. Body is covered with a thick chitinous cuticle forming an exoskeleton.
3. Body segments usually bear paired lateral and jointed appendages.
4. Musculature is not continuous but comprises separate striped muscles.
5. Body cavity is haemocoel. The true coelom is reduced to the spaces of the organs.
6. Digestive system complete; mouthparts adapted for various modes of feeding.
7. Circulatory system open with a dorsal heart, arteries and blood sinuses.
8. Respiration by general body surface, gills, tracheae or book-lungs.
9. Excretory organs are *green glands* or *Malpighian tubules.*
10. Nervous system with a dorsal nerve ring and a double ventral nerve cord.
11. Cilia are entirely absent from all parts of the body.
12. Sexes are generally separate and sexual dimorphism is often exhibited by several forms.
13. Fertilization is internal. Development is usually indirect through larval stages.

14. Parental care is also often well marked in many arthropods.

Phylum Arthropoda is divided into following classes:

Class-I. Crustacea

Class-II. Myripoda

Class-III. Insecta

Class-IV. Arachnida

CLASS-I. CRUSTACEA

1. The body is divisible into head, thorax and abdomen, when head and thorax are fused they form cephalothorax.
2. The exoskeleton is chitinous.
3. The body is metamerically segmented, bilaterally symmetrical and triploblastic.
4. Each segment bears paired and jointed appendages.
5. The body cavity is filled with *haemolymph* and known as *haemocoel.*
6. Respiration takes places either by the general body surface or by gills.
7. Excretion occurs by modified coelomoducts, which may take the form either of maxillary glands or antennary (green) glands.
8. These are unisexual except many Cirripedia and some Isopods.
9. The development is indirect.

Class Crustacea is divided into 7 sub-classes

Sub-class-1. Branchiopoda

1. Body has varying number of segments.
2. Antennules are small, uniramous and usually unsegmented.
3. Compound eyes are present.
4. Thorax bears four or more flat, leaf-like respiratory appendages also help in filter feeding and locomotion.
5. Abdomen without appendages.
6. Parthenogenesis is common.

Sub-class-Branchiopoda is divided in three orders:

Order (i) Anostraca

1. The carapace is absent.
2. Eyes are stalked.
3. Antennae are uniramous, prehensile in male and reduced in female.
4. Trunk elongated with 20 or more segments. Anterior 11-19 segments bear appendages.

5. The caudal styles are unjointed.
   Examples : *Branchipus, Streptocephalus, Artemia.*

Order (ii) Notostraca

1. A large dorsal shield-shaped carapace is present.
2. Eyes are sessile i.e. stalkless.
3. Antennae are reduced.
4. Anterior half of the trunk bears 35-70 pairs of appendages.
5. The caudal styles are many jointed.
   Example : *Apus, Lepidurus.*

Order (iii) Conchostraca

1. The carapace is present but divided in two portions or valves.
2. Antennae biramous and large.
3. Eyes are sessile.
4. Caudal styles are in the form of unjointed, curved claws.
   Examples : *Lynceus, Estheria.*

Order (iv) Cladocera

1. The carapace bivalved, encloses the trunk but not the head and usually ends posteriorly in an apical spine.
2. Eyes are unstalked and united together.
3. Antennae biramous used for swimming.
4. Trunk segments 10-32, each with a pair of appendages.
5. Caudal styles are unjointed claw-like.
   Examples : *Daphnia, Leptodora.*

Sub-class-2. Ostracoda

1. Compound eyes may or may not be present.
2. Carapace is bivalved and compressed which encloses entire body.
3. Head forms about half of the body.
4. Thorax usually bears only two pairs of cylindrical appendages.
5. Abdomen is reduced, without appendages and ends in a pair of caudal styles.
6. Antennules and antennae locomotory.
7. Mandibles have a palp.
8. Larva is *Nauplius.*
9. Parthenogenesis is common.
   Sub-class Ostracoda is divided into following orders:

Order (i) Myodocopa

1. Marine crustaceans.
2. Carapace with antennal notches.
3. Antennae biramous.
4. Trunk appendages two pair.

   Examples : *Cypridina, Conchoecia.*

Order (ii) Cladocopa

1. Marine forms.
2. Carapace unnotched.
3. Both pairs of antennae used in swimming.
4. Second pair of antennae with 2 branches.

   Example : *Polycope.*

Order (iii) Podocopa

1. Marine or freshwater.
2. Carapace unnotched.
3. Second antennae leg-like, clawed at tips.
4. Trunk appendages 2 pairs.

   Examples : *Cypris, Entocythere, Cythereis.*

Order (iv) Platycopa

1. Marine forms
2. Carapace unnotched.
3. Both pairs of antennae large but not used for swimming, second pair flattened.
4. Trunk appendages single pair.

   Example : *Cytherella.*

Sub-class-3. Mystacocardia

1. Primitive marine forms.
2. Size microscopic, cylindrical.
3. Antennules and antennae long, prominent.
4. Eyespots, compound eyes absent.
5. Thorax 4-segments, each with one pair of simple appendages.
6. Excretion by both antennal and maxillary glands.
7. Ventral nerve cord paired.
8. Sexes separate, earliest larva is *metanauplius.*

   Example : *Derocheilocarus.*

Sub-class-4. Copepoda

1. Fresh water or marine, free-living, commensal or parasitic.
2. Body small, elongated, composed of head, thorax and abdomen.
3. Typically 10 free trunk somites, with 4 lacking appendages. Appendages mostly biramous.
4. Three ocelli often fused as median eye, no compound eyes.
5. Antennules and antennae well developed and used for swimming.
6. Abdomen without appendages, ends in a pair of caudal styles.
7. Excretion by maxillary gland.
8. Eggs on abdomen of female in 1 or 2 egg sacs.

Subclass Copepoda is divided into 7 orders.

Examples : *Cyclops, Salmincola, Caligus, Doropygus, Calanus, Monstrilla.*

Sub-class-5. Branchiura

1. Ectoparasites of fresh water and marine fishes and some amphibians.
2. Body flat, carapace large, disc-like, covering head and thorax.
3. Abdomen small, bilobed.
4. First maxillae modified as suckers.
5. Compound eyes present, sessile.
6. Thoracic appendages 4 pairs, large, biramous, some of them are provided with flagella.

Examples : *Argulus, Dolops.*

Sub-class-6. Cirripedia

1. Marine forms.
2. Adults sessile, attached or parasitic; attached by cement gland on first antenna.
3. Carapace becomes calcareous and mantle surrounding body.
4. Six or fewer pairs of slender and bristly biramous appendages behind mouth used in food gathering.
5. Abdomen vestigeal, without appendages, ending in a pair of caudal styles.
6. Antennae and compound eyes disappear in adult.
7. Monoecious. Development includes free swimming *nauplius* which passes through a *cypris* stage which is enclosed in bivalve shell.

This subclass is divided into following orders :

Order (i) Thoracica

1. Non-parasitic sedentary marine barnacles.

2. Body surrounded by calcareous shell.
3. Six pairs of biramous thoracic limbs.
4. Alimentary canal is present.
5. Hermaphrodite.

Examples : *Balanus, Lepas.*

Order (ii) Acrothoracica

1. Parasitic, boring in the shells of mollusc.
2. Body is covered by large mantle. Calcareous shell is absent.
3. Less than 6 pairs of thoracic appendages.
4. Alimentary canal is present.
5. Chitinous disc is present for attachment.

Example : *Alcippe.*

Order (iii) Ascothoracica

1. Parasitizing corals and echinoderms.
2. Mouth appendages are modified into piercing and sucking organs.
3. Digestive tract with branches into mantle.
4. 6 pairs of trunk appendages.

Examples : *Laura, Petrarca.*

Order (iv) Apoda

1. Parasitic in mantle cavity of *Alepas.*
2. Body without mantle and thoracic appendages.
3. Body maggot-like.
4. Anus absent.
5. Hermaphrodite.

Example : *Proteolepas.*

Order (v) Rhizocephala

1. Parasitic on crabs.
2. A mantle but no shell.
3. Appendages, gut are absent.
4. Body sac-like, with absorptive roots penetrating host.

Example : *Sacculina.*

Subclass-7. Malacostraca

1. Mostly large sized crustaceans, marine and fresh water.
2. Body typically of 19 segments (5 head, 8 thorax and 6 abdomen).
3. Exoskeleton of head united with few or more thoracic segments to form cephalothoracic carapace.

4. Stalked compound eyes.
5. Abdominal caudal style absent.
6. Development through zoaea stage. Nauplius stage rarely occurs.

Series (A) Phyllocardia

1. Abdomen of eight somites.

Order (i) Leptostraca

1. Marine forms.
2. Body made up of 21 segments. Abdomen of 7 segments.
3. Thoracic appendages similar and foliaceous.
4. Telson has a pair of caudal styles.
5. Carapace bivaled. Eyes stalked.
   Example : *Nebalia.*

Series (B) Eumalacostraca

1. Abdomen of 6 or fewer somites.

Super-order (A) Syncarida

1. Carapace is absent.

Order (i) Anaspidacea

1. Fresh waters of Australia.
2. First thoracic segment fused with head.
3. Eyes stalked, sessile or absent.
   Example : *Anaspides.*

Order (ii) Bathynellacea

1. First thoracic segment is not fused with head.
2. Eyes absent.
3. Telson fused with the sixth abdominal segment.
   Example : *Bathynella.*

Super-order (B) Peracarida

1. Carapace, when present, leaving last 4 or more thoracic somites distinct.
2. Females with thoracic brood pouch where youngs develop.
3. Development direct without larval stage.

Order (i) Mysidacea

1. Marine forms.
2. Carapace covers almost entire thorax.
3. Eyes stalked.

4. Thoracic appendages are biramous.
5. Uropods form broad fan-like tail fin.

   Examples : *Mysis.*

Order (ii) Cumacea

1. Small, mostly marine, burrowing in surface sand or mud.
2. Carapace with 2 anterior extensions often joined over head.
3. Head and thorax greatly enlarged.
4. Abdomen slender, mobile.
5. Uropods slender.

   Example : *Diastyles, Cumopsis.*

Order (iii) Tanaidacea

1. Minute, marine, upto depths of 3650 meters in mud or tubes.
2. Carapace is small and fused to first 2 thoracic segments.
3. Second thoracic appendage with chela.
4. Telson unjointed fused with last abdominal segment.
5. Uropods slender.

   Examples : *Tanais, Apseudes.*

Order (iv) Isopoda

1. Mostly marine, some freshwater and few parasitic.
2. Body usually depressed dorso-ventrally.
3. Carapace is absent. Abdomen short, partly or all fused.
4. Gills and heart abdominal.

   Examples : *Ligia, Asellus, Hemioniscus, Oniscus.*

Order (v) Amphipoda

1. Mostly marine, some freshwater and terrestrial.
2. Body often laterally compressed.
3. Carapace is absent.
4. Telson usually distinct.
5. Gills and heart in thoracic region.

   Example : *Gammarus, Caprella, Cyamus.*

Super-order (C) Hoplocarida

1. Head with two movable anterior segment bearing eyes and antennules.

Order (i) Stomatopoda

1. Marine on bottom, in sand or cervices.

2. Antennal scales enlarged.
3. Gills on abdominal appendages modified for raptorial feeding.
4. Carapace small.

   Examples : *Squilla, Pseudosquilla, Chloridella.*

Super-order (D) Eucarida

1. Carapace is present, fused to and covering all the thorax.
2. Eyes stalked.
3. Gills and heart in thoracic region.
4. Brood pouch absent.
5. Development usually indirect with a *zoea* larva.

Order (i) Euphausiacea

1. Marine, small, pelagic forms.
2. Carapace does not cover sides of body and the gills.
3. Thoracic appendages all biramous and similar, none modified as maxillipedes.
4. Pleopods have long setae, used for swimming.
5. Larva is *nauplius.*

   Examples : *Euphausia, Thysanopoda.*

Order (ii) Decapoda

1. Mostly marine forms.
2. Carapace covers the entire thorax.
3. Thoracic appendages are modified as 5 pairs of walking legs and three pairs of maxillipedes.
4. Thoracic appendages mostly uniramous.
5. Statocyst present.
6. Gills usually present in three series on the thorax.

   Decapoda is divided into 3 suborders:

Sub-order (a) Macrura

1. Abdomen is well developed and elongated.
2. The abdomen terminates in an expanded fan-like tail fin composed of telson and the uropods.
3. Eyes are not enclosed in orbits.

   Examples : *Lucifer, Palaemon, Homarus, Astacus, Cambarus.*

Sub-order (b) Anomura

1. Abdomen is more or less reduced and flexed.

2. Tail fin is also reduced.

   Examples : *Eupagurus, Hippa, Birgus.*

Sub-order (c) Brachyura

1. Abdomen greatly reduced, hard, permanently flexed under cephalothorax.
2. Antennules and the eyes are both capable of being retracted into cavities.
3. Rostrum and uropods absent.
4. First legs usually large and chelate.

   Examples : *Cancer, Inachus, Uca, Libinia.*

CLASS-II. MYRIPODA

1. Terrestrial and air breathing arthropods.
2. Body is usually elongated with numerous segments, each bearing one or two pairs of appendages.
3. Head is distinct, bearing a pair of antennae, mandibles, eyes and maxillae.
4. Respiration by tracheae.
5. Excretory organs are one or two pairs of Malpighian tubules associated with hindgut.
6. Sexes are separate.

Subclass-1. Progeneata

1. The genital apertures are situated near the anterior end of the body.

Order (i) Pauropoda

1. The trunk segments are eleven and anal segment is also present.
2. There are 9 pairs of legs.
3. The head bears one pair of antennae, one pair of mandibles and one pair of maxillae.
4. These are without blood vascular system and trachea.

   Example : *Pauropus.*

Order (ii) Diplopoda

1. Body is usually cylindrical.
2. Body bears numerous segments each with two pairs of legs (expect first four).
3. Genital opening on the third segment behind the head.

   Example : *Julus, Spirobolus.*

Order (iii) Symphyla

1. There are not more than 12 segments in the trunk.
2. The head bears one pair of antennae, one pair of mandibles and 2 pairs of maxillae.
3. There is a single pair of tracheae.

   Example : *Scolopendrella, Scutigerella.*

Subclass-2. Opisthogoneata

1. Genital apertures are situated at the posterior extremity of the body.

Order (i) Chilopoda

1. Body is usually dorso-ventrally flattened.
2. Head bears many jointed antennae, a pair of mandibles and two pairs of maxillae.
3. Poisonous claws are present.
4. Trunk bears 15-117 segments each with a single pair of legs.

   Examples : *Scoloplendra, Scutigera.*

CLASS-III. INSECTA

1. These are air breathing, mostly terrestrial and rarely aquatic arthropods.
2. Body is divisible into head, thorax and abdomen.
3. Head bears a pair of antennae, compound eyes, mandibles and two pairs of maxillae.
4. Thorax consists of three segments bearing 3 pairs of legs and 2 pairs of wings.
5. Abdomen 7-11 segments which are without appendages in adult.
6. Liver absent but salivary glands are present.
7. Respiration by tracheae.
8. Excretion by Malpighian tubules.
9. Sexes separate.
10. Development is usually indirect, completed by complicated metamorphosis. Sometimes direct without larval stage.

    Class Insecta is divided into two sub-classes:

    Sub-class-I. Apterygota

    Sub-class-II. Pterygota.

Sub-class-I. Apterygota

1. Wingless, primitive insects.
2. Metamorphosis is absent.

Order (i) Protura

1. Minute, elongated whitish insects.
2. Antennae, eyes and ocelli are lacking.
3. Wings are absent.
4. Abdomen short consists of 12 segments. Bilateral styli on first three segments.
5. Tracheae lacking.
6. Malpighian tubules are small papillae.

   Example : *Acerentulus.*

Order (ii) Collembola

1. Minute, somewhat tubular or globose forms.
2. Eye patches consist of one to several lateral ocelli, dorsal ocelli lacking. Antennae short usually 4-segmented.
3. Abdomen consists of 6 segments, lobe-like organ on first segment. 4th segment bears furcula.
4. Mouthparts entognathus, chewing type.

   Examples : *Podura, Isotoma Sminthurus.*

Order (iii) Archeognatha

1. Small, elongate, with many scales on parts of body.
2. Mouthparts are ectognathus, chewing type.
3. Antennae filiform, large compound eyes and 3 ocelli are present.
4. Abdomen small bilateral styli on venter of several segments.
5. Females with elongated jointed ovipositor.

   Example : *Machilis.*

Order (iv) Thysanura

1. Small, elongated insects. Usually covered with scales.
2. Compoud eyes usually present 2-3 ocelli.
3. Abdomen small, bilateral style on venter of several segments.
4. Abdomen consists of 10 or 11 segments.

   Example : *Lepisma, Japyx.*

Sub-class-II. Pterygota

1. Adults are winged, some are wing-less.
2. Metamorphosis complete or incomplete.

Division (A) Exopterygota (Heterometabola)

1. Wings develop externally as buds.

2. Metamorphosis simple without plupal stage.
3. Young stage, called *nymphs*, develop gradually.

Order (i) Odonata

1. Medium to very large sized insects.
2. Head large, mobile bearing short antennae, large compound eyes and 3 dorsal ocelli.
3. Two pairs of membranous wings, venation net-like.
4. Abdomen elongate. Adult males with gonopores on 9th segment and complex penis on second segment.
5. Metamorphosis incomplete. The nymphs are aquatic, predatory with caudal tracheal gills.
   Examples : Dragon flies (*Lestes*), Damsel flies.

Order (ii) Ephemeroptera

1. Soft bodied, fragile, small to medium sized.
2. Mouthparts chewing type in nymph and vestigial in adults.
3. Antennae short. Compound eyes and 3 ocelli are present.
4. Two pairs of membranous wings. Fore-wings larger than hind-wings.
5. Abdomen with a pair of long cerci and often with a median caudal filament.
6. Metamorphoses incomplete. Nymph bears gills on the first 4-7 segments.
   Example : Mayflies (*Ephemera*).

Order (iii) Plecoptera

1. Small to medium sized, soft-bodies, elongated, flattened insects.
2. Mouthparts chewing type, usually vestigial in adults.
3. Antennae long and tapering. Well developed compound eyes, usually 3 dorsal ocelli are present.
4. Two pairs of membranous wings, males of some species apterous or brachypterous.
5. Metamorphosis incomplete. Nymphs aquatic with tracheal gills.
   Examples : Stoneflies (*Isoperla*, *Pteronarcys*).

Order (iv) Embioptera

1. Minute to small, elongated insects.
2. Head large bearing filiform antennae, compound eyes present, ocelli absent.
3. Mouthparts chewing type.
4. Thorax nearly as long as abdomen.

5. Some males with 2 pairs of membranous wings of nearly equal size and shape, females and some males wingless.
6. Metamorphosis gradual in male but absent in female.
   Examples : Web-spinners (*Oligotoma*), *Embia*.

Order (v) Phasmida

1. Large size insects. Body elongate, cylindrical with short prothorax, large meso-and meta-thorax.
2. Head broad and bears a pair of long filiform or moniliform antennae. Compound eyes are present 2-3 or no ocelli.
3. Mouthparts biting type.
4. Wings present or absent.
5. Abdomen 11 segmental. Ovipositor small and somewhat concealed. Cerci short, unsegmented.
   Examples : *Phyllium, Carausius* (stick insect).

Order (vi) Orthoptera

1. Minute to very large sized insects.
2. Mouthparts chewing type. All gradation between hypognathous and opisthognathous.
3. A pair of antennae of variable structure.
4. Compound eyes and usually 2 or 3 ocelli are present.
5. Two pairs of wings are present. Fore-wings are hard, narrow, parchment like opaque with indistinct venation forming tegmina (wing covers).
6. Abdomen 11 segmented. Females in several species with well developed ovipositors. Cerci variable, short, long, clasper-like, segmented or unsegmented.
   Examples : Grasshoppers (*Poecilocerus*), Locusts (*Schistocerca*), Crickets (*Gryllus*), Mole cricket (*Gryllotalpa*).

Order (vi) Dermaptera

1. Small to medium sized insects.
2. Mouthparts chewing type.
3. Antennae long slender. Compound eyes present in most of the forms may be vestigial or absent in some forms.
4. Fore-wings short, leathery and veinless, hind-wings membraneous, broad, fan-like.
5. Abdomen 11 segmented bearing a pair of forceps-like cerci at the end.

6. Metamorphosis gradual.
   Example : Earwigs (*Forficula*).

Order (vii) Grylloblattoidea

1. Medium to large, somewhat elongated and slender.
2. Mouthparts biting and chewing type.
3. Antennae filiform, compound eyes reduced or absent. Ocelli also absent.
4. Wings absent.
5. Abdomen well developed with sword-shaped ovipositor. Cerci long.
6. Metamorphosis simple.
   Example : *Grylloblattà*.

Order (viii) Isoptera

1. Social and polymorphic insects living in colonies under a caste system.
2. Mouthparts chewing type.
3. Compound eyes are present in all winged forms; present or absent in apterous forms. O or 2 dorsal ocelli are present.
4. Antennae short, moniliform or filiform.
5. Wings when present, two pairs, similar in size, shape and venation.
6. Cerci short and simple.
7. Metamorphosis simple.
   Examples : Termites (*Zootermopsis nasutitermes*).

Order (ix) Blattaria

1. Medium to large sized insects, some what dorso-ventrally flattened.
2. Mouthparts chewing type.
3. Compound eyes well developed, 2 ocelli in some. Antennae long filiform.
4. Wings present or absent, when present fore-wings tegmina, hind-wing membranous.
5. Pair of styli on 9th sternum of males. Ovipositor reduced and concealed by 7th abdominal sternum.
6. Metamorphosis simple.
   Example : *Periplaneta, Blatta*.

Order (x) Mantodea

1. Medium to large sized insects. Body usually elongated, somewhat cylindrical, with elongated prothorax.
2. Head small, triangular and bears large compound eyes with 3 or more ocelli.

3. Mouthparts chewing type.
4. Raptorial forelegs at anterior end of elongated prothorax.
5. Wings usually present in males, reduced or absent in females.
6. Abdomen comparatively short, multisegmented cerci, pair of styli usually associated with 9th sternum in males.
7. Metamorphosis simple.

   Example : Praying mantids (*Tenodera*).

Order (xi) Zoraptera

1. Minute sized insects.
2. Mouthparts chewing type.
3. Compound eyes and ocelli present in winged forms but absent in wingless forms.
4. Wings present or absent when present fore-wings larger than hind-wings.
5. Abdomen short with one segmented cerci.
6. Metamorphosis simple.

   Example : *Zorotypous*.

Order (xii) Psocoptera

1. Minute sized insects. Head capsule large as compared to rest of body.
2. Mouthparts chewing type. Labial silk-glands are present.
3. Compound eyes weakly or strongly developed. Ocelli 3 in winged forms absent in wingless forms.
4. Wings when present, two pairs, membranous.
5. Cerci absent.
6. Metamorphosis simple.

   Examples : Book-lice (*Liposcellis*), *Peripsocus*.

Order (xiii) Mallophaga

1. Small, apterous, ectoparasitic insects.
2. Body is broad or elongated, flattened dorsally.
3. Head generally large. Compound eyes reduced. Ocelli absent.
4. Mouthparts biting type.
5. Cerci absent.
6. Metamorphosis absent.

   Examples : Chewing lice or biting lice (*Trichodectes*, *Menacanthus*).

Order (xiv) Anoplura

1. Minute, dorso-ventrally flattened, ectoparasitic.

2. Head narrower than thorax.
3. Mouthparts piercing and sucking, retracted into head when not feeding.
4. Compound eyes weakly developed or lacking, ocelli also absent.
5. Antennae short 3-5 segments, setiform.
6. Wings are absent.
7. Legs adapted for clinging to the hairs of the host and each terminating in a curved claw.
8. Metamorphosis absent.

   Examples : Human body-louse (*Pediculus*), *Phthirtr.*

Order (xv) Thysanoptera

1. Minute to small sized insects, Body slender.
2. Mouthparts rasping and sucking type : Asymmetrical, the right mandible reduced or absent.
3. Small, compound eyes; 3 dorsal ocelli in winged forms absent in wingles forms.
4. Antennae short 6-10 segments.
5. Wings two pairs, narrow fringed; coupled by basal hooks.
6. Abdomen 10-11 segments with an ovipositor. Cerci lacking.
7. Metamorphosis simple.

   Example : Thrips (*Heliothrips*).

Order (xvi) Hemiptera

1. Medium to very large sized insects.
2. Mouthparts piercing and sucking type.
3. Antennae 2-10 or rarely 25 segmented.
4. Eyes large with or without ocelli.
5. Wings when present, two pairs, fore-wings usually thickened basally and membranous apically (sub-order *Heteroptera*) and wholly membranous (sub-order *Homoptera*).
6. A triangular plate is located between the bases of the wings.
7. Anal respiratory filaments are present in some aquatic forms and anal filaments in male coccids.
8. Metamorphosis gradual.

   Examples : Giant water bug (*Belostoma*), Bedbug (*Cimex*), Water strider (*Gerris*), Squash bug (*Anasa*), Leaf hopper (*Paraphlepsius*), Cicada (*Magicicada*).

Division (B) Endopterygota (Holometabola)

1. Wings develop internally.
2. Metamorphosis complete.

Order (i) Coleoptera

1. Minute to large sized insects.
2. Antennae variable in shape and size usually 11 segmented.
3. Compound eyes present or absent. Ocelli usually absent but present in some groups. Larva without compound eyes but lateral ocelli present.
4. Fore-wings sclerotized *elytra* which protect the membranous hind-wings at rest.
5. Legs typically cursorial, some adapted for swimming, jumping or digging.
6. Abdomen usually 10 segmented. Cerci absent.
7. Mouthparts biting and chewing types.
8. Larvae diverse usually catterpillar-like or maggot-like.
9. Metamorphosis complex.

   Examples : Ground beetle (*Calosoma*), Blister beetle (*Epicauta*), Alfalfa beetle (*Hypnera*).

Order (ii) Strepsiptera

1. Minute sized mostly endoparasitic insects.
2. Mouthparts with reduced mandibles.
3. Compound eyes present in males and free-living females.
4. Antennae flabellate in males, absent in females.
5. Wings two pairs in males, fore-wings clublike, hind-wings membranous and fan-like.
6. 3-5 genital pores in the abdomen of females

   Example : *Stylops.*

Order (iii) Neuroptera

1. Small to very large-sized, soft-bodied insects.
2. Mouthparts chewings type in adult and grasping-sucking type in larvae.
3. Compound eyes present. Ocelli present or absent.
4. Antennae filiform.
5. Wings two pairs, similar in size and appearance.
6. Cerci absent. variously modified ovipositors are present.

7. Metamorphosis complete.
   Examples : Green lacewing (*Chrysopa*), Snake fly (*Agulla*), Dobson fly (*Corydalis*).

Order (iv) Mecoptera

1. Small to medium sized, slender, predaceous insects.
2. Head usually elongated. Mouthparts chewing type.
3. Compound eyes are present, Ocelli 2-3. Antennae long and filiform.
4. Wings 2 pairs, membranous, commonly with dark spots.
5. Cerci short, simple or two segmented.
6. Metamorphosis complete.
   Examples : Scorpion fly *(Panorpa)*, Hanging Scorpion fly (*Bittacus*).

Order (v) Diptera

1. Minute to medium sized, diurnal or nocturnal, terrestial or aquatic insects.
2. Mouthparts variously modified for sucking, lapping, piercing and sponging.
3. Compound eyes present. Most species with 3 ocelli.
4. Wings are present, fore-wings developed for flight while hind-wings are reduced to *halters*.
5. Cerci present or absent.
6. Metamorphosis complete.
   Examples : Mosquitoes (*Aedes, Anopheles, Culex*), Black fly (*Simulium*), House fly (*Musca*), Fruit fly (*Drosophila*).

Order (vi) Siphonoptera

1. Minute, hard-bodied bilaterally compressed insects.
2. Mouthparts piercing and sucking type. In larvae chewing type.
3. Compound eyes lacking. Generally 2 ocelli.
4. Antennae short contained within antennal grooves.
5. Wing absent.
6. Male terminalia highly modified with one pair of 2 segemented claspers. Cerci lacking.
7. Metamorphosis complete.
   Examples : Human fleas (*Pulex*), Poultry flea (*Echinophaga*), Rat flea (*Xenopsylla*), Dog and cat flea (*Ctenocephalides*).

Order (vii) Trichoptera

1. Small to medium sized insects.

2. Monthparts are non-feeding in adult stage. Adapted for inbibling fluid. Mandibles weakly developed or lacking. Larvae with chewing type.
3. Compound eyes present, 2 or 3 ocelli.
4. Antennae range from setaceous to filiform.
5. Wings 2 pairs, membranous, hind-wings broader then fore-wings; covered with modified hair-like setae.
6. Larvae aquatic
7. Cerci one or two segmented.
8. Metamorphosis holometabolic.

   Example : Caddis fly (*Philopotamus*).

Order (viii) Lepidoptera

1. Medium to large sized insects.
2. Body and wings are completely covered by flat over-lapping scales and hairs.
3. Mouthparts usually long, sucking proboscis.
4. Large compound eyes, 2 or 2 dorsal ocelli.
5. Antennae always promient, vary in form.
6. Wings two pairs, fore-wings often largest.
7. Metamorphosis complete.
8. Larvae called *catepillars* with 3 pairs of thoracic legs and 2-4 pairs of abdominal prolegs.

   Example : Butterflies (*Danaus*), Moths (*Bombyx*).

Order (ix) Lepidoptera

1. Medium to large sized, social or parasitic insects.
2. Antennae 12 segmented in male and 13 segmented in females.
3. Mouthparts variable, chewing to lapping or sucking.
4. Compound eyes usually well developed. Commonly three dorsal ocelli.
5. Two pairs of membranous wings. Hind-wings smaller than fore-wings, many wingless.
6. Abdomen with basal pedical.
7. Female with conspicuous ovipositor modified into saw, drell or sting.
8. Metamorphosis complete.

   Examples : Honey bees (*Apis*), Thief ant (*Solenopsis*), Wasp (*Vespa*), Ants etc

CLASS-IV. ARACHNIDA

1. These are air breathing, mostly terrestrial arthropods.
2. Body is divisible into cephalothorax and abdomen.
3. Cephalothorax bears two pairs of jointed appendages, the chelicerae and pedipalpi and four pairs of legs.
4. Antennae and true jaws are absent.
5. Eyes are sessile and simple.
6. Respiration by tracheae or book-lungs or book-gills.
7. Heart is dorsal and tubular.
8. Excretion by coxal glands or Malpighian tubules or both.
9. Sexes are separate with conspicuous sexual dimorphosim.
10. Development direct.

Order (i) Eurypterida

1. Extinct forms.
2. Exo-skeleton is characteristically sculptured.
3. Cephalothorax is relatively small and the abdomen bears 12 free segments and a telson.
4. Cephalothorax bears 6 pairs of appendages.

   Example : *Eurypterus*

Order (ii) Xiphosura

1. Marine arachnids burrowing in the sand.
2. These animals have large sub-circular, apparently unsegmented body with two body divisions-anterior large cephalothorax (*prosoma*) and posterior small abdomen (*opisthosoma*), the two are joined by a hinge.
3. The cephalothorax bears a pair of short chelate appendages and 5 pairs of legs.
4. Abdomen bears operculum.
5. Respiration takes place by lamelliform gills attached to the abdominal appendages.
6. Sexes separate. Development includes a *trilobite* larva.

   Example : King crab (*Limulus*).

Order (iii) Scorpionida

1. Terrestrial animals found under stones in tropical and sub-tropical regions.
2. Body is divisible into cephalothorax and abdomen.
3. Prosoma bears a pair of chelicerae, a pair of pedipalpi and four pairs of walking legs.

4. Respiration takes place by 4 pairs of book-lungs.
   Examples : *Palamnaeus, Buthus.*

Order (iv) Pedipalpida

1. Body consists of cephalothorax and abdomen.
2. The chelicerae are simple and non-chelate.
3. Respiratory organs are two pairs of book-lungs.
   Example : *Phrynichus.*

Order (v) Araneida

1. Found under stones, bark, leaves of trees etc.
2. Body consists of an undivided cephalothorax and an unsegmented abdomen.
3. Cephalothorax bears 6 pairs of appendages.
4. Chelicerae are sub-chelate with poison glands. Pedipalpi are simple and 6 jointed.
5. Eight eyes are arranged dorsally on the head.
6. Respiration by book-lungs or book-lung with tracheae.
   Example : Spider (*Aranea*), *Argiope.*

Order (vi) Palpigradi

1. Small arachnids.
2. Body consists of cephalothorax and an abdomen of 10 segments.
3. Chelicerae are chelate and pedipalpi leg-like.
4. Respiratory organs are 3 pair of book-lungs.
   Example : *Koenenia.*

Order (vii) Solifugae

1. Body divisible into prosoma and opisthosoma.
2. Prosoma consists of three fused and last three free segments.
3. Opisthosoma, composed of 10 segments, is not marked off from prosoma.
4. Chelicerae are large and chelate. Pedipalpi are elongated and leg-like.
5. Respiration by tracheae. Poison glands are absent.
   Example : *Galeodes.*

Order (viii) Pseudoscorpionida

1. Minute arachnids having the appearance of scorpion, found under the tree barks.
2. Body consists of cephalothorax and an abdomen of 12 segments.

3. A pair of spinning glands are present.
4. Respiration by tracheae.
   Examples : *Chelifer.*

Order (ix) Ricinulei

1. Body consists of cephalothorax and an abdomen of 9 segments.
2. Cephalothorax and pedipalpi bear movable plates or cucullus.
3. Respiratory organs are tracheae.
   Example : *Cryptocellus.*

Order (x) Phalangida

1. Body consists of unsegmented cephalothorax and an abdomen of 10 segments.
2. Chelicerae are chelate and pedipalpi leg - like.
3. Spinning glands are absent.
4. Respiratory organs are tracheae.
   Example : *Obligolophus.*

Order (xi) Acarina

1. Body is unsegmented.
2. Chelicerae and pedipalpi are usually small and associated with the mouth parts.
3. Mouthparts are piercing and sucking type.
4. Respiration by tracheae or skin.
   Example : Ticks (*Ixodes*), Mites (*Chorioptes*).

PHYLUM-ONYCHOPHORA

Aberrant forms as its several characters are of Phylum Annelida and Arthropoda combined. Body soft, worm-like covered by papillae and hairy spines. Cuticle is free, having transverse rings, metamerism not marked externally. Jaws are only one pair. Series of short, fleshy leg pairs provided with two claws. Respiratory organs are tracheae. Sexes are separate. Regarded as living fossil.

Example : *Peripatus.*

BRANCHIPUS

Phylum	-	Arthropoda
Class	-	Crustacea
Subclass	-	Branchiopoda
Order	-	Anostraca
Genus	-	*Branchipus*

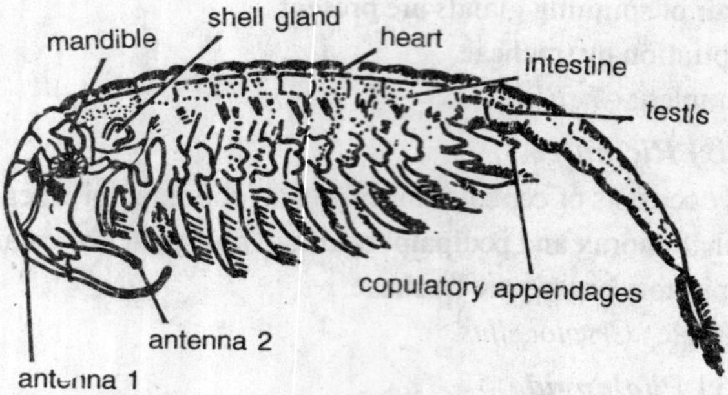

Fig. 1.1. *Branchipus.*

1. It lives in fresh water lakes, ponds etc.
2. It is commonly known as *brine shrimp*.
3. Body is elongated pinkish, semitransparent and diferentiated into head, thorax and abdomen.
4. Carapace is absent.
5. Eyes are situated on unjointed stalkes.
6. Antennae many and are modified into prehensile organs.
7. The caudal styles are unjointed.
8. Eggs are laid in mud during summer and hatch in the next summer.

STREPTOCEPHALUS

The systematic position is same as that of *Branchipus.*

1. *Streptocephalus* has the peculiar habit of swimming on its back near the surface of water with its leaf-like appendages and occurs in fresh-water ponds and ditches.
2. Body is long and sub-cylindrical, beautiful orange red in colour.
3. Carpace is absent.
4. Body can be distinguished into head, thorax and abdomen.
5. Sexual dimorphism is common i.e., male and female are easily distinguishable.
6. Head bears a pair of antennules, a pair of antennae, a pair of mandibles and two pairs of maxillae.
7. In male, antennae are long and modified for holding female during copulation.
8. On the dorsal side of the head a *nuchal sense organ* and a *median eye* are present.

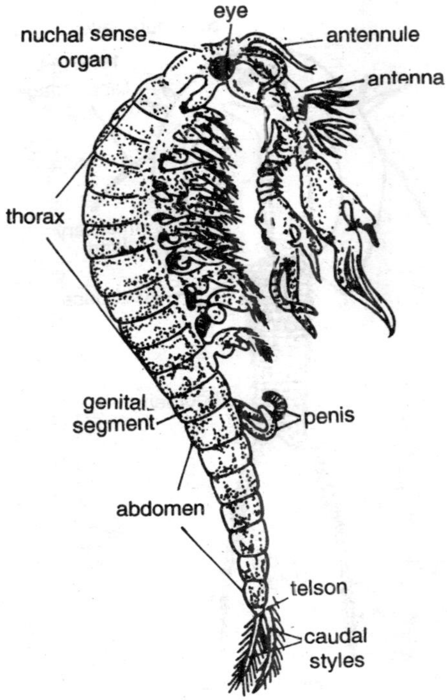

Fig. 1.2. *Streptocephalus* (Male).

9. **Trunk consists of nineteen segments, of these the first eleven segments, each bears a pair of broad foliaceous appendages.**
10. The twelth trunk segment carries the genital opening.
11. The segments from 13 to 19 constitute the abdomen which is devoid of appendages.
12. Anus lies ventrally in front of the telson.

APUS

Phylum	-	Arthropoda
Class	-	Crustacea
Subclass	-	Branchiopoda
Order	-	Notostraca
Genus	-	*Apus*

1. It is a fresh water crustacean found in most part of the world especially in U.S.A., Kansas, Texas, Mexico etc.

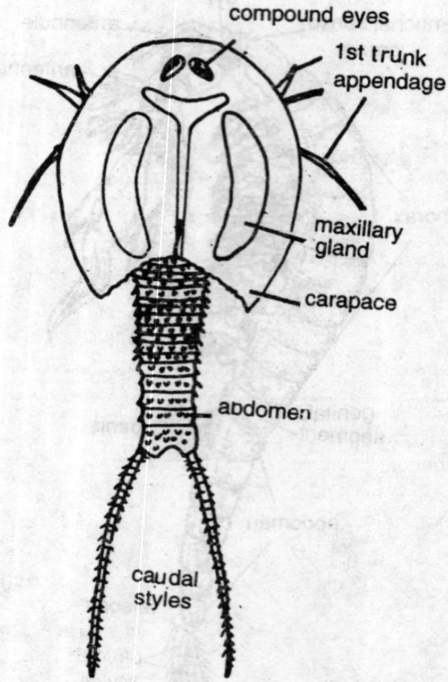

Fig. 1.3. *Apus.*

2. It is commonly called *tadpole shrimp*.
3. Body is elongated with broad shield-shaped carapace like horse shoe.
4. Head is broad and depressed bears paired eyes, a median eye and a dorsal organ above the antennules and antennae below.
5. Shell glands are present on the lateral surface of carapace.
6. The original segments of the body seem replaced by secondary annulation.
7. Caudal styles are long, filamentous and joined.
8. Males are rare, hence reproduction commonly occurs by *parthenogenesis*.

DAPHNIA

Phylum	-	Arthropoda
Class	-	Crustacea
Subclass	-	Branchiopoda
Order	-	Cladocera
Genus	-	*Daphnia*

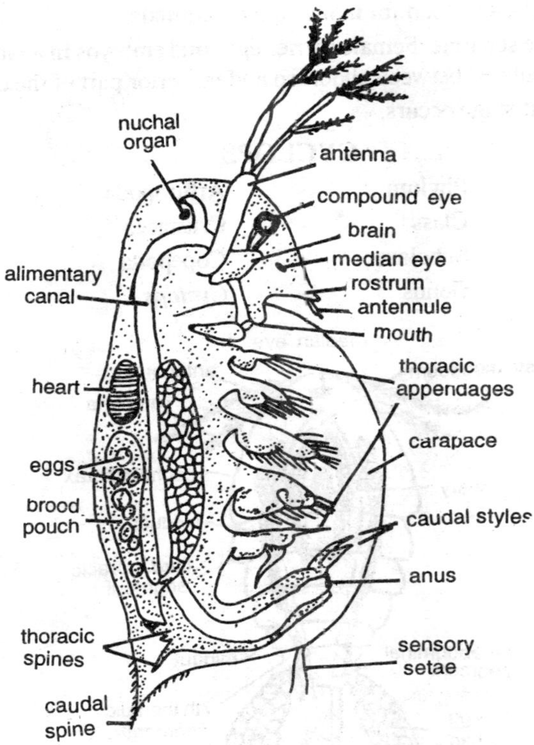

Fig. 1.4. *Daphnia*.

1. It is a fresh water crustacean found in ponds, pools and ditches.
2. It is commonly known as *water flea*.
3. The body, except head, is covered by folded carapace.
4. Head and abdomen are turned downward.
5. Head bears sessile eyes, nuchal sense organs in front of it, a pair of antennules, a pair of biramous antennae, large mandibles and two pairs of maxillae.
6. Large biramous antennae are the chief organs of locomotion.
7. Thorax bears usually five pairs of leaf-like appendages.
8. Abdomen is devoid of appendages.
9. The female develops a brood pouch between posterior part of carapace and abdomen in which eggs are stored.
10. The alimentary canal is simple.
11. Heart is a little sac.

12. It is a delicious food for many aquatic animals.
13. Sexes are separate. Female carries eggs and embryos in a large brood pouch situated between abdomen and posterior part of the carapace.
14. No larval stage occurs.

CYCLOPS

Phylum	-	Arthropoda
Class	-	Crustacea
Subclass	-	Copepoda
Genus	-	*Cyclops*

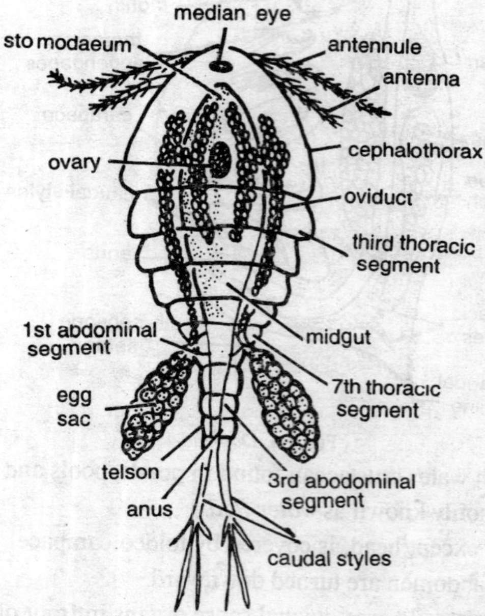

Fig. 1.5. *Cyclops.*

1. *Cyclops* is found in fresh water ponds, ditches and also in brackish water.
2. The body is elongated, somewhat oval in shape.
3. The cephalothrax is formed by the fusion of head and the first thoracic segments.
4. Carapace is absent.
5. The cephalothorax is followed by the five free segments of the thorax and an abdomen of four segments.

6. The last segment of the thorax is fused in the female with first abdominal segment and bears the genital openings.
7. The antennules are large, entire and segmented and they are meant for copulation in male.
8. The antennae are short and the mandibles have palps.
9. There are two pairs of maxillae following with are maxillipedes and which are the first thoracic appendages. Rest are used for swimming.
10. Abdomen is devoid of appendages except the caudal styles.
11. Sexes are separate. Mature female carries two egg sacs attached to the abdomen.
12. *Cyclops* serves as the intermediate host for the broad tapewarm (*Diphyllobothrium*) and the guinea worm (*Dracunculus*) of man.

CYPRIS

Phylum	-	Arthropoda
Class	-	Crustacea
Subclass	-	Ostracoda
Genus	-	*Cypris*

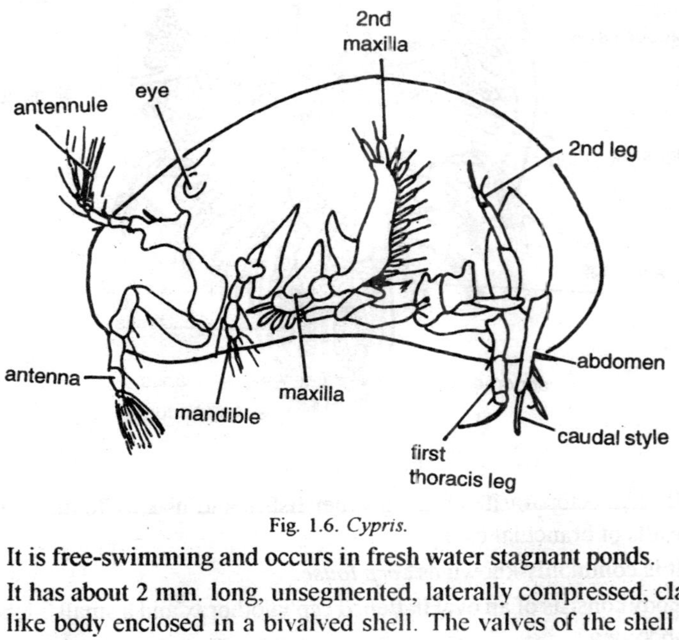

Fig. 1.6. *Cypris.*

1. It is free-swimming and occurs in fresh water stagnant ponds.
2. It has about 2 mm. long, unsegmented, laterally compressed, clam-like body enclosed in a bivalved shell. The valves of the shell are closed by a transverse adductor muscle.

3. Compound eyes are absent but there is a prominant median eye. Antennule and antennae are large.
4. It is an omnivorous filter feeder.
5. Respiration by integument.
7. Development by *parthenogenesis*.

ARGULUS

Phylum	-	Arthropoda
Class	-	Crustacea
Subclass	-	Branchura
Genus	-	*Argulus*

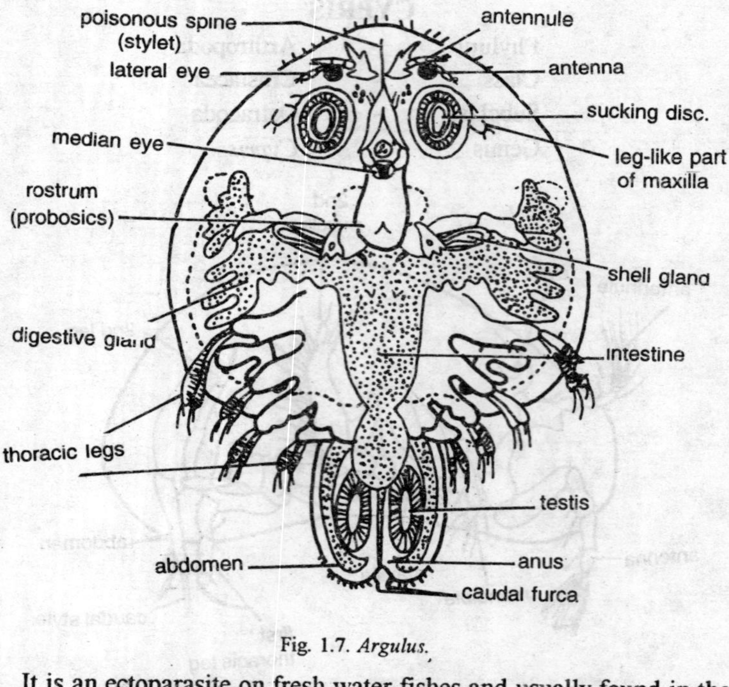

Fig. 1.7. *Argulus.*

1. It is an ectoparasite on fresh water fishes and usually found in the walls of branchial cavity.
2. It is commonly known as *carp louse*.
3. Body consists of an oval flattened cephalothorax and a small bilobed abdomen.

4. Mandibles and maxillae form piercing organs and are surrounded by a sucking tube called *proboscis.*
5. A poisonous spine, the *stylet*, lies in front of the proboscis.
6. Second maxillae are divided into two parts, an anterior sucking disc for attachment and a posterior leg-like part.
7. A pair of large compound eyes and a median eye are present on the head.
8. Thorax bears 4 pairs of swimming appendages.
9. Gills are absent, respiration probably through the carapace surface.
10. Fertilization is internal.
11. Development is direct.

LUCIFER

Phylum	-	Arthropoda
Class	-	Crustacea
Subclass	-	Malacostraca
Order	-	Decapoda
Genus	-	*Lucifer*

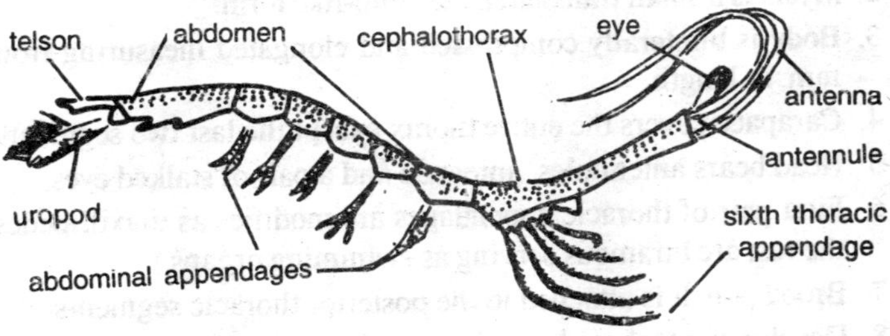

Fig. 1.8. *Lucifer.*

1. It is a pelagic crustacean found in India, Asia and Europe.
2. The animal is minute, delicate and slender.
3. Body is divisible into cephalothorax and abdomen.
4. Cephalothorax is smaller than abdomen.
5. Head is elongated and bears long antennae and stalked eyes.
6. The thoracic appendages are non-chelate and the last two thoracic somites are without appendages.
7. Gills are absent.
8. Development indirect through *nauplius* larva which moults through *zoaea* and a *mysis* stages.

MYSIS

Phylum	-	Arthropoda
Class	-	Crustacea
Subclass	-	Malacostraca
Order	-	Mysidacea
Genus	-	*Mysis*

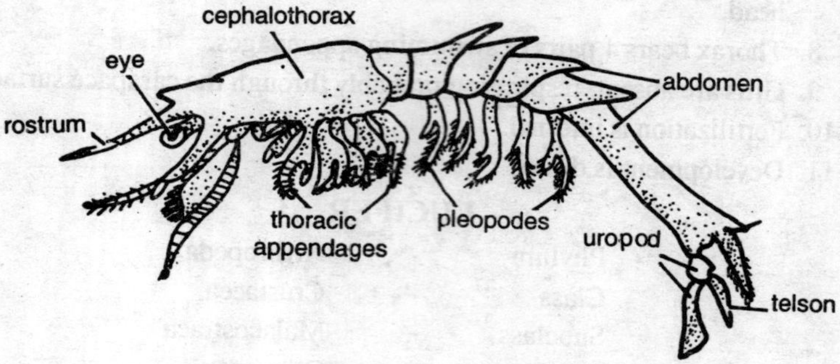

Fig. 1.9. *Mysis.*

1. It is found sometimes in sea water and sometimes in fresh water.
2. *Mysis* is a small transparent, shrimp-like form.
3. Body is bilaterally compressed and elongated measuring from 2-6 mm. in length.
4. Carapace covers the entire thorax except the last two segments.
5. Head bears antennules, antennae and a pair of stalked eyes.
6. First pair of thoracic appendages are modifies as maxillipedes and the rest are biramous serving as swimming organs.
7. Brood pouch is attached to the posterior thoracic segments.
8. Development takes place within the brood pouch so there is no larval stage.

GAMMARUS

Phylum	-	Arthropoda
Class	-	Crustacea
Subclass	-	Malacostraca
Order	-	Amphipoda
Genus	-	*Gammarus*

1. It is marine as well as found in fresh water. It is commonly called *sand-flea*.
2. Body is elongated and laterally compressed.

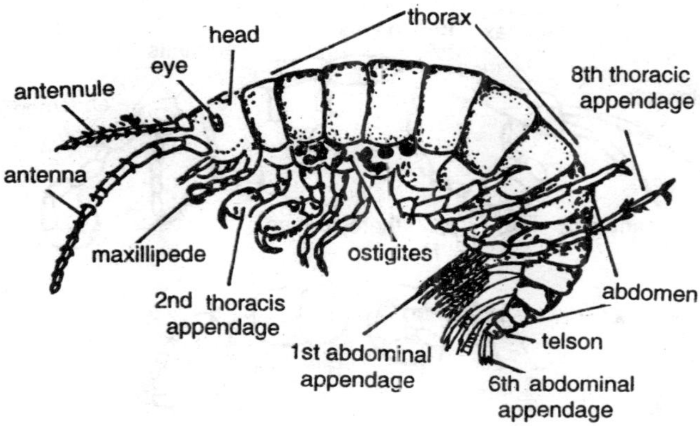

Fig. 1.10. *Grammarus.*

3. Cephalothorax is formed by the fusion of head and first thoracic segment.
4. Carapace lacking.
5. Head bears a pair of antennule, antennae and sessile eyes.
6. First thoracic segment bears maxillipeds and the remaining seven segments bear walking legs. First two pairs of legs are chelate.
7. The abdomen is of a considerabale size and flexed ventrally, bears a bifid telson.
8. First three pairs of abdominal appendages are used for swimming and rest for jumping on the ground.
9. It feeds upon dead and decaying organisms.

CAPRELLA

The systematic position is same as that of *Gammarus.*

1. *Caprella* liver in association of seaweeds, hydroid and polyzoan colonies.
2. It has cosmopolitan distribution. Specially found in U.S.A., Alaska to Santa Barbara.
3. It is *no-body crab* with peculiar and extremely slender body and limbs.
4. Body has elongated segments, cosisting cephalothorax and abdomen.
5. Two thoracic segments fuse with head, forming cephalothorax containing eyes, antennules and antennae. Antennae are shorter than antennules.
6. First antennae longer than the second.
7 Thorax has six free segments and contains heart.

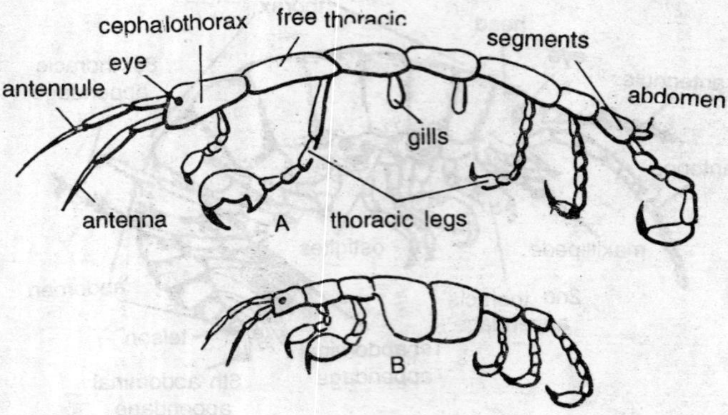

Fig. 1.11. *Caprella.*

8. Gills confined to third and fourth free thoracic segments on which the legs may be wanting while the rest have sub-chelate walking legs.
9. Abdomen reduced to a small knob bearing a pair of rudimentary legs in the male.
10. Male is smaller than the female.
11. It resembles in colour and form with polyzoan hydroid to which it mimics. It holds by 3 pairs of long posterior prehensile thoracic legs.

LEPAS

Phylum	-	Arthropoda
Class	-	Crustacea
Subclass	-	Cirripedia
Order	-	Thoracica
Genus	-	*Lepas*

1. It is sessile in habit and found attached in the clusters of floating objects like logs and the bottom of ships in the sea.
2. It has cosmopolitan distribution.
3. It is commonly called the *goose barnacle*.
4. Body consists of a long stalk or pedicel and the body proper or capitulum.
5. Pedicle is covered by wrinkled skin.
6. The capitulum is enclosed in a bivalve carapace strengthened by 5 calcareous plates, 2 proximal scuta, 2 distal terga and a single dorsal carina.

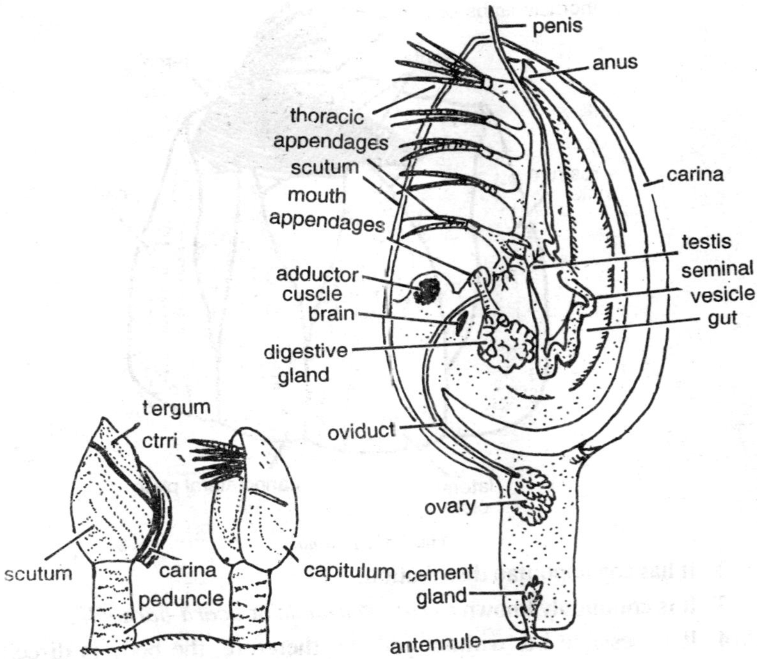

Fig. 1.12. *Lepas.*

7. The pedicle is really the highly modified anteriormost part of the head contains the vestige of the antennules, ovaries and the cement gland which provide the stickly secretion for attachment.
8. Antennae and paired eyes are absent.
9. Mouth is provided with a pair of mandibles and two pairs of maxillae.
10. Thorax bears 6 pairs of appendages.
11. Behind the appendages is a penis.
12. Abdomen is reduced to a process without any appendages.
13. It is bisexual, paired testes lie at the side of digestive tract and ovaries in the pedicle.
14. Development indirect through *nauplius* which moults into *cypris* stage which metamorphoses into adult.

BALANUS

The systematic position is the same as that the *Lepas.*

1. It is found attached to rocks and floating woods, shell of molluscs, carapace of turtle etc. in sea water.

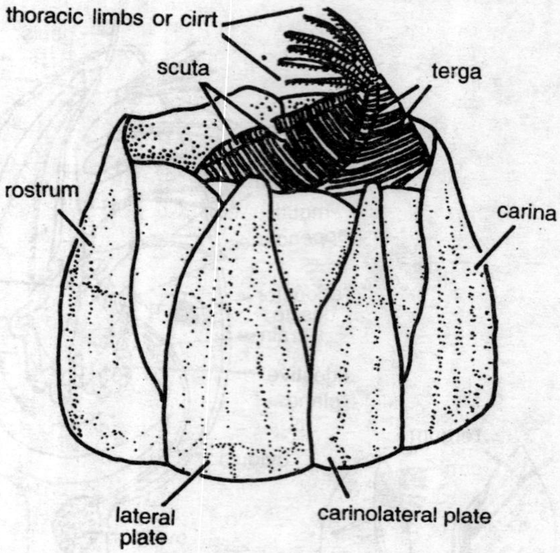

Fig. 1.13. *Balanus.*

2. It has cosmopolitan distribution.
3. It is commonly known as *rock-barnacle* or *acord-barnacle*.
4. It is sessible i.e. without pedicle, therefore, the body is directly attached to the rock or substratum.
5. Body is enclosed in a cylindrical case formed of six calcareous plates - unpaired carina and rostrum and two pairs of carinolaterals.
6. At the top of the shell there is an opening which is closed by two pairs of valves called *opercular valves*.
7. From the opening 6 pairs of thoracic feet which sweep in food particles of animal.
8. Eyes and antennae are absent.
9. Bisexual or hermaphrodite.
10. Development indirect and a *nauplius* larva is set free which moults into *cypris* stage and finally metamorphoses into adult.

SACCULINA

Phylum	-	Arthropoda
Class	-	Crustacea
Subclass	-	Cirripedia
Order	-	Rhizocephala
Genus	-	*Sacculina*

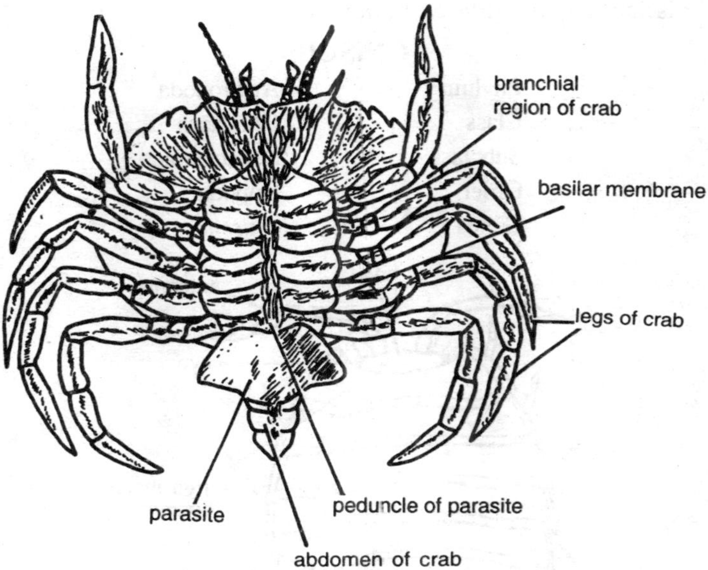

Fig. 1.14. *Sacculina* on the abdomen of crab.

1. *Sacculina* is parasite on carbs found attached between thorax and abdomen.
2. It has world-wide distribution.
3. It is commonly called *root-headed barnacle*.
4. Adult is extremely degenerate and appears as a fleshy tumour on the host.
5. Body without carapace, segementation, appendages and alimentary canal.
6. The peduncle gives out many delicate root-like process into the tissue of host through which it absorbs nourishment.
7. Opening of mantle cavity at the posterior end of the body is known as cloacal aperture
8. These are bisexual.
9. Development is indirect through cirripide *nauplius*, which transform into *cypris*. Cypris attaches itself to the crab by its antennules. The cypris moults and called *kentrogen* larva. It travels within the blood and attached to ventral side of intestine. Rootlets grow out from it and larva metamorphoses into adult.
10. It causes changes in the secondary sexual characters of host. Male

crab develops female characters while female ovary degenerates. This is known as *parasitic castration.*

ONISCUS

Phylum	-	Arthropoda
Class		Crustacea
Subclass	-	Malacostraca
Order	-	Isopoda
Genus	-	*Oniscus*

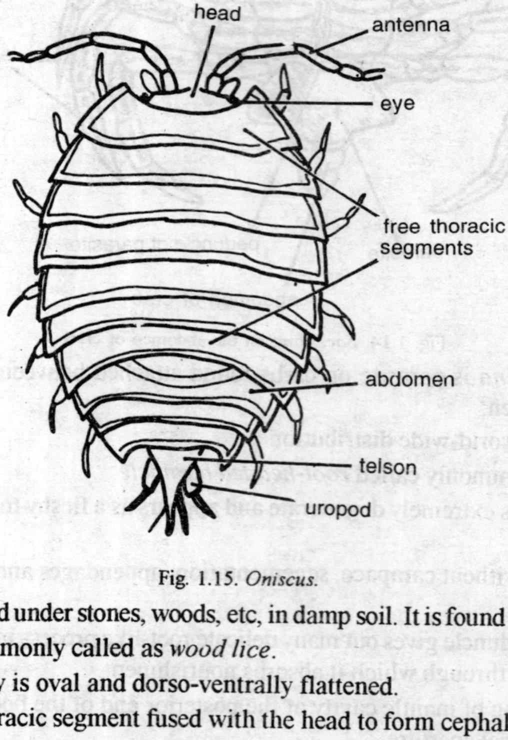

Fig. 1.15. *Oniscus.*

1. It is found under stones, woods, etc, in damp soil. It is found in Europe.
2. It is commonly called as *wood lice*.
3. The body is oval and dorso-ventrally flattened.
4. First thoracic segment fused with the head to form cephalothorax.
5. Unstalked compound eyes are present.
6. Antennules and antennae are uniramous.
7. The abdominal legs are adapted for aerial respiration and their endopodiates bear delicate branchiae traversed by minute tube called *pseudotracheae*.
8. Abdomen short.
9. Younges develop in a brood pouch, present on ventral side.

BOPYRUS

Systematic position is same as that of *Oniscus.*

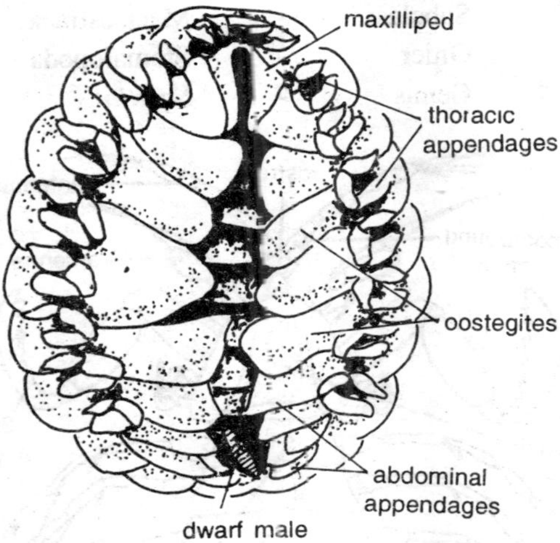

Fig. 1.16. *Bopyrus.* Ventral view.

1. *Bopyrus* belonging to family Bopyridae inhabits the gill chambers (branchial cavity) of the prawns.
2. It has cosmopolitan distribution.
3. Parasitic isopods exhibt extraordinary sexual dimorphism.
4. Adult female is greatly deformed bopyrid with an asymmetric body and without distinct segmentation.
5. Body is more or less oval.
6. It has suctorial mouth parts with simple piercing mandibles.
7. Maxillipedes are widely expanded covering the head appendages in front.
8. Large lamellae or oostegites develop from the bases of the thoracic limbs and form a brood pouch. Seven pairs of thoracic legs are short and hooked for attachment.
9. The abdominal appendages are plate-like functioning as gills.
10. Dwarf male is usually found attached to the female beneath the last pair of oostegites. It is more isopodian since it normally possesses developed thoracic and abdominal appendages.

SQUILLA

Phylum	-	Arthropoda
Class	-	Crustacea
Subclass	-	Malacostraca
Order	-	Stomatopoda
Genus	-	*Squilla*

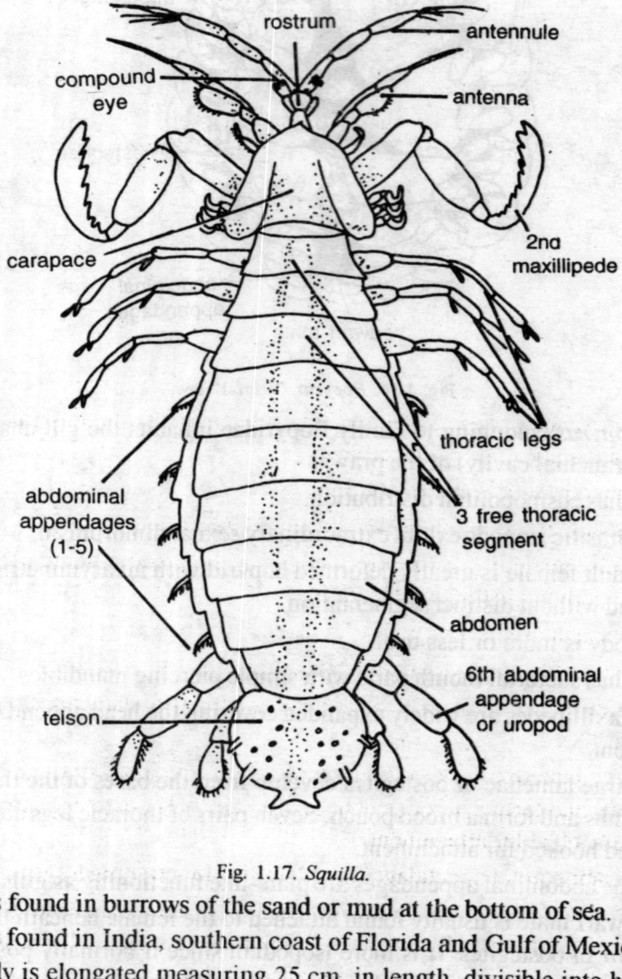

Fig. 1.17. *Squilla.*

1. It is found in burrows of the sand or mud at the bottom of sea.
2. It is found in India, southern coast of Florida and Gulf of Mexico.
3. Body is elongated measuring 25 cm. in length, divisible into head, thorax and abdomen.

4. Carapace is thin and uncalcified. It is produced anteriorly into a movable rostrum.
5. Compound eyes are stalked.
6. Anterior 5 pairs of thoracic appendages are uniramous and directed forwards to act as maxillipedes. The remaining thoracic appendages are biramous and act as legs.
7. The 2nd pair of maxillipedes are very large and form powerful prehensile weapons.
8. Abdomen is relatively large and has a broad telson bearing 3 pairs of marginal spines.
9. Pleopods are biramous and large. They bear gills on their exopodities.
10. Sixth appendages form large uropods which are flattened and form the tail-fan along with the telson.
11. The eggs are carried about by the last three pairs of maxillipedes.
12. The larvae are transparent pelagic *alima*, which metamorphoses into adult *Squilla*.

ASTACUS

Phylum	-	Arthropoda
Class	-	Crustacea
Subclass	-	Malacostraca
Order	-	Decapoda
Genus	-	*Astacus*

1. *Astacus* is found in freshwater ponds, streams and lakes all over the world.
2. It is commonly called *cray-fish*.
3. The body is long and robust, somewhat compressed.
4. Body is divisible into cephelothorax and abdomen.
5. The line of demarcation between the head and thorax is possibly indicated by cervical groove.
6. Rostrum is short and without serrated margins.
7. Head appendages 5 pairs. Eyes, antennules and antennae are present.
8. Thoracic appendages 8 pairs. First three pairs of thoracic legs are chelate and the first pair is largest.
9. Abdomen 6 segmented, with 6 pairs of appendages.
10. In females, eggs are carries on by the pleopods.
11. The development is direct.

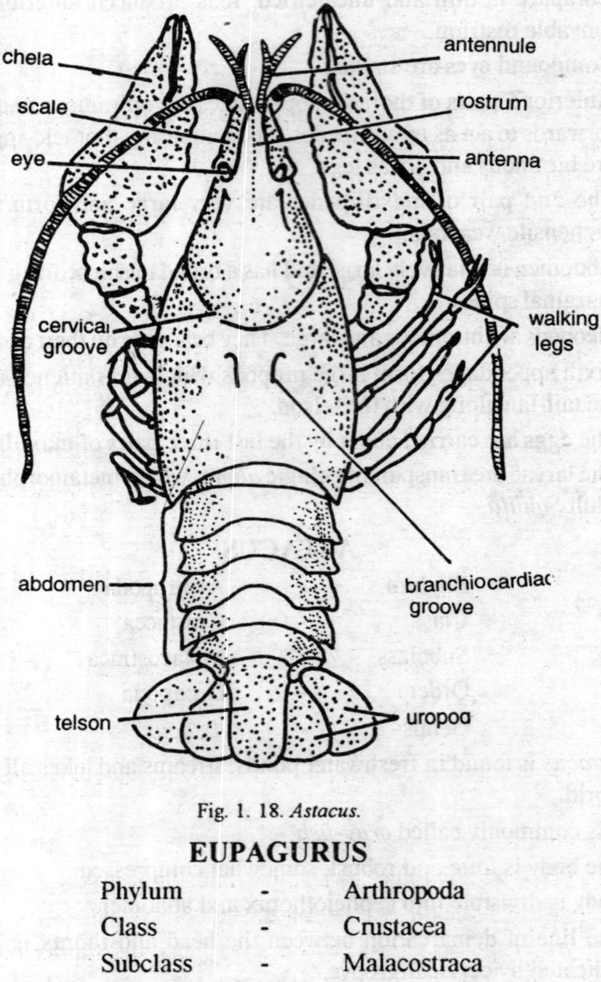

Fig. 1. 18. *Astacus.*

EUPAGURUS

Phylum	-	Arthropoda
Class	-	Crustacea
Subclass	-	Malacostraca
Order	-	Decapoda
Genus	-	*Eupagurus*

1. It is found in the empty shells of gastropod molluscs.
2. It has cosmopolitan distribution.
3. It is commonly called *hermit crab*.
4. The body is highly asymmetrical and consists of cephalothorax and abdomen.

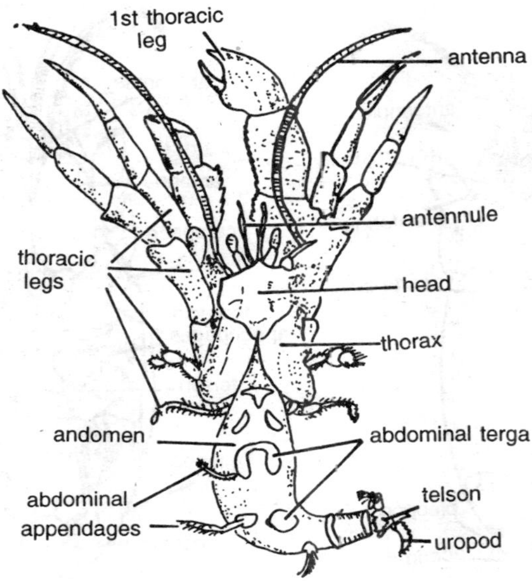

Fig. 1.19. *Eupagurus.*

5. Cephelothorax is broad and flattened bearings short antennules, antennae and stalked eyes.
6. Of the 5 pairs of thoracic legs, first are chelate, 2nd and 3rd are non-chelate, 4th and 5th are chelate but reduced.
7. Abdomen is soft, cylindrical, coiled and unsegmented. It bears reduced appendages on the left side and none on right side. Uropods are, however, well developed.
8. Swimmerets are found only in females for carrying eggs.
9. The close association between the hermit crab and sea-anemone in which both are benefetted by each other called *commensalism.*

HIPPA

Phylum	-	Arthropoda
Class	-	Crustacea
Subclass	-	Malacostraca
Order	-	Decapoda
Genus	-	*Hippa*

1. It is found in the sand near the sea. It is commonly found on Pacific Coast, U.S.A. and Oregon to Maxixo.
2. It is commonly known as *mole-crab.*

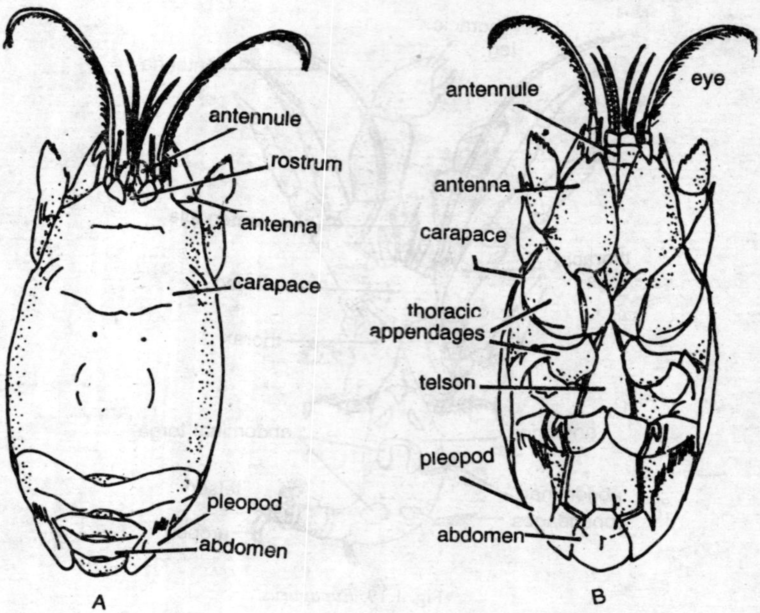

Fig. 1.20. *Hippa.* A—Dorsal view, B—Ventral view.

3. Body is ovate with large cephalothorax.
4. Carapace is smooth.
5. Mouthparts are poorly developed.
6. Head bears long stalked eyes, a pair of biramous antennules and a pair of uniramous antennae.
7. Rostrum is simple and pointed.
8. Thoracic appendages 7 pairs, walking legs 5 pairs adapted for digging. First two pairs of walking legs partially chelate.
9. Abdomen 6 segmented first three pairs of abdominal appendages modifies for swimming and last three pairs are adapted for darting.
10. Feeds like an earthworm, swallowing sand.

CARCINUS

Phylum	-	Arthropoda
Class	-	Crustacea
Subclass	-	Malacostraca
Order	-	Decapoda
Genus	-	*Carcinus*

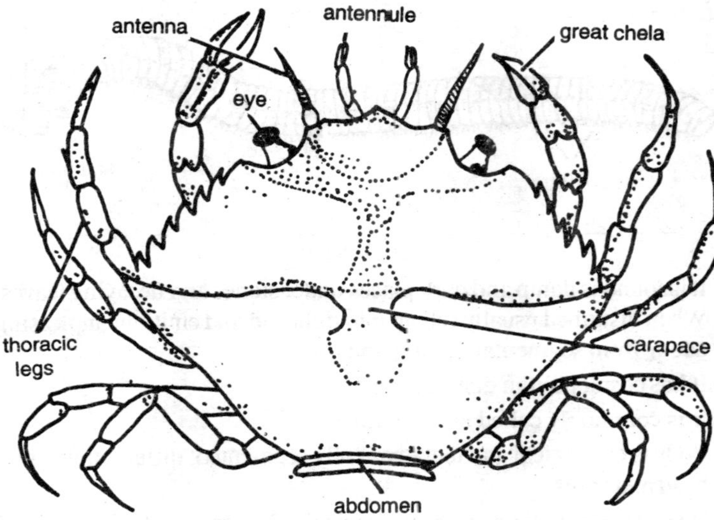

Fig. 1.21. *Carcinus.*

1. It is commonly found burried among rocks and in sand near about the water. It has cosmopolitan in distribution.
2. Cephelophorax broad and somewhat oval with a broad carapace.
3. The antennae and antennules are short.
4. The eyes are stalked and lodged in a pit of the carapace.
5. Third maxillipedes broad, flat and plate-like.
6. Walking legs 5 pairs, the first pair is large chelate.
7. Abdomen reduced and is permanently flexed below cephalothorax.
8. Males with 2 pairs of pleopods, whereas female possesses 4 pairs of them. Uropods absent.
9. Eggs are carries by the females which are attached to the pleopods.
10. Development indirect. Egg develop into *zoaea* which passes into *megalopa* and finally metamorphoses into adult.

JULUS

Phylum	-	Arthorpoda
Class	-	Myripoda
Subclass	-	Progoneata
Order	-	Diplopoda
Genus	-	*Julus*

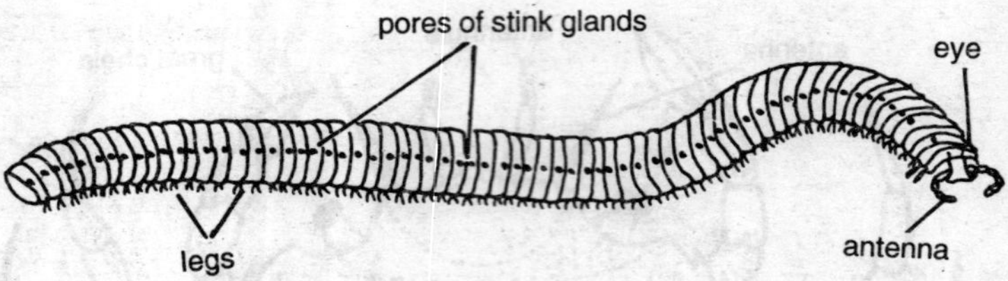

Fig. 1.22. *Julus.*

1. It is found is damp and dark places under stones and decaying leaves. When disturbed usually roll up into tight coil and emits an unpleasing odour from segmental scent glands.
2. It has cosmopolitan distribution.
3. It is commonly called as *wire worm.*
4. Body is cylindrical and metamerically segmented; differentiated into head and trunk.
5. Head small; antennae short generally 7 segmented. One pair of maxillae and mandibles forming *gnathochilarium.*
6. Eyes are minute.
7. Trunk cylindrical consisting of very closely ring-like segments from 25 to more than 100 segments.
8. In each segment there are two pairs of legs. Since it has numerous legs it is called *millipede.*
9. Poison glands and poison claws absent.
10. Vascular system well developed.
11. Respiration by tracheal tubes.
12. Sexes are separate.
13. Genital openings situated at the anterior end of body.
14. In male one or both pairs of legs of 7th segment are modified to form copulatory organs.
15. Herbivorous and even scavanger.

SCOLOPENDRA

Phylum	-	Arthropoda
Class	-	Myripoda
Subclass	-	Opisthogoneata
Order	-	Chilopoda
Genus	-	*Scolopendra*

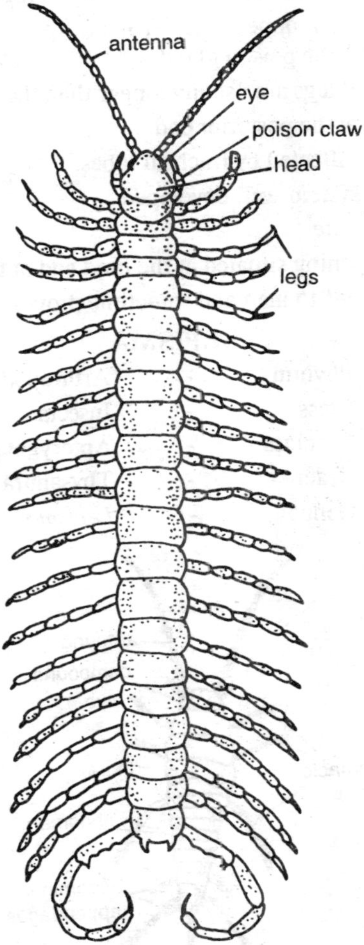

Fig. 1.23. *Scolopendra*—the centipede.

1. It is found under stones, in cervices and other dark places. It is found in India and America.
2. It is commonly called *Centipede*.
3. Body is elongated and dorso-ventrally flattened with numerous segments.
4. Head is distinct and bears a pair of antennae, a pair of mandibles and two pairs of maxillae.

5. Trunk segments numerous, each bearing a single pair of legs.
6. First pair of trunk appendages or maxillipedes bears a sharp claw connected with the poison gland.
7. The last pair of legs are usually longer than the rest.
8. The anus lies at the posterior end.
9. Respiration is affected by tracheal tubes.
10. The vascular system well developed.
11. Sexes are separate.
12. The genital opening situated at the hind end of the body.
13. Bite is dangerous to man and sometime proves to be fatal.

LEPISMA

Phylum	-	Arthropoda
Class	-	Insecta
Subclass	-	Apterygota
Order	-	Thysanura
Genus	-	*Lepisma*

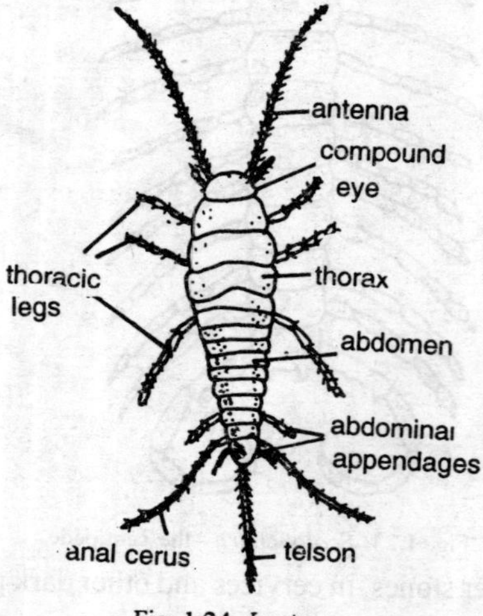

Fig. 1.24. *Lepisma.*

1. It is found in houses, in books, papers, clothes, under pictures in warm countries. It feeds on starch.
2. It is commonly known as *silver fish.*
3. Body is flattened and covered with silvery scales.
4. Body divisible into head, thorax and abdomen.

5. Head small with a pair of filiform antennae and eyes.
6. Mouthparts chewing type.
7. Thorax three segmented and bears 3 pairs of legs, wings are absent.
8. Abdomen 11 segmented with 3 anal cerci and a number of small delicate appendages.
9. Development direct without metamorphosis.

ISOTOMA

Phylum	-	Arthropoda
Class	-	Insectsa
Subclass	-	Apterygota
Order	-	Collembola
Genus	-	*Isotoma*

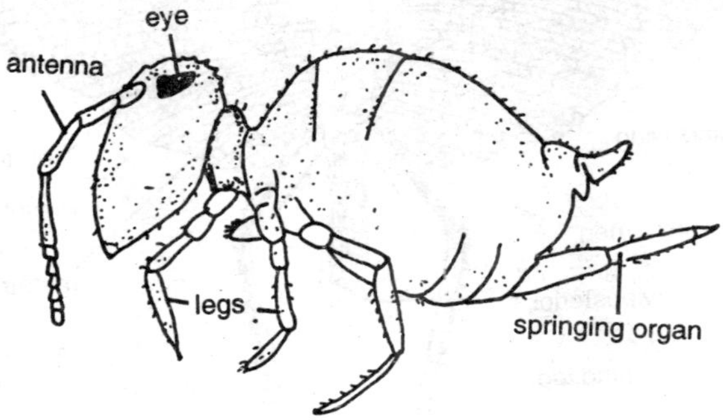

Fig. 1.25. *Isotoma.*

1. These are found in damp places, under leaves, cerives of bark etc.
2. It is commonly called *Spring tail*.
3. Body is divisible into head, thorax and abdomen.
4. Head with 4 jointed antennae.
5. Compound eyes are absent, Ocelli scattered.
6. Thorax 3 segmented with 3 pairs of legs. Wings absent.
7. Abdomen 6 segmented. First segment with a tubular adhesive organs, *collophore*. 3rd segment with *tentaculum* and 4th with *furcula* (spring organ).
8. It secretes glue-like substance used for sticking body to the object.

SCHISTOCERCA

Phylum	-	Arthropoda
Class	-	Insecta
Subclass	-	Pterygota
Division	-	Exopterygota
Order	-	Orthoptera
Genus	-	*Schistocerca*

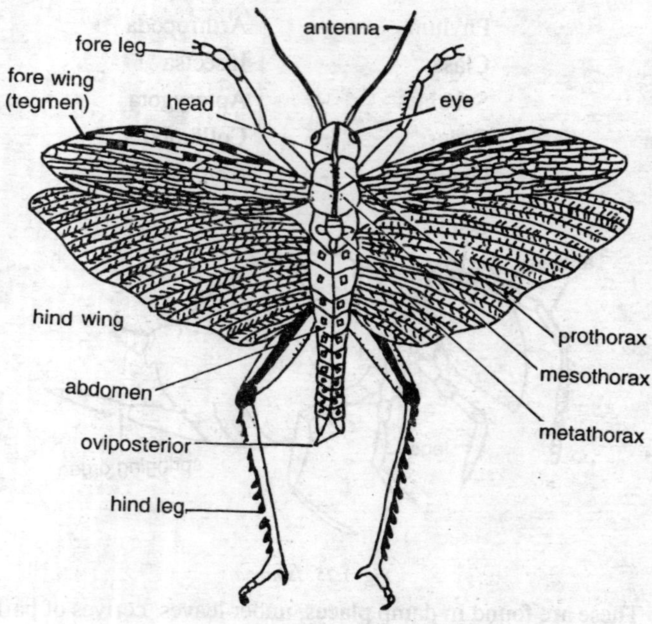

Fig. 1.26. *Schistocerca.*

1. Gregarious herbivorous insect found in green vegetation.
2. Commonly found in Africa, Arabia, India, Japan, Mexico, U.S.A. and Afghanisthan.
3. Body is divisible into head, thorax and abdomen.
4. Head with one pair of long antennae and well developed compound eyes.
5. Thorax 3 segmented. 3 pairs of legs and two pairs of wings are present. Wings are dissimilar and unequal.
6. The fore-wings elongated, straight, thick, leathery with numerous brownish black sports.

7. The membranous hind-wings are large, plaietd anal area.
8. Legs well developed.
9. The hind-legs modifies for jumping having long basally stout femora.
10. The abdomen has 10 evident and vestigial eleventh segments.
11. Mouthparts biting and chewing type.
12. Male with genital lamina incised posteriorly.
13. This is the most destructive of all locusts and its ravages were known in Biblical times.
14. They occur in two phases : (a) *Solitary* or non-migratory phase having yellow colour with dark green and brown patches (b) *Gregareous* or migratory phase have pink body with dark brown spots.

GRYLLUS

Phylum	-	Arthropoda
Class	-	Insecta
Subclass	-	Pterygota
Division	-	Exopterygota
Order	-	Orthoptera
Genus	-	*Gryllus*

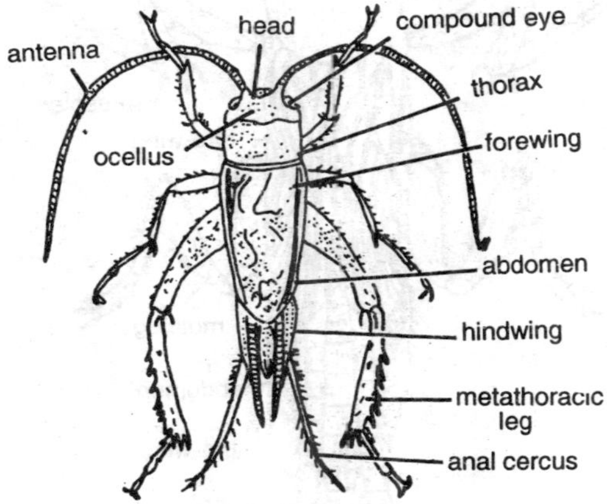

Fig. 1.27. *Gryllus*.

1. It is a noctural and omnivorous insect found under various object, stones etc. during day time.
2. It is found in India, Sri Lanka, Asia, Burma, U.S.A., Europe and Canada.

3. It is commonly known as *circket*.
4. Body divisible into head, thorax and abdomen.
5. Head bears compound eyes and a pair of filiform antennae.
6. Mouthparts chewing type.
7. Fore-wings hard, narrow and parchment like with pigmented spots (tegmina). Hind-wings are membranous and folded.
8. Tibia of the fore-legs bear tympanic organs and the hind-legs are modified for jumping.
9. Well developed ovipositor in females.
10. Male circket chirps by rubbing from wings together during courtship.

GRYLLOTALPA

The systematic position is same as that of *Gryllus*.

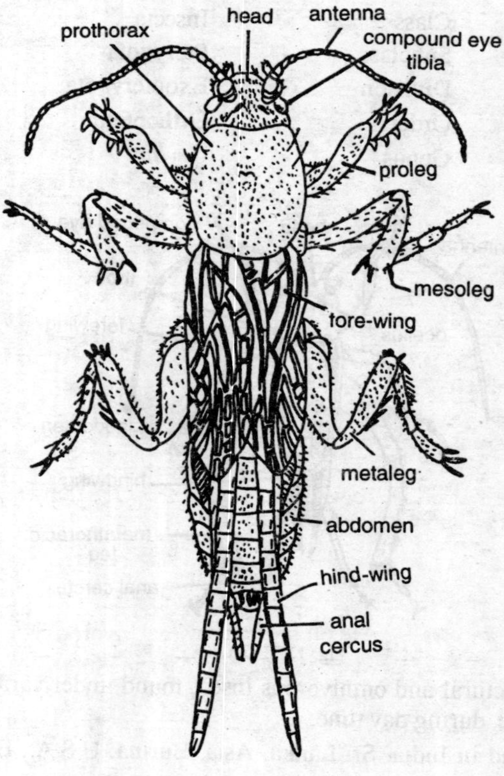

Fig. 1.28. *Gryllotalpa*.

1. *Gryllotalpa* is a nocturnal, burrowing insect.
2. It is found in Indian plains, warmer parts of Europe and Asia.
3. It is commonly known as *mole-circket*.
4. Body is covered with short, fine hairs and is devisible into head, thorax and abdomen.
5. Head is well developed, bears a pair of antennae and commpound eyes.
6. Mouth-parts are biting and chewing type.
7. Tibiae of the first pairs of legs are especially modified with stout spines for digging purpose.
8. Fore-wings are short and hard membranous.
9. Ovipositor is not protruded.
10. Cerci are unjointed and long.

MANTIS

Phylum	-	Arthropoda
Class	-	Insecta
Subclass	-	Pterygota
Division	-	Exopterygota
Order	-	Mantodea
Genus	-	*Mantis*

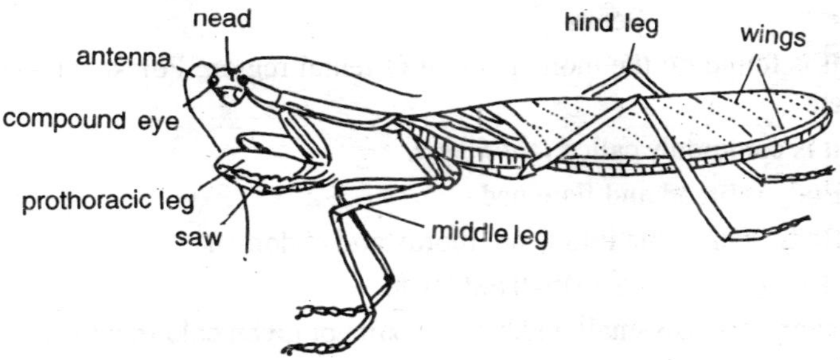

Fig. 1.29. *Mantis*.

1. These are arboreal feeding on living prey (carnivorous).
2. It is found in North America, U.S.A. Africa, South Europe and Western Asia.
3. It is commonly known as *preying mantis*.
4. Body is long, slender, usually green in colour.
5. Body is divisible into head, thorax and abdomen.

6. Head is small, triangular with vertical face and freely movable on slender neck.
7. Large compound eyes, three ocelli and long filiform antennae are present on head.
8. Mouthparts are chewing type.
9. Prothorax is much elongated and prothoracic legs are modifies for grasping and holding the prey.
10. Wings are well developed. These are folded flat and overlapping the sides of the body.
11. Abdomen 10 segmented. Ovipositor is no exected.
12. Nineth segment in males bear a pair of styli.
13. Short flattened and segmental cerci present.
14. Female eats the male after nuptial. 200 or more eggs are enclosed in the ootheca formed of frothy quick drying material.

PHYLLIUM

Phylum	-	Arthropoda
Class	-	Insecta
Subclass	-	Pterygota
Division	-	Exopterygota
Order	-	Phasmida
Genus	-	*Phyllium*

1. It is found on the moist parts of Oriental region, Forests of Assam etc.
2. It is commonly called *leaf-insect*.
3. Body is broad and flattened.
4. Body is divisible into head, thorax and abdomen.
5. Head is small bears small antennae.
6. Thorax is also small and bears 3-pairs of green coloured legs.
7. Tegmen or first pairs of wings are small than the other pair, having leaf-like colouration and veination.
8. Sexes distinct. Male with small tegmina, well developed wings and narrow bodies. Female with extensive tegmina produced into leaf like expansions.
9. It is an example of *mimicry* or *camouflage* in which protective resemblance in both form and colouration, is similar to the objects in their surroundings.

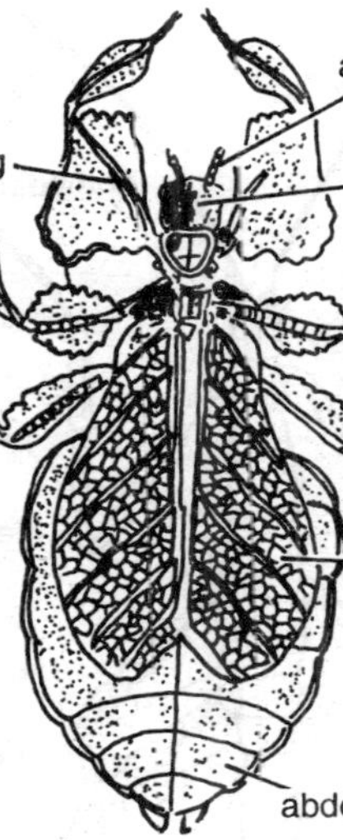

Fig. 1.30. *Phyllium.*

CARAUSIUS

Phylum	-	Arthropoda
Class	-	Insecta
Subclass	-	Pterygota
Division	-	Exopterygota
Order	-	Phasmida
Genus	-	*Carausius*

1. It is found in tropical forests and is arboreal. It is found in India, Sri Lanka, Burma, Asia and Europe.
2. It is commonly known as *stick-insect*.
3. Body is elongated and slender with long legs and antennae resembling a twing or stick or any other foliage on which they feed.
4. Head is small with a pair of filiform antennae and compound eyes. Mounthparts chewing type.
5. Thorax is elongated bearing 3 pairs of legs and the wings are absent.
6. Abdomen 10 segmented.
7. Cerci small and unsegmented.
8. Sexes are separate, Male is small. active and winged. Female is large. sluggish and apterous.

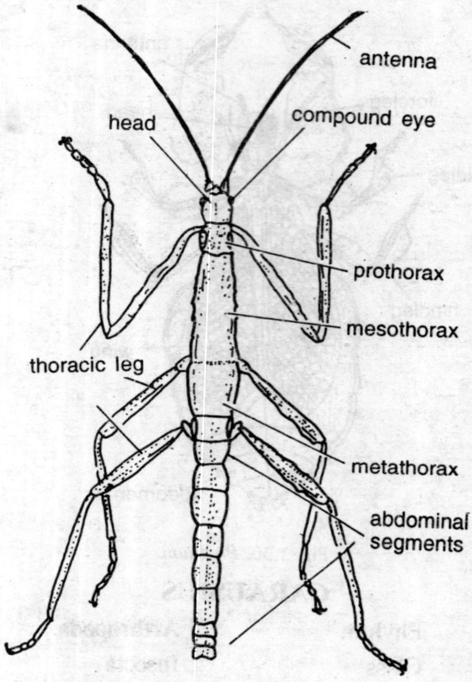

Fig. 1.31. *Carausius.*

9. Exclusively parthenogenetic.
10. Power of regeneration or *autotomy* is very great.

FORFICULA

Phylum	-	Arthropoda
Class	-	Insecta
Subclass	-	Pteygota
Division	-	Exopterygota
Order	-	Dermaptera
Genus	-	*Forficula*

1. Found under stones, moist places under stones in decaying vegetable matter etc.
2. It is found in Europe, N. America, Mexico, Asia, Africa and on the coasts of Indian ocean.
3. It is commonly known as *earwing*.
4. Small insect with elongate body.

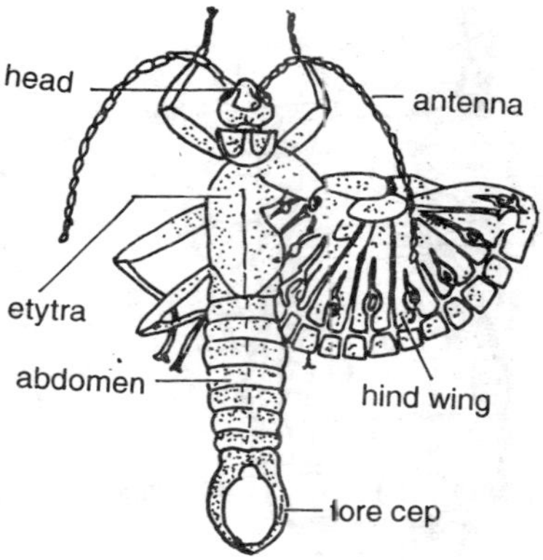

Fig. 1.32. *Forficula.*

5. Body is divisible into head, thorax and abdomen.
6. Head contains a pair of eyes and a pair of long filiform antennae.
7. Mouthparts chewing type.
8. Thorax contains 3 pairs of walking legs and peculiar hind-wings.
9. The fore-wings form hard veinless truncated wing cover below which the hind-wings are present.
10. Anal cerci are modified into a pair of forceps which is raised upwards and is opened widely in a threatening manner, when the animal is disturbed.
11. Ovipositor absent.
12. Metamorphosis gradual and it exhibits parental care for eggs and young ones.

PEDICULUS

Phylum	-	Arthropoda
Class	-	Insecta
Suborder	-	Pterygota
Division	-	Exopterygota
Order	-	Anoplura
Genus	-	*Pediculus*
Species	-	*humanus*

1. These are minute, flat, wingless and ectoparasite of man. It has cosmopolitan distribution.
2. It is commonly known as *human louse*.
3. Body is dorso-ventrally flattened and pale in colour with dark marking along the sides.

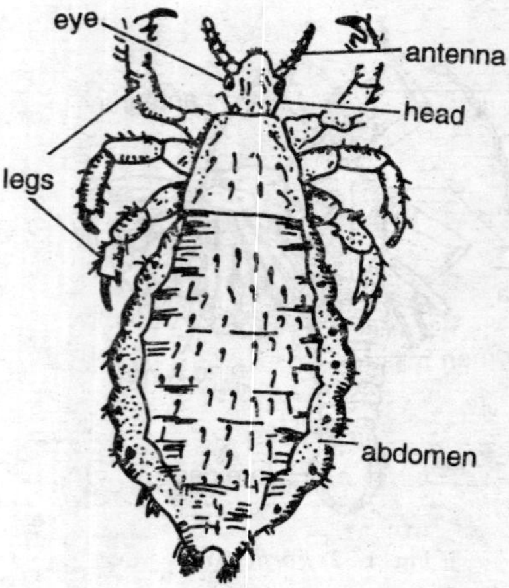

Fig. 1.33. *Pediculus.*

4. Head bears a pairs of antennae which is 5 segmented, and a pair of compound eyes.
5. Mouthparts piercing and sucking type.
6. Thoracic segments are fused in one mass and contain 3 pairs of clawed legs, which form clinging organs with the tissue of the host.
7. Abdomen is 9 segmented.
8. Males are smaller than females and their posterior ends turned upward.
9. Metamorphosis is absent.
10. It is a major factor in transmitting *relapsing fever*, *typhus* and *trench fever*. It sucks the blood of man.

EPHEMERA

Phylum	-	Arthropoda
Class	-	Insecta
Subclass	-	Pterygota
Division	-	Exopterygota
Order	-	Ephemerida
Genus	-	*Ephemera*

1. It is a soft-bodies and most familiar insect found on the margins of streams, lacks etc.
2. It is found in U.S.A., India, Sri Lanka and Burma.
3. It is commonly called as *Mayfly*.
4. Body is divisible into head thorax and abdomen.

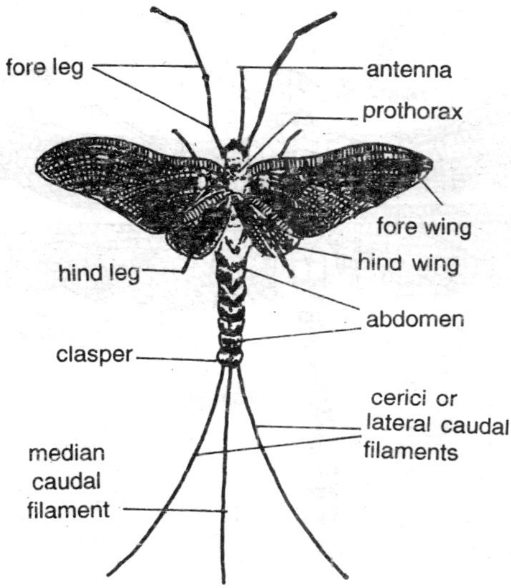

Fig. 1.34. *Ephemera.*

5. Head is small contain small setose antennae, compound eyes and 3 ocelli.
6. Mouthparts vestigial.
7. 3 pairs of inconstant walking legs.
8. Two pairs of membranous wings, forewings larger than hind-wings.
9. Abdomen is 10 segment with a reduced tergum. Last segment bears a pair of long many jointed cerci. Tergum is produced into a long caudal filament.
10. Males has 3 jointed, clasping organ and the female has no ovipositor.
11. Nymph or naiad lives for a long time upto 3 years in water, feeding in aquatic plants and animals. The imago moults prior to mating and is extremely short lived.

DRAGON FLY

Phylum	-	Arthoptera
Class	-	Insecta
Subclass	-	Pterygota
Division	-	Exopterygota
Order	-	Odonata
Genus	-	*Libellula*

1. It has world wide distribution and generally found darting balancing in the air.

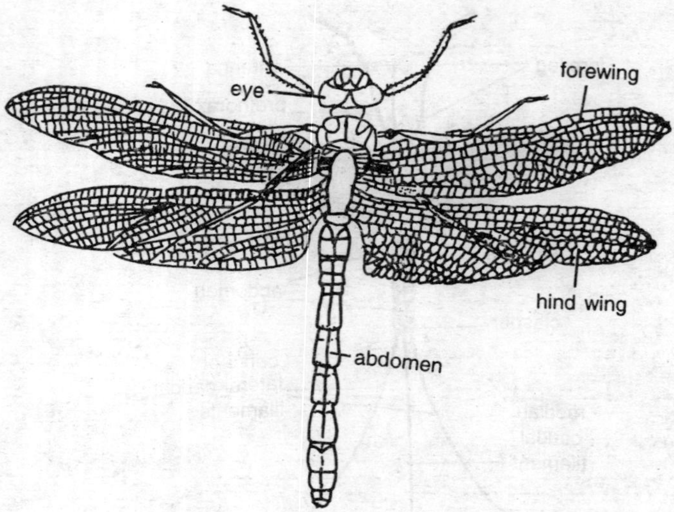

Fig. 1.35. *Libellula.*

2. Body is slender and divisbile into head, thorax and abdomen.
3. Head bears a pair of large compound eyes and small inconspicuous antennae.
4. Mouthparts are of chewing type.
5. Prothorax is small. Meso-and metathorax are fused. Legs are weak and used for resting.
6. Both pairs of wings are well developed and membranous. These are held in a horizontal position.
7. The abdomen in elongated often slender made up of 10 segments. The genital openings in both, male and female, lies at the hind end. Male copulatory organs are present on the 2nd and 3rd sternite.
8. Nymphs are aquatic, breath by rectal gills and feeding upon aquatic insects and organisms.

NEPA

Phylum	-	Arthropoda
Class	-	Insecta
Subclass	-	Pterygota
Division	-	Exopterygota
Order	-	Hemiptera
Genus	-	*Nepa*

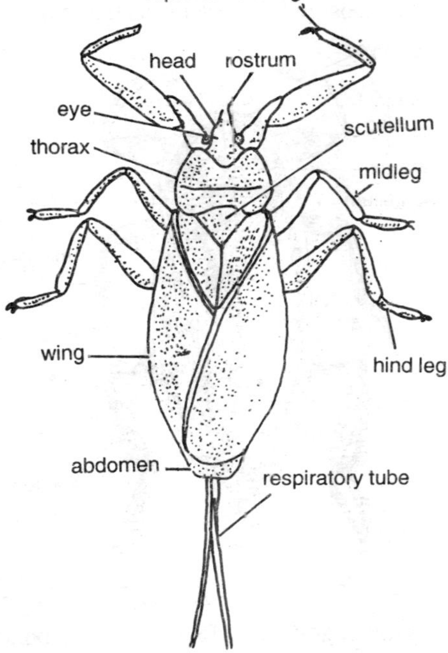

Fig. 1.36. *Nepa.*

1. It is commonly found in shallow water. It has world wide distribution.
2. It is commonly called *water scorpion*.
3. The body is dorso-ventrally flattened and divisible into head, thorax and abdomen.
4. Head comprises a pair of 3 jointed antennae and compound eyes.
5. Mouthparts form a long rostrum adapted for piercing and sucking.
6. The respiratory tube is most peculiar. It consists of two spine-like processes.
7. There are three pairs of false spiracles situated on 3rd, 4th, and 5th ventral abdominal segments respectively.
8. The false spiracles comprise sieve-like structures with the perforations occuleded by a delicate membrane.
9. Sexes are separate.
10. The female is provided with a pointed toothed ovipositor.
11. The eggs are deposited in chains. The ova adhere to one another by means of sevan long filaments radiating from one extremity.

BELOSTOMA

The systematic position is the same as that of *Nepa.*

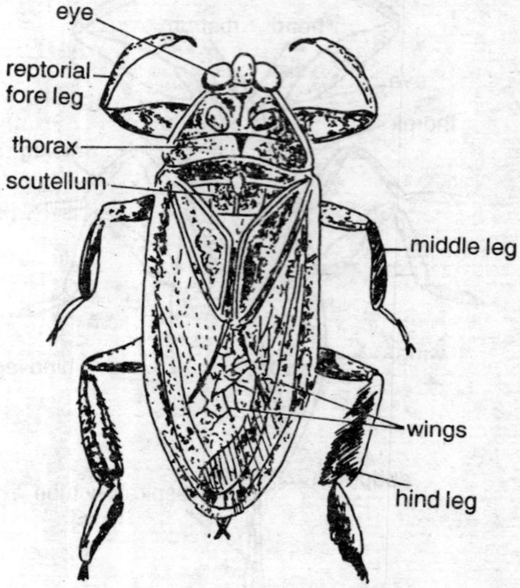

Fig. 1.37. *Belostoma.*

1. *Belostoma* is the largest insect found in rivers, ponds, lakes, feeding on insects, tadpoles, snails etc.
2. It is found in India, U.K., North America and South Africa.
3. It is commonly called *giant water bug.*
4. Body is dorso-ventrally flattened and divisible into three parts, head, thorax and abdomen.
5. Head is small with a pair of large eyes and 4 jointed antennae. Mouthparts piercing and sucking type.
6. The thorax bears 3 pairs of legs. The mid-legs and hind-legs are flattened and adapted for swimming, the fore-legs *raptorial.*
7. First pair of wings *hemielytra* and second pair membranous.
8. The abdomen is provided with two retractile apical appendages.
9. Ovipositor well developed. Females of some species deposit their eggs on the back of male (parental care).

DYSDERCUS

The systematic position is the same as that of *Nepa.*

1. It is a common pest of cotton and commonly known as *Red cotton-bug.*

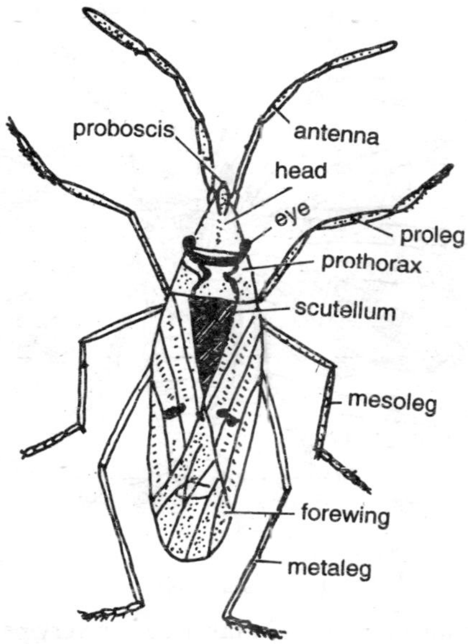

Fig. 1.38. *Dysdercus.*

2. Body is black and red in colour.
3. Body divisible into head thorax and abdomen.
4. Head bears a pair of compound eyes and 4 jointed antennae.
5. Mouth parts piercing and sucking type.
6. Wings two pairs, basal half of the forewings hard and leathery (hemelytra), Hindwings soft and membranous.
7. Triangular scutellum present between the leases of wings.
8. Pronotum large.

CICADA

The systematic position is same as that of *Nepa.*

1. It is found in the forests and moist places, abundant in tropical, subtropical and temperate regions.
2. Body is divisible into head, thorax and abdomen.
3. The head is broad possessing two lateral compound eyes and two very short antennae.
4. The thorax comprises prothorax, mesothorax and metathorax.
5. There are two pairs of wings : forewings are larger and somewhat triangular and the hindwings are oval.
6. The abdomen is elongated and consists of ten segments.

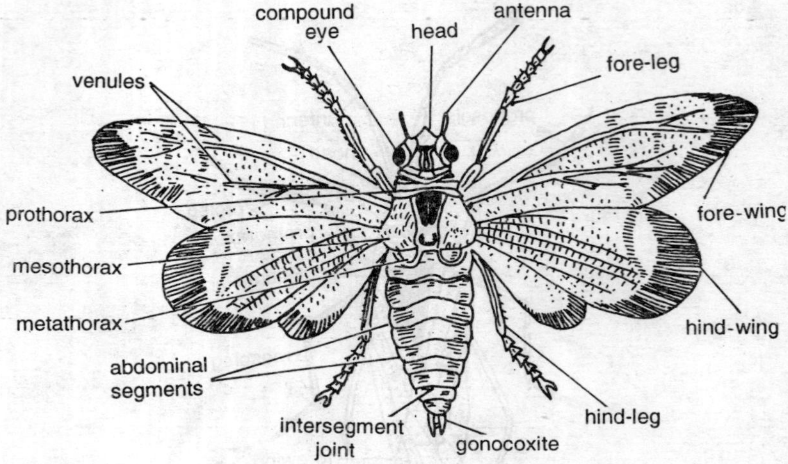

Fig. 1.39. *Cicada.*

7. The sound production habit of this insect is very peculiar.
8. The sound is produced only by the male which possesses a drum-like apparatus present ventrally on the basal abdominal segments, with the help of special muscular arrangement.
9. Metamorphosis incomplete, egg hatches into nymph.
10. The life-history is very interesting as it takes about 13 to 17 years to complete i.e. the nymph moults 25 times in about 13-17 years to become adult.
11. Adults soon dies after mating.

APHID

The systematic position is the same as that of *Nepa*.

1. It is ectoparasite on plants sucking their juice. It is found specially on the plants of mustard, wheat and cotton.
2. It is commonly called *plant-lice*.
3. Body is short divisible into head, thorax and abdomen.
4. Head is distinct with long straight antennae. Eyes are small.
5. Mouth parts are piecrcing and sucking type.
6. Thorax and abdomen robust. Abdomen is elongated.
7. Exhibit alternation of sexual and parthenogenetic phases, females always develop from parthenogenesis.
8. Both apterous are winged forms.

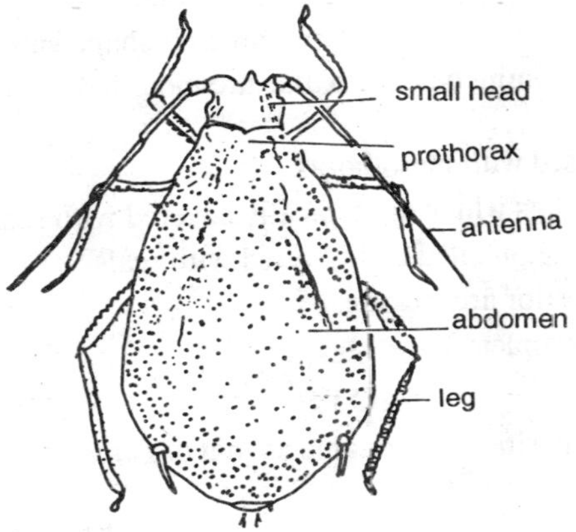

Fig. 1.40. Apple-grain Aphid.

DANAUS

Phylum	-	Arthropoda
Class	-	Insecta
Subclass	-	Pterygota
Division	-	Endopterygota
Order	-	Lepidoptera
Genus	-	*Danaus*

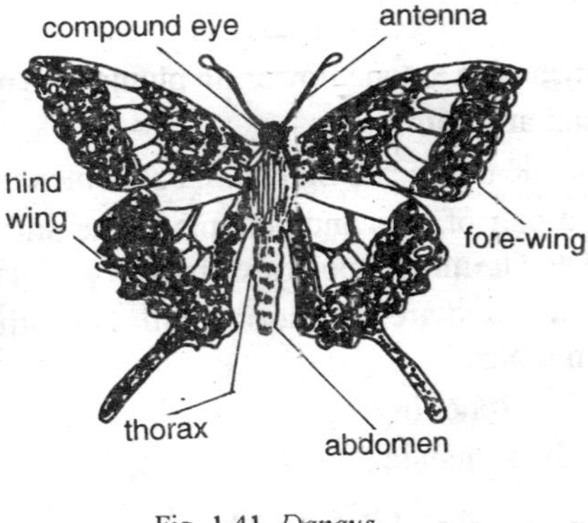

Fig. 1.41. *Danaus.*

1. It is a flying, terrestrial, diurnel insect found in gardens.
2. Commonly called *butter-flies.*
3. Beautifully and brightly coloured insects.

4. Body is divisible into head, thorax and abdomen.
5. Head with a pair of compound eyes and club-shaped antennae.
6. Mouthparts siphoning type. Maxillae are modified to form sucking tube.
7. Prothorax is fused with mesothorax.
8. Two pairs of wings which are typically covered with scales.
9. Abdomen is 10 segmented and covered with scales.
10. Cerci and ovipositor are lacking.
11. Development complete.

MOTH

The systematic position is the same as that of *Danaus.*

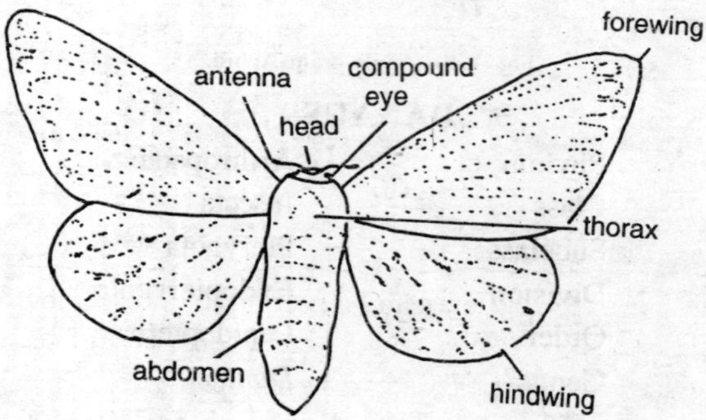

Fig. 1.42. Moth.

1. It is found in garden among flowering plants. It is nocternal in habit.
2. Body is broad and stumpy.
3. Body is divisible into head, thorax and abdomen.
4. Head bears a pair of eyes and antennae. The antennae are usually feather-like and terminal knobs are absent.
5. Wings two pairs and are held horizontally in resting condition and covered with scales.
6. Mouthparts are siphoning type.
7. Abdomen is 10 segmented.

COCCINELLA

Phylum	-	Arthropoda
Class	-	Insecta
Subclass	-	Pterygota

Division	-	Endopterygota
Order	-	Coleoptera
Genus	-	*Coccinella*

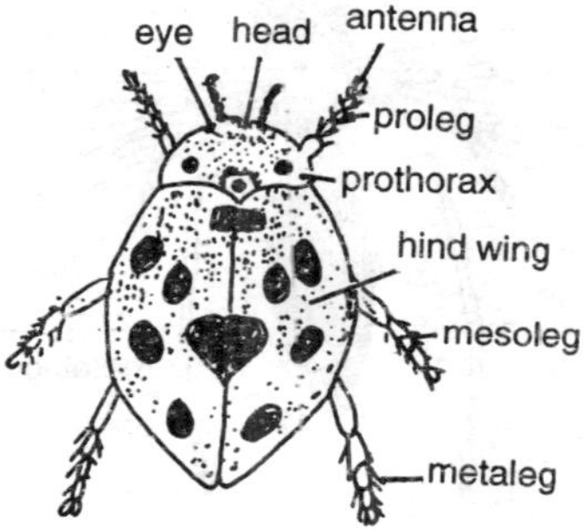

Fig. 1.43. *Coccinella.*

1. These are commonly called, *Lady Bird-Beetle*. These are terrestrial animal.
2. Body divisible into head, thorax and abdomen.
3. Head incompletely hidden by pronotum.
4. Head bears compound eyes and a pair of antennae. Mouthparts are chewing type.
5. Large number of species are carnivorous and predacious. Feeds on aphids, coccids and other soft bodies insects.

VESPA

Phylum	-	Arthropoda
Class	-	Insecta
Subclass	-	Pterygota
Division	-	Endoptezygota
Order	-	Hymenoptera
Genus	-	*Vespa*

1. It is an Indian hornet, which usually makes large nests of mud over walls and trees.
2. Body strongly built, usually chestnut red in colour with yellow, transverse bars.
3. Body is divisible into head, thorax and abdomen.
4. Head is globular concave posteriorly with small ocelli and compound eyes.

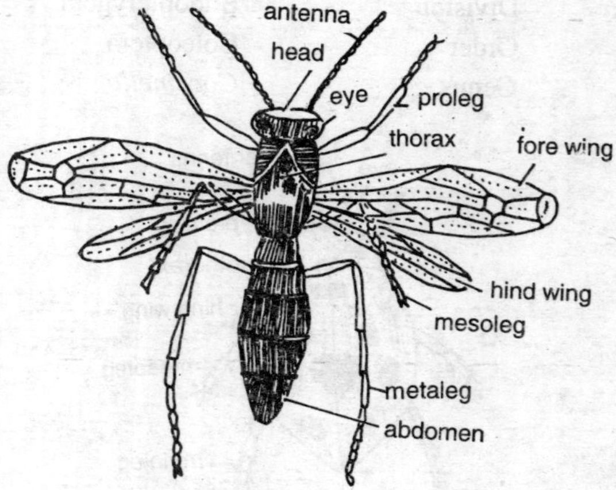

Fig. 1.44. *Vespa.*

5. Thorax deep and cubical wings are membranous. Hind-wings short and bear hooks on their anterior margin which fit into a groove on the posterior margin of fore-wings.
6. The abdomen petiolate, large with pointed apex.
7. Modified ovipositor as sting is present. It stings painfully and causes local pain and swelling.
8. *Vespa orientalis* is a social colonial insect. Colonly consists of queen, males and workers. The impregnated female hibernates in winter and forms a new colony at the next spring.

CAMPONOTUS

The systematic position is the same as that of *Vespa.*

1. Usually social insects lives in burrows, holes, cervices etc.
2. Body divisible into head, thorax and abdomen.
3. Head varies in shape and mandibles are variable in forms.
4. Antennae consists of 4-13 segments. Male has one more segment than the female.
5. Compound eyes and ocelli are well developed in males where as reduced or vestigial in females and workers.
6. *Stridulatory* (sound producing) *organ* is present is abdomen.
7. Large and well developed stings are present in females and workers.
8. Mating takes place in winged stage during naptial flight wings are then lost.

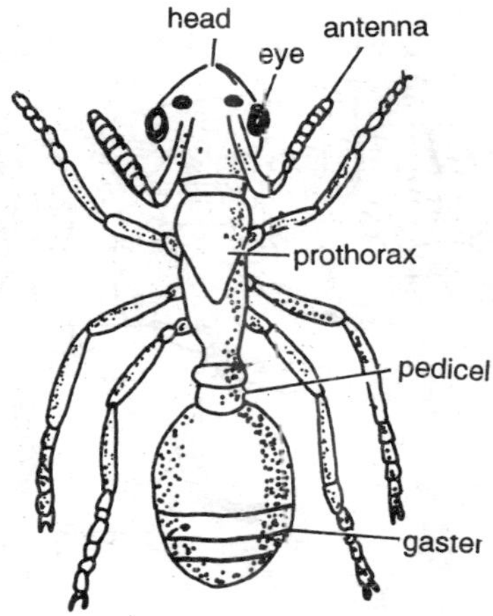

Fig. 1.45. *Camponotus.*

9. These are scavanger but are also harmful to man by destroying grain and seeds.
10. Other commen Indian ants are : *Ocephylla smaragetina* (Weaver), *Holeomyrmex* (harvester) and *Formica* (red ant).

XENOPSYLLA

Phylum	-	Arthropoda
Class	-	Insecta
Subclass	-	Pterygota
Division	-	Endopterygota
Order	-	Siphonoptera
Genus	-	*Xenopyslla*

1. It is a sanguivorous ectoparasite of rats, birds and man.
2. It is commonly found in India, Sri Lanka and Malaysia.
3. It is commonly called *rat-flea*.
4. Body is laterally compressed without wings.
5. Integument is covered with spines and bristles.
6. Head broad jointed with small thorax. Abdomen is very large.
7. Antennae are clavate, short and concealed in grooves.
8. Compound eyes and ocelli are absent.
9. Mouthparts piercing and sucking type.
10. Walking legs are elongated, clawed and modified for jumping.
11. It transmits '*plague*' from rat to man.

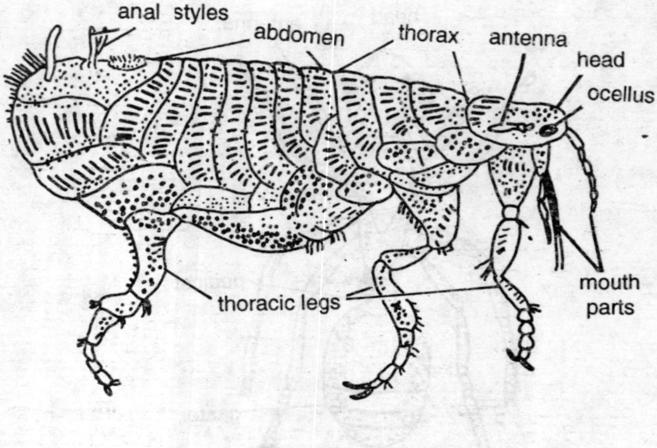

Fig. 1.46. *Xenopsylla.*

ARANEA

Phylum	-	Arthropoda
Class	-	Arachnida
Order	-	Araneida
Genus	-	*Aranea*

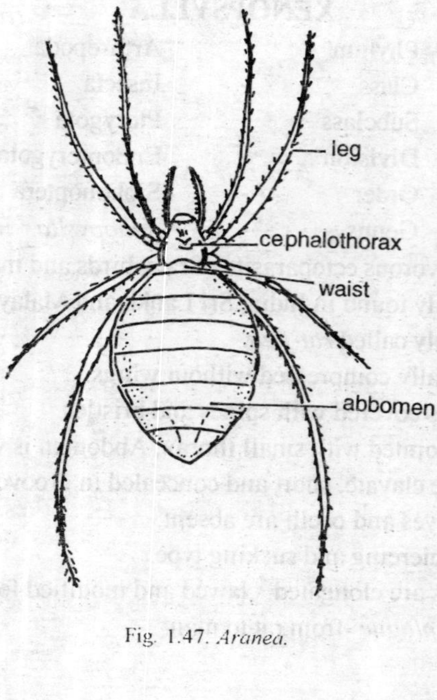

Fig. 1.47. *Aranea.*

1. It is found in houses and gardens.
2. It is commonly known as *orb-webbed spider*.
3. Body consists of cephalothorax or *prosoma* and abdomen or *opisthosoma*, the two being connected by a narrow *pedical*.
4. Prosoma with simple 8 eyes and 6 pairs of appendages. A pair of sub-chelate chelicerae, a pair of non-chelate pedipalps and 4 pairs of walking legs.
5. Opisthosoma is unsegmented, bears three pairs of spinnerets which produce thread for making web.
6. Poison glands open on the fangs of chelicerae.
7. Respiration by book-lungs and tracheae.
8. Excretion by coxal gland and Malpighian tubules.
9. Sexes are separete. Female is larger than male and eats up after copulation.

SARCOPTES

Phylum	-	Arthropoda
Class	-	Arachnida
Order	-	Acarina
Genus	-	*Sarcoptes*

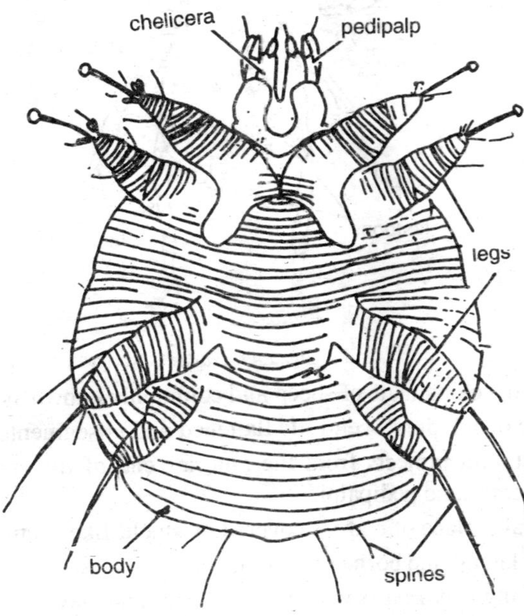

Fig. 1.48. *Sarcoptes*.

1. It is an ectoparasite on human skin and causes horizontal burrows in region between fingers, elbow etc. Leads to echzema. Commonly known as *itch-mite of human*.
2. It is rounded or oval and dorso-ventrally flattened with transverse striations and short spines.
3. Prosoma and opisthosoma not differentiated.
4. Four pairs of legs. Anterior two legs are stronger and project beyond the body and provided with terminal stalked suckers. Posterior two legs are shorter and attached more ventrally and carry long bristles.
5. Mouthparts borne on capitulum; chelicerae chelate.

IXODES

The systematic position is the same as that of *Sarcoptes*.

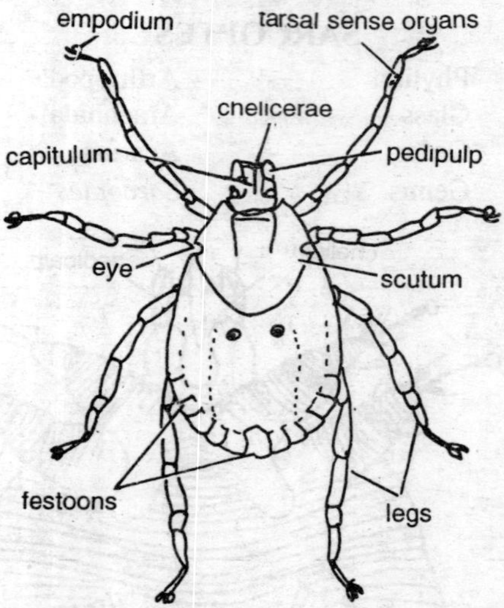

Fig. 1.49. *Ixodes.*

1. It is an ectoparasite of sheep and commonly known as *tick*.
2. Body ovoid, dorso-ventrally flattened and unsegmented.
3. Capitulum projects from the anterior end of the body and bears chelicerae and pedipalpi.
4. Dorsal surface of body is covered by shield-like scutum.
5. Eyes lateral and borne on scutum.
6. Walking legs 4 pairs with adhesive pads and claws.
7. Anus with ad-anal shield.

8. Female is larger than male.
9. They produce *ixodin* which acts as anesthetic, so that the host does not feel pains.

CHELIFER

Phylum	-	Arthropoda
Class	-	Arachnida
Order	-	Pseudoscorpionida
Genus	-	*Chelifer*

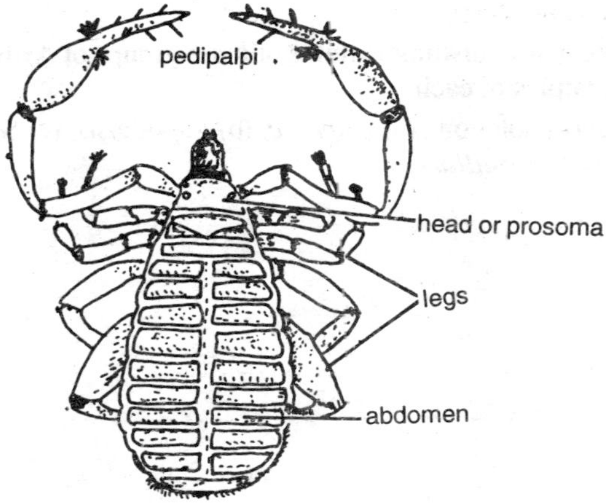

Fig. 1.50. *Chelifer.*

1. It is found commonly under the bark of trees.
2. Minute animals which have the appearance of scorpion but without tail hence commonly called *pseudoscorpion*.
3. Body consists of *prosoma* and an *opisthosoma*.
4. Prosoma is broad and continuous bearing 6 pairs of appendages.
5. The chelicerae small 2 jointed and chelate.
6. Pedipalpi 5 jointed large and claw-like.
7. A pair of spining glands open on the tips of chelicerae.
8. Abdomen is 12 segmented.
9. Respiration by tracheae.
10. Males have a pair of reversible organ *ram's horns*, supposed to be copulatory in function.

Revision Questions

1. Give characteristics of Phylum Arthropoda. Name the classes and bring out differences between them in the form of a table.
2. Classify Arthropoda up to classes giving their diagnostic characters and familiar examples.
3. Place the following animals in their proper position in the classification, giving their distinctive characters : (a) *Sacculina,* (b) *Limulus,* (c) *Millipede.*
4. Classify the class Crustacea. Give characters and examples of its various sub-classes.
5. Give the distinguishing features of main groups of Arthropoda with two examples of each.
6. Write short notes on : (a) *Daphnia,* (b) *Eupagurus,* (c) *Sacculina,* (d) *Cyclops,* (e) *Squilla.*

2

Palaemon

Palaemon, commonly known as *prawn*, is found in fresh water streams, ponds, rivers etc. It serves as a good example to exhibit the characteristics of class Crustacea

SYSTEMATIC POSITION

Phylum	-	Arthropoda
Class	-	Crustacea
Subclass	-	Malacostraca
Order	-	Decapoda
Suborder	-	Macrura
Genus	-	*Palaemon*

Habits and Habitat

A number of species of *Palaemon*, such as *P.carcinus, P.malcolmsonii* etc., are found in the fresh water streams, ponds etc. of India, Bangla Desh and distant countries such as West Indies, Mexico and Australia. These animals generally live in slow-running waters and are bottom-feeders. They are nocturnal in habit and come out in the night in search of food. The food is generally algae or other water plants and sometimes, insects. *P.malcolmsonii* breeds in May, June and July and the females carry eggs attached to their abdominal appendages.

EXTERNAL MORPHOLOGY

Shape and Size

Palaemon is a bilaterally symmetrical animal. The body is elongated and spindle-shaped and is well-adapted for swimming in water. The size of the adult *Palaemon* varies considerably in different species. *P.carcinus* measures about 75 cm in length (found in Travancore). *P.malcolmsonii* from Central India, Tamilnadu and Bengal is about 25-30 cm in length *P.lamarrei,* common in lakes and ponds is quite small about 2.5-5 cm. in length. *P.vulgaris,* a common prawn of America, measures 30 mm. in length.

Colouration

The embryonic stages of prawn are usually translucent and whitish in colour. Adult animals of different species are of different colours. It is pale-yellow or greenish with a brown tinge or with orange-red patches or dull pale-blue.

Division of body

The body is divisible into two distinct regions:

I. Anterior, immovable and unjointed *cephalothorax*.

II. Posterior, movable and jointed *abdomen*.

I. Cephalothorax

It is rigid and immovable. It is straight and oval in cross section. It is composed of 13 segments. In the embryonic condition, cephalothorax consists of first 14 fused segments and a pre-segmental region. Later on during development, first segment of the embryo disappears in the adult, while 13 segments remain. There are 13 pairs of appendages in these thirteen segments. The presegmental region is also found in adults and it possesses a pair of stalked compound eyes.

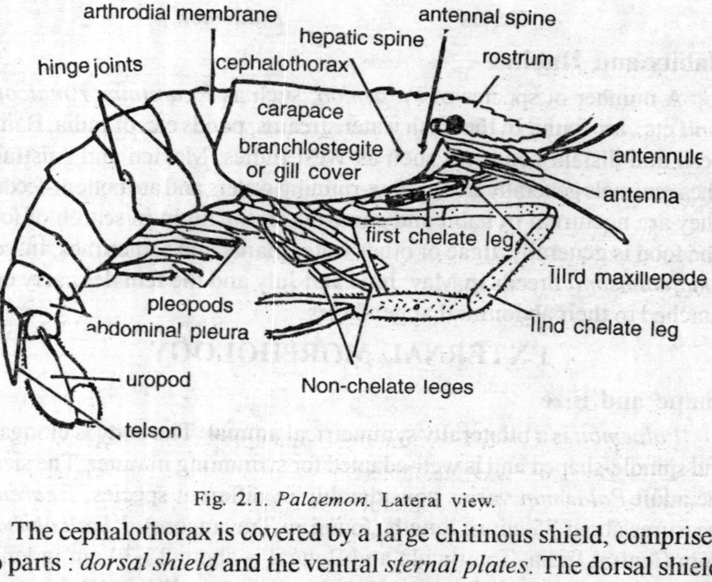

Fig. 2.1. *Palaemon*. Lateral view.

The cephalothorax is covered by a large chitinous shield, comprises two parts : *dorsal shield* and the ventral *sternal plates*. The dorsal shield further consists of anterior *dorsal plate* and posterior *carapace* but there is no line of demarcation between two. The dorsal plate is triangular and it extends forward as a laterally compressed process, the *rostrum*. The upper and lower margins of the rostrum are serrated and densely fringed with setae. At the base of rostrum, on either side, there is an *orbital notch* to which movable jointed stalk of the compound eye is attached. Just below each orbital notch, the dorsal plate bears a prominant *antennal spine*. Behind each antennal spine there is a relatively small *hepatic spine*.

The carapace covers the thorax dorsally and laterally, and represents the fused thoracic terga and pleura. On either side of the body, the carapace hangs freely as a convex *gill cover* or *branchiostegite*. It encloses a narrow *gill-chamber*, to house the gills.

II. Abdomen

It consists of 6 posterior movable segments and a terminal conical structure, the *telson*. The telson articulates with the 6th abdominal segment by a pair of hinge-joints.

Each abdominal segment is covered dorsally by a convex sclerite, the *tergum* and ventrally by the *sternum*. The tergum extends downwards on either side as a free flap, the *pleuron*. The pleuron covers the base of the appendage of its side, and is connected with it by a small sclerite, the *epimeron*. Pleura of the 2nd abdominal segment are largest and differ from others in over-lapping the pleura of both the Ist and the 3rd segments. Pleura of the 6th segment are small and point backwards. The adjacent terga and sterna are connected by flexible *arthrodial membranes* of uncalcified cuticle. The adjacent terga articulate by a hinge-joint on either side. However, the hinge joint between the 3rd and 4th segments are lacking. These joints allow only up and down movement of the segments.

External apertures

Mouth is silt-like aperture opening mid-ventrally at the anterior end of the cephalothorax. It is surrounded by maxillipedes and maxillae. *Anus* is in the form of a longitudinal aperture, located ventrally at the base of telson. One pair of *renal apertures* open on raised papillae on the inner surface of the antennae. Sexes are separate. Paired *female genital apertures* situated at the inner surface of the coxae of third walking legs in female. The paired *male genital apertures* are located on the inner surface of coxae of the 5th pair of walking legs in male. The openings of the statocysts are two lying one in a deep depression dorsally on the basal segment of each antennule.

APPENDAGES

The appendages of prawn show considerable diversity in structure. In different regions of the body the appendages are well-adopted for different functions. These appendages are homologous as all have the same typical structure and development.

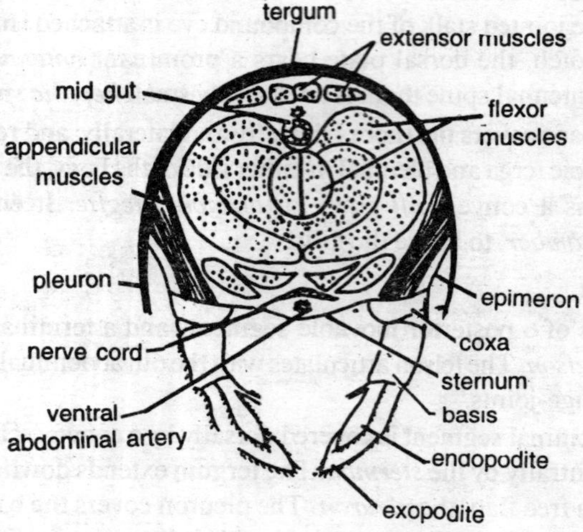

Fig. 2.2. *Palaemon*. Section through abdominal region showing sclerites.

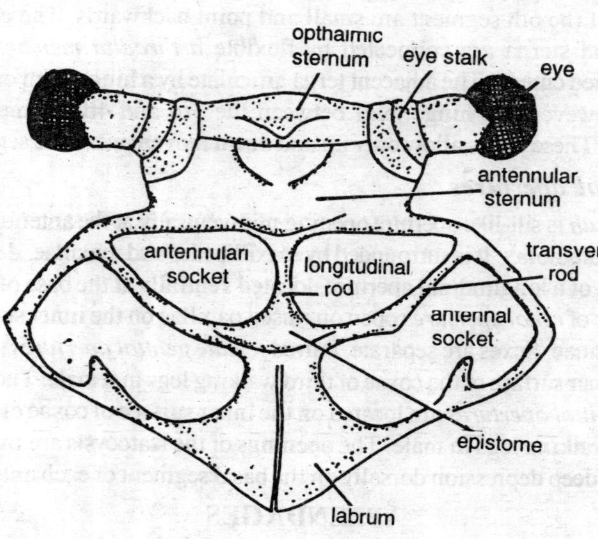

Fig. 2.3. *Palaemon*. Front view of preoral region.

Appendages appear very different from one another and can be reduced to a general type of appendage called, *stenopodium*. A stenopodium consists of two parts (i) *protopodite* or *sympod* or *stalk* and (ii) *rami* supported on the protopodite. The protopodite is a two jointed structure : the proximal podomere with which the appendage is attached to the body is *coxopodite* or *coxa,* and the distal joint is known as *basipodite* or *basis*. The rami are leaf like structures, the outer one is called *exopodite,* and inner one is *endopodite*. When there is only one ramus the appendage is called *uniramous* and when both are present it is a *biramous appendage.*

There are 19 pairs of appendages which can be studies under 3 headings.

I. Cephalic appendages.
II. Thoracic appendages.
III. Abdominal appendages.

I. Cephalic appendages

As in all arthropods, there are 6 segments in the head. On the first, there is a pair of compound eyes. The cephalic appendages are not serially homologous with the other appendages, since they arise in a different way. Thus there are 5 pairs of appendages in cephalic region. These are antennules, antennae, mandibles, maxillulae and maxillae.

1. Antennules

These lie just beneath the eye stalks. Each consists of a protopodite, which is made up of three podomeres called *precoxa, coxa* and *basis*. The basal podomere, the precoxa is very large and hollow. The cavity of precoxa lodges the *statocyst,* whose opening is situated near its proximal end. From the outer margin of precoxa arises a spiny lobe, the *stylocerite*. The coxa is short and cylindrical. It bears setae on its inner and anterior borders. The basis is longer than the coxa and lacks setae. It bears a pair of long, slender, many-jointed, tapering processes, the *feelers*, with swollen bases. The inner feeler is unbranched but the outer one is divided into two unequal branches, the inner smaller branch bearing special *olfactory setae*. It is wrong to regard the two feelers as the two rami of biramous limb, perhaps they represent only the endopodite subdivided. The antennules are tactile in function.

2. Antennae

The antennae belong to third segment, each with one feeler, situated immediately below and behind the antennules. The protopodite has two podomeres, the *coxa* and *basis*. Each one is swollen at the base as it lodges the *renal gland* within. The coxa is small in size and bears the renal

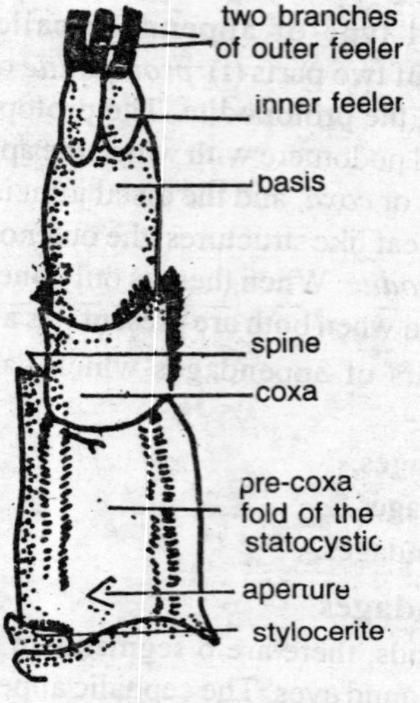

Fig. 2.4. *Palaemon*. Antennule.

aperture on its inner margin. The basis bears two rami: a long, slender, many jointed *feeler* with a swollen base and a broad leaf-like *squama*. Of these the feeler corresponds to the endopodite and the squama the exopodite. Squama acts as a balancer in swimming, is fringed with setae along its inner border. Its outer border is smooth terminating in a small spine in front.

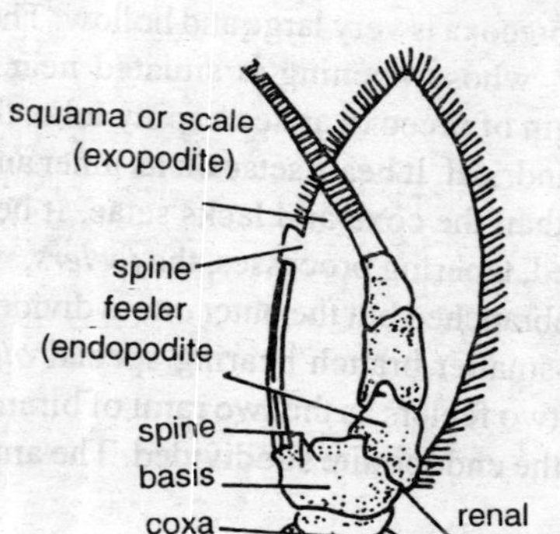

Fig. 2.5. *Palaemon*. Antenna.

3. Mandibles

The mandibles belong to the fourth segment, lie on the sides of the mouth. They are very stout and highly chitinized. The protopodite has two podomeres *coxa* and *basis*. The coxa is very large and can be divided into two parts: proximal, spoon-shaped hollow *apophysis,* and the distal solid, *head*. The head is formed of two processes - a *molar process* having 5-6 yellow dental plates and a plate like *incisor process* ending in three teeth. Each mandible bears three segmented *mandibular palp* that arises from the outer border of the head. The basal segment of the palp represents the basis of protopodite, the outer two segments represent the endopodite, and the exopodite is absent. The mandibles are masticatory in function.

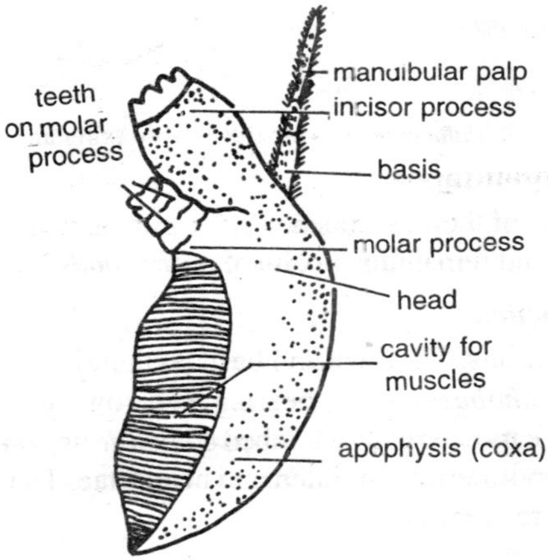

Fig. 2.6 *Palaemon*. Mandible.

4. *Maxillulae*

These are small thin, and leaf-like appendages of 5th segment. Coxa and basis are broad, and project inward as *gnathobases* or *jaws*, fringed with bristles. Endopodite forms a curved process bifurcated at the apex. The exopodite is absent.The maxillulae help in the manipulation of food.

5. *Maxillae*

These are thin, flattened and delicate leaf-like structures present in the last cephalic segment. Coxa is small and incompletely divided into two representing a fusion of precoxa and coxa. The basis is large and produced into a forked and strongly setose gnathobase, the limbs of the fork are called *endites*. Endopodite is small and unjointed but the exopodite is large, flattened and fan-like called *scaphognathite*.The movement of scaphognathite creates a water current passing over the gills. It also bears sensory setae along its free margin. The maxillae help in manipulation and respiration.

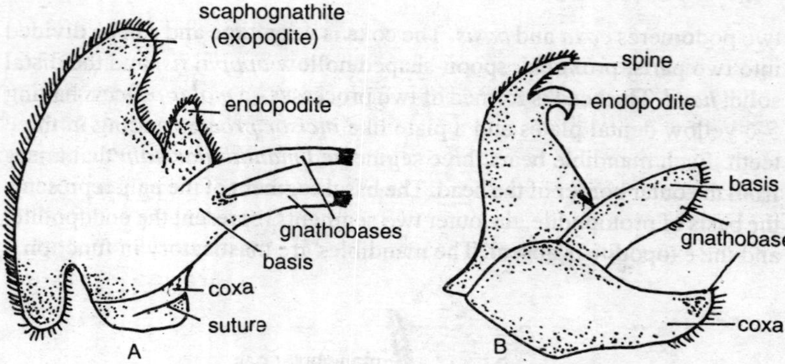

Fig. 2.7. *Palaemon*. A—Maxilla, B—Maxillula.

II. Thoracic Appendages

There are 8 pairs of thoracic appendages in prawn. Anterior 3 pairs are called *maxillipedes* and remaining 5 pairs of *paraeopods* or *walking legs*.

1. First maxillipedes

These are foliaceous. The coxa and basis of their protopodite extend inwards as two *gnathobases*, each bearing two rows of spines on the inner border. Coxa bears externally a bilobed *epipodite* or *primitive gill*.The endopodite and exopodite are unjointed and bear setae. The maxillipedes help in feeding and respiration.

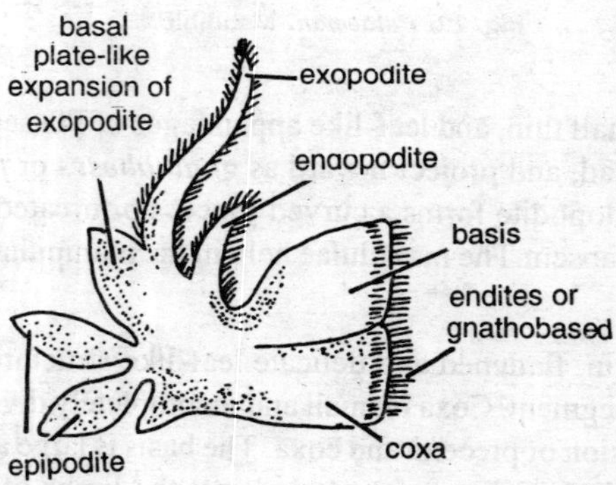

Fig. 2.8. *Palaemon.* I Maxillipede.

2. Second maxillipedes

The protopodite is represented by a coxa, which is short and bears setae on the inner side and a small epipodite and a gill on the outer border.

Basis is short and is immovably joined with the endopodite which consists of five podomeres: (i) *ischium*, (ii) *merus*, (iii) *carpus*, (iv) *propodus* and (v) *dactylus*. Last two are bent backwards and inwards. Exopodite is long, slender, unjointed and is fringed with setae. The functions of second maxillipedes are also the same as those of the first maxillipedes i.e. to help feeding as well as respiration.

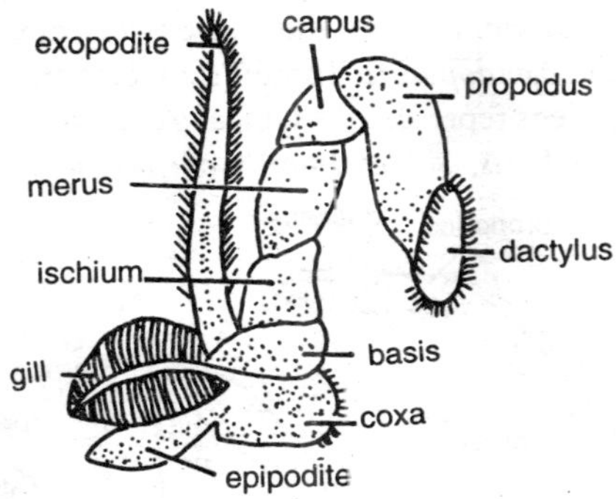

Fig. 2.9. *Palaemon.* II Maxillipede.

3. *Third maxillipedes*

These are leg-like in appearance but possess same parts as 2nd maxillipedes. The coxa bears an epipodite on the outer side and setae

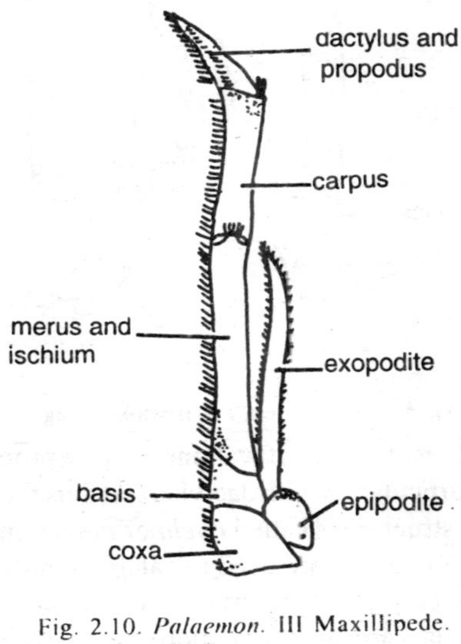

Fig. 2.10. *Palaemon.* III Maxillipede.

along the inner border. The basis carries outer unsegmented *exopodite* and an inner three-jointed *endopodite*. The first podomere represents fused *ischium* and *merus*, the 2nd *carpus* and the third represents fused *propodus* and *dactylus*.

4. Walking legs

The 5 pairs of walking legs differ from the maxillipedes in the absence of *exopodite* and *epipodites*. Thus they have two segmented *protopodite* and a five segmented *endopodite*. These 7 segments are arranged in a row. The fourth pair of legs represent the typical structure. It contains all the 7 podomeres as *coxa, basis, ischium, merus, carpus, propodus* and *dactylus*.

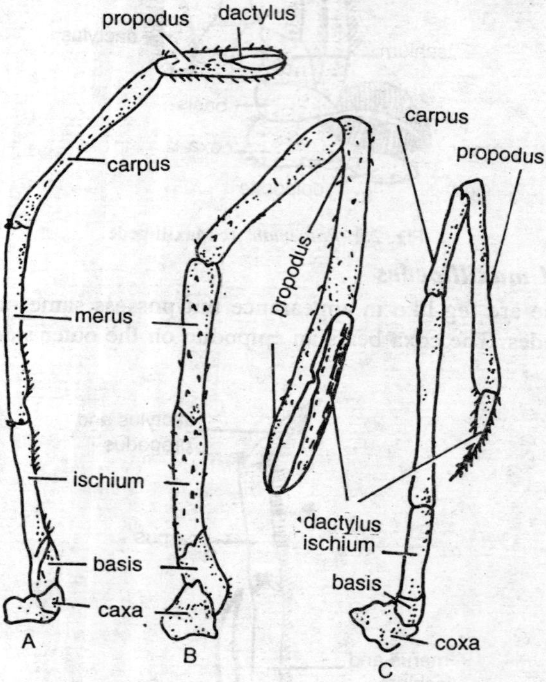

Fig. 2.11. *Palaemon*. A—walking leg, B—II walking leg, C—III walking leg.

In the first and second pairs of walking legs the propodus is prolonged distally beyond its articulation with dactylus and thus forms a long forceps-like structure. The structure is called *chela* or *pincer* and the leg is called *chelate leg*. The chelate legs are used to catch the prey and transfer it to the mouth and also serve as the organs of offence and defence. In male the 2nd pair of chelate legs are larger and more stronger than that of female.

Descriptive Table of the Appendages of *Palaemon*

Appendage	*Protopodite*	*Exopodite*	*Endopodite*	*Function*
I. Antennule	3 segments : statocyst in basal segment	many-jointed filament	many-jointed filament	tactile; chemical; equilibration
II. Antenna	2 segments; excretory pore in basal segment	broad, thin, dagger-like lateral projection	a long many-jointed "feeler"	tactile; chemical
III. Mandible	2 segments; a heavy jaw and basal segment of palp	absent	small; 2 distal segments of palp	crushing food
IV. 1st Maxilla	2 thin lamellae extending inward	absent	a small outer lamella	
V. 2nd Maxilla	2 bilobed lamellae	dorsal half of plate, the scaphognathite	1 segment; small, pointed	creates current of water in gill chamber
VI. 1st Maxillipede	2 thin segments extending inward; a broad plate, the epipodite extending outward	a long basal segment bearing a many-jointed filament	small; 2 segments	chemical; tactile; holds food
VII. 2nd Maxillipede	2 segments; a basal coxopodite bearing a gill, and a basipodite bearing the exopodite and endopodite	similar to VI	5 segments; the basal one long and fused with the basipodite	similar to VI

Appendage	*Protopodite*	*Exopodite*	*Endopodite*	*Function*
VIII. 3nd Maxillipede	similar to VII	smilar to VI	similar to VII, but larger	similar to VI
IX. 1st Walking Leg (Chela Cheliped, or Pincher)	2 segments; coxopodite, and basipodite	absent	5 segments, the terminal two forming a powerful pincher	offense and defense; aids in walking; tactile
X. 2nd Walking Leg (Pereiopod)	similar to IX	absent	as in IX, but not so heavy	walking; prehension; toilet implements
XI. 3rd Walking Leg	similar to IX; coxopodite of female contains genital pore	absent	similar to X	similar to X
XII. 4th Walking Leg	similar to IX	absent	similar to X, but no pincher at end	walking
XIII. 5th Walking Leg	similar to IX; coxopodite of male bears genital pore	absent	similar to XII	walking; cleaning abdomen and eggs
XIV. 1st Abdominal (1st Pleopod or Swimmeret)				reduced in female; in male, protopodite and endopodite fused together forming an organ for transferring sperm

Appendage	*Protopodite*	*Exopodite*	*Endopodite*	*Function*
XV. 2nd Abdominal (2nd pleopod or Swimmeret)	in female 2 segments	in female many-jointed filament	in female like exopodite but longer	in female as in XVI, in male modified for transferring sperm to female
XVI. 3rd Abdominal (3rd pleopod or Swimmeret)	2 segments	many-jointed filament	like exopodite but longer	creates current of water; in female used for attachment of eggs and young
XVII. 4th Abdominal (4th Pleopod or Swimmeret)	2 segments	as in XVI	as in XVI	as in XVI
XVIII. 5th abdominal (5th Pleopod or Swimmeret)	as in XVII	as in XVI	as in XVI	as in XVI
XIX. 6th Abdominal (Uropod)	1 short, broad segment	flat oval plate divided by transverse groove into two parts	flat oval plate	swimming

The third, fourth and fifth pairs of walking legs are typical and non-chelate. In female, each of the third leg bears a crescentic female genital aperture on the inner side of its coxa, while in male, each of the fifth leg bears a slit-like male genital aperture on its inner side of the arthrodial membrane connecting the leg and thorax.

III. Abdominal appendages

There are 6 pairs of abdominal appendages. These are collectively called *pleopods* or *swimmerets.* These are biramous appendages and help in swimming.

In a typical swimmeret (the 3rd abdominal appendage) the protopodite consists of two podomeres, a short ring-like segment devoid of setae is known as the coxa and the basis which is cylindrical, bears setae on its surface and carries exopodite and endopodite. The endopodite is a leaf-like structure and gives to a small, rod-like structure, the *appendix interna*, with a knob-like head bearing many hook-like processes.

Differences from the typical swimmeret described above, are to be seen in the 1st, 2nd and 6th abdominal appendages of male and 1st and 6th of the female.

First abdominal appendages

In these appendages the endopodite is reduced and *appendix interna* is absent.

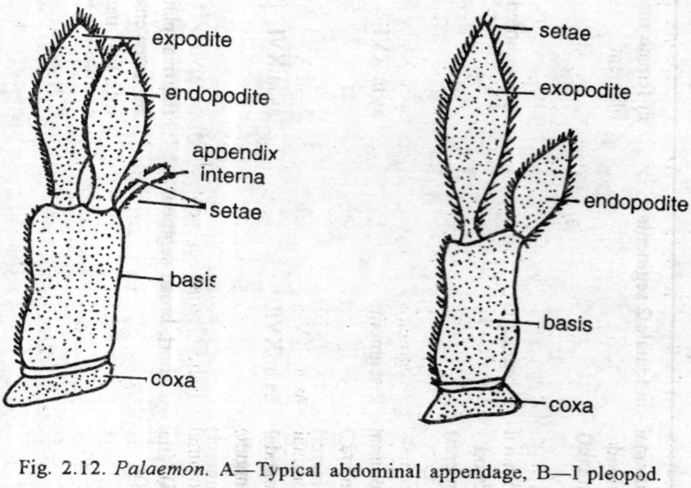

Fig. 2.12. *Palaemon.* A—Typical abdominal appendage, B—I pleopod.

Second abdominal appendages of male

These are typical in female but in male the appendix interna gives out a rod-shaped process on its inner side called *appendix masculina.* It is covered with setae.

Sixth abdominal appendages

The sixth pair of abdominal appendages, called *uropods*, are large and lie one on either side of the telson forming tail-fin which enables the prawn to take a backward stearing in water. The coxa and basis fuse together to form a triangular *sympod*, bearing the oar-shaped endopodite and exopodite. The exopodite is bigger than the endopodite and incompletely divided in the middle by a transverse suture. Their margins, except the outer border of the exopodite, are fringed with numerous setae.

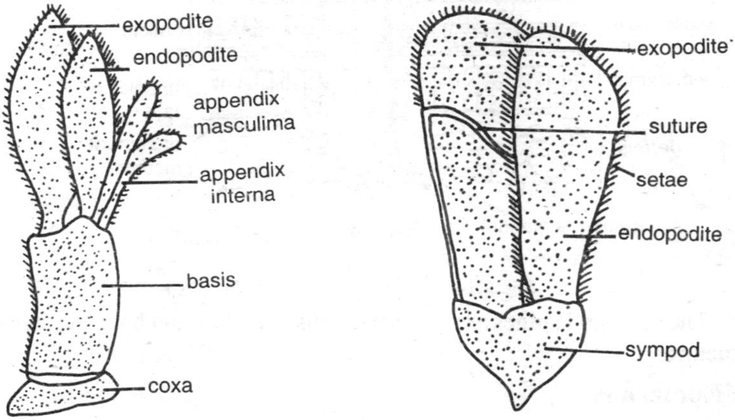

Fig. 2.13. *Palaemon*. A—II pleopod, B—Uropod.

BODY WALL

The body wall of *Palaemon* consists of an outer layer of *cuticle*, a middle-single layered *epidermis* and an inner *dermis*.

1. Cuticle

The cuticle of the body wall has become transformed into the exoskeleton which is thick and rigid at some places and thin and membranous at other. Cuticle is composed of two layers: (i) The outer layer is a thin non-chitinous *epicuticle*. It is yellow hyaline layer. It is further made up of two layers. Outer extremely thin made up of lipids and inner relatively thick protein layer. The protein layer is hardened and pigmented by its combination with an oxidised phenol. The epicuticle is secreted by the tegumental gland cells present is dermis. (ii) The *endocuticle* contains layers of chitin and has three successive layers : the outermost layer is a thin pigmented layer, below it is thick calcified layer and under it is an uncalcified layer. The pigmented layer is dark due to protein deposits and it becomes hardened by a process of tanning or sclerotization. In the calcified layers are deposits of carbonates and phosphates of calcium which make the integument hard. The chitin is an acetate of a polysaccharide containing glycosamine secreted by the underlying epidermis. The cuticle is relatively impermeable, except where it is thin and allows the passage of gases or absorption of water.

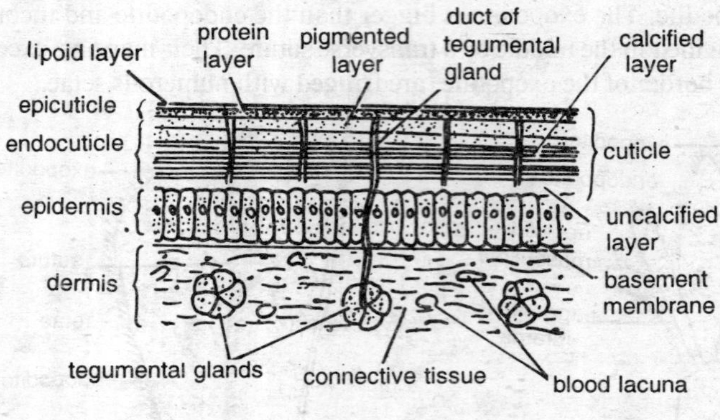

Fig. 2.14. *Palaemon.* V. S. of Skin.

The seate and spines are the outgrowths of cuticle and have the same structure.

2. Epidermis

It comprises a single layer of glandular columnar epithelium resting on a thin basement membrane which is secreted by the underlying dermis. The epidermis is often called *chitogenous* epithelium as it secretes chitin.

3. Dermis

The dermis is thin and is made up of loose connective tissue beast with blood lacunae. It contains three types of tegumental glands that open out by fine ductules passing through overlying epidermis and the cuticle.

BODY CAVITY

In *Palaemon* the body cavity is mostly occupied by musculature and organs (having interspaces containing blood) and all these together form a body cavity or *haemocoel*. However, true coelom is restricted to kidneys and gonads.

MUSCULATURE

The bulk of the body of the *Palaemon* consists mainly of muscles which are enclosed within the exoskeleton. These muscles are voluntary or striated type. The unstriated muscles are found in the organs such as blood vascular system and digestive system. There are two sets of muscles in prawn, the *appendicular muscles* that are segmentally and the *abdominal muscles* for straightening and flexure of the abdomen. The muscles are arranged in opposing series. By the contraction of one set the limbs are flexed (*flexor muscles*) and by the contraction of other, the limbs return to their normal position (*extensor muscles*). In the abdomen too such muscles are found. In the terminal sclerite of chelate leg, the flexor muscle become *adductor*, closing the chela, and the extensor, *abductor* to open the chela.

The extensor of abdominal muscles include a pair of *dorsal extensor*, four pairs of inter-tergal *abdominal muscles*, a pair of *intertergal muscles* and two pairs of *extensor* of the telson.

The flexors of abdominal muscles include 5 pairs of main flexor (first 5-segments), four pairs of *accessory flexor* muscles (first 4 segments), a pair of flexor (between first abdominal segment and thorax), two pairs of *lateral thoraco-abdominal* muscles, two pairs of *flexor muscles* of telson and a pair of *ventral superficial flexor* muscles.

All the appendages have their own muscles.

ENDOSKELETON

Palaemon has an endoskeleton in the cephalothorax formed of a series of ingrowths of the exoskeleton called *apodemes* or *endophragma*.

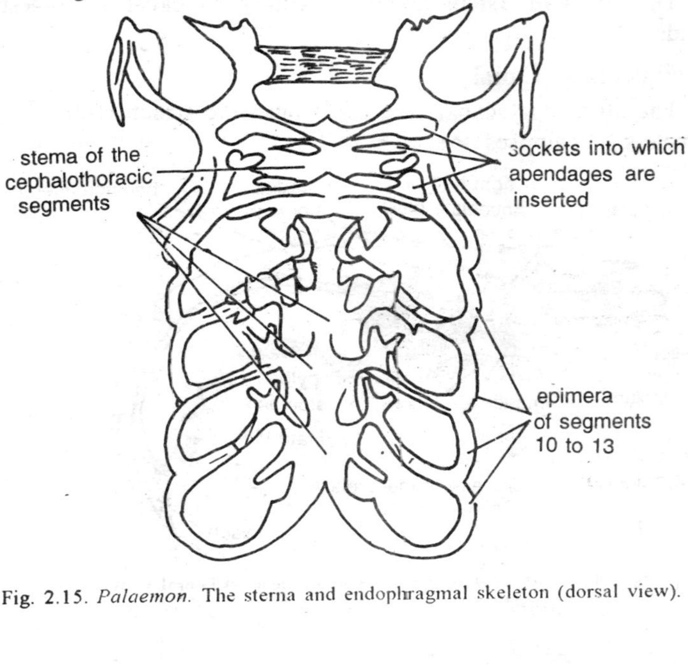

Fig. 2.15. *Palaemon*. The sterna and endophragmal skeleton (dorsal view).

These ingrowths are connected together to form a regular framework that is known as *endophragmal skeleton.*

The endophragmal skeleton is well developed in the segments 11-13, where, in each segment, it consists of an outgrowth from the antero-lateral end of epimeron of the same sternal segment called *endopleurite posterior.*

The endopleurite posterior meets an ingrowth from the posterior-lateral end of the epimeron of the segment in front, called *endopleurite anterior.* These in turn meet respective ingrowths of the sternum of the same segment an *endosternite posterior* and an *endosternite anterior*. From the mesosternite, on each side arises an upwardly directed 'Y' shaped plate. The inner limb of the Y is called *mesophragm* and outer *paraphragm*. These apodemes provide space for attachment of abdominal flexor muscles while other apodemes provide space for the attachment of appendicular muscles. A typical apodeme with complete number of endosternites, endopleurites and Y shaped structure occurs only between 11th, 12th and 13th segments. In other segments the apodeme is incomplete.

In mandibular region there is a H-shaped structure, the *cephalic apodeme*, for the insertion of muscles of branchiostegites, small gastric muscles and the muscles of mandibles.

DIGESTIVE SYSTEM

The digestive system comprises, alimentary canal and digestive glands.

1. Alimentary Canal

The alimentary canal is straight tube-like structure of varying diameters. It comprises :

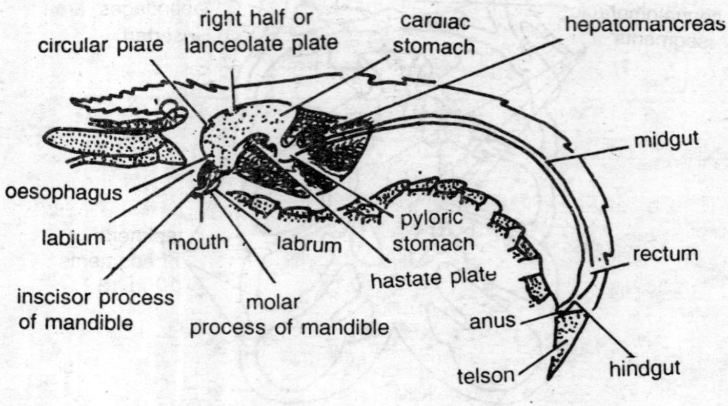

Fig. 2.16. *Palaemon.* Alimentary canal in lateral view.

1. Foregut or stomodaeum.
2. Midgut or mesenteron
3. Hindgut or proctodaeum.

1. Foregut

The foregut comprises the mouth, buccal cavity, oesophagus and stomach. Foregut is ectodermal in origin hence it is internally lined by ectoderm and cuticle.

(i) *Mouth.* The mouth is large slit-like aperture situated on the ventral side of the head below the anterior end. It is bounded anteriorly by a shield shaped *labrum* or upper lip, laterally by the *incisor processes* of mandibles and posteriorly by a thin *labium* or lower lip which is bilobed. Mouth leads into the buccal cavity.

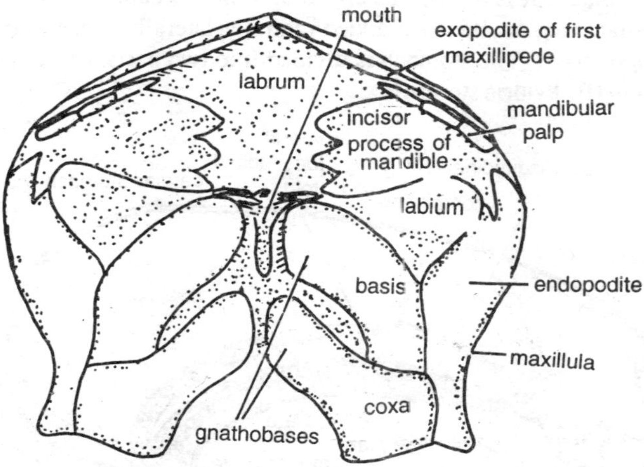

Fig. 2.17. *Palaemon.* Ventral view of oral region.

(ii) *Buccal cavity.* The buccal cavity is an antero-posteriorly compressed small chamber with irregularly folded thick chitinous lining. The molar processes of mandibles project into it and lie opposite each other.

(iii) *Oesophagus.* The oesophagus is a short, wide, vertical tube joined with the buccal cavity and stomach. Its inner wall is muscular and is produced into four longitudinal folds : an anterior fold which is short, a posterior fold and two lateral folds, which are large and more prominent.

The internal cuticle of oesophagus is covered with various bristles. Oesophagus opens into the stomach; the opening is provided with a valve like structure.

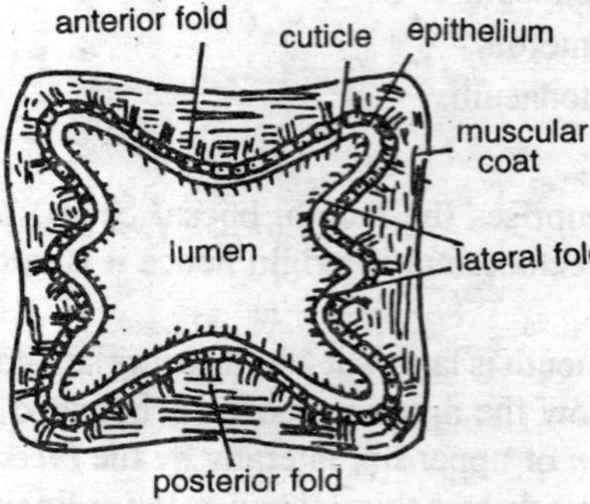

Fig. 2.18. *Palaemon*. T.S. of oesophagus.

(iv) *Stomach*. It is a spacious chamber which occupies most of the cephalothoracic region. It is surrounded ventrally, laterally and posteriorly by *hepatopancreas*. The stomach is divided into two parts. (A) Cardiac stomach and (B) Pyloric stomach.

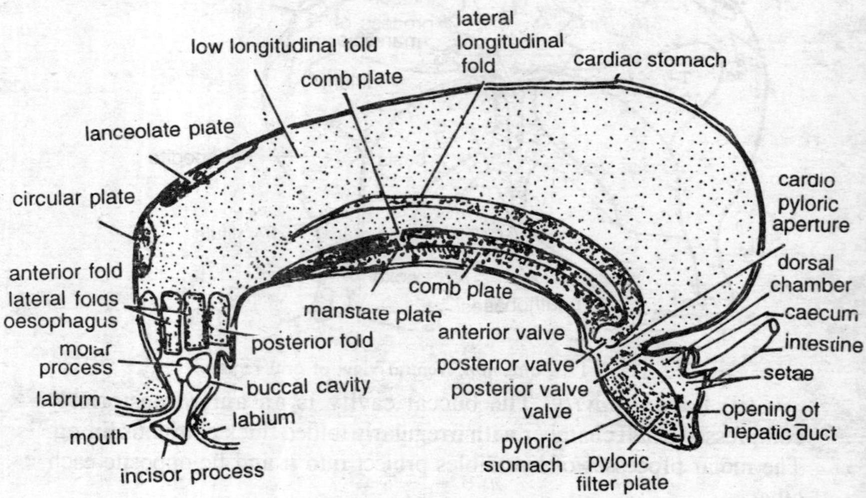

Fig. 2.19. *Palaemon*. V.L.S. of foregut.

(A) Cardiac stomach. It is the anterior sac-like part of the stomach in which the layer of cuticle on the inner surface is raised into many fine and indistinct longitudinal folds which bear fine bristles. The wall of stomach is supported by a number of cuticular plates. These plates are as follows :

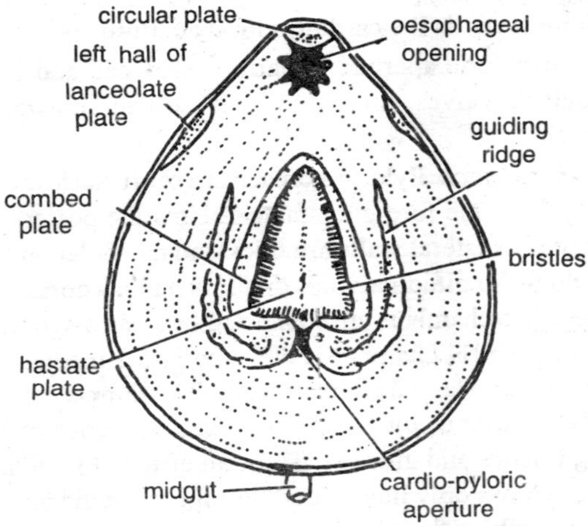

Fig. 2.20. *Palaemon.* Floor of cardiac stomach.

(a) *Circular plate* is present in front of oesophageal aperture forming its anterior wall.

(b) *Lanceolate plate* is embedded mesially in the roof of the stomach just behind the circular plate.

(c) *Hastate plate* is a triangular plate embedded in the middle of floor of stomach. It is raised in the middle with sloping sides. Its upper surface is covered with a thick growth of delicate setae. The posterior broad margin is depressed and fringed with setae to form the anterior valve of the cardio-pyloric aperture. The lateral margins of the hastate plate are supported by two circular supporting rods. A narrow lateral groove runs along the outer border of each supporting rod.

(d) *Groove plates* form the floor of lateral grooves and are in the form of open drain pipes.

(e) *Combed plate.* Each lateral groove on its outer side is supported by a long ridge plate or combed plate. The two combed plates are united anteriorly but remain separate posteriorly by cardio-pyloric apertures. The inner border of each combed plate is beset with a row of long delicate bristles which bridge over the lateral groove and partially overlap the lateral margins of the hastate plate. The bristles remain in constant movement.

Just outside the combed plates, the wall of cardiac stomach is folded on each lateral side to form guiding *ridges* or *lateral longitudinal folds*. These ridges are low in front but gradually increase in high behind, but posteriorly the high walls of these folds bend inwards over the cardiopyloric aperture, forming lateral valves. Thus these valves guard the passage of food towards the cardio-pyloric aperture.

The cardio-pyloric aperture is narrow x-shaped and is guarded by four cardiopyloric valves, through which cardiac stomach opens into pyloric stomach.

(B) Pyloric stomach. Pyloric stomach is a short, narrow chamber. It is the posterior part of the fore-gut situated below the posterior end of the cardiac stomach. Its lateral walls are thick and muscular and project into the lumen as thick, longitudinal folds dividing it into a dorsal and a ventral chamber. These two chambers are connected by a narrow vertical passage. The floor of the ventral chamber is elevated into a median longitudinal ridge so as to form two lateral compartments. The floor is provided with an 'Λ' shaped *filter plate* or *pyloric filter*. It bears a series of longitudinal ridges having bristles and grooves. Thus, an efficient *strainer* or *filter* is formed, which allows only liquid food through it. At the posterior end of the chamber just behind the pyloric filter, there is a pair of small openings by which digestive juice from the hepatopancreatic duct is poured into the pyloric stomach. These openings are provided by posteriorly directed setae, which act as valves to control the direction of the flow of the secretion.

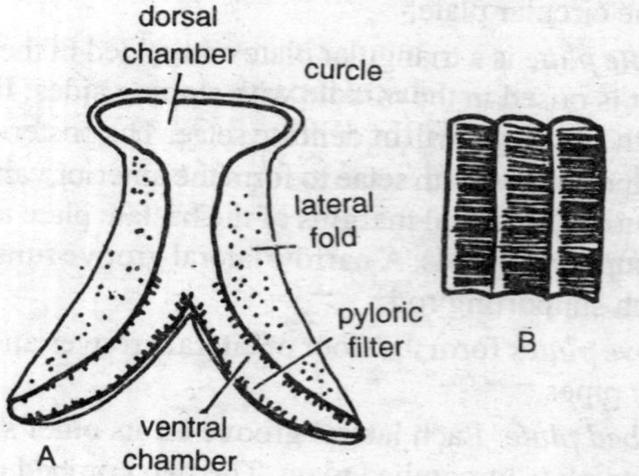

Fig. 2.21. *Palaemon.* A—T.S. pyloric stomach B—A part of pyloric stomach.

The dorsal chamber gives out a small blind caecum dorsally and opens into the mid-gut. The junction of dorsal chamber and midgut is guarded by the median dorsal and two lateral groups of elongated setae. These setae control the flow of food material.

2. Midgut or mesenteron

The midgut is a long, narrow, straight and slender tube running back along the median line above the mass of the ventral abdominal muscles upto the sixth abdominal segment. The internal lining of midgut is formed by epithelium which in the posterior part is thrown into many longitudinal folds thus greatly reducing its lumen.

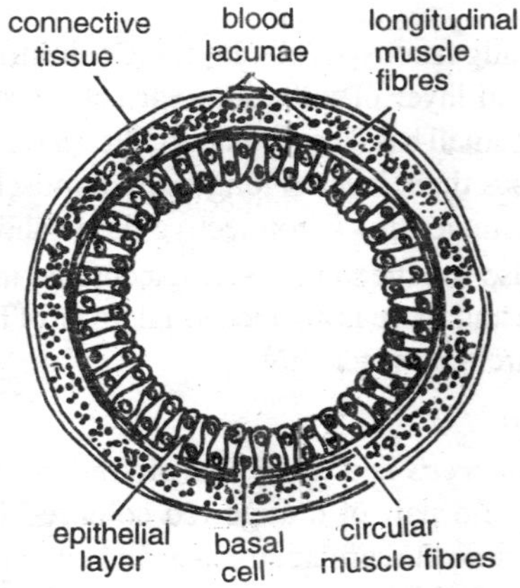

Fig. 2.22. *Palaemon*. T.S. of mesenteron.

3. Hindgut or proctodaeum

It is the last part of alimentary canal which is internally lined with cuticle. Its anterior enlarged muscular part is called the *intestinal bulb* and the posterior less wide part is the *rectum*. It opens out through the anus which is placed at the base of the telson.

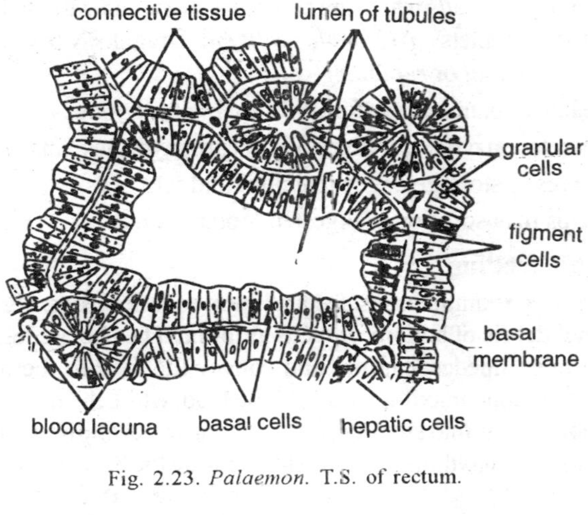

Fig. 2.23. *Palaemon*. T.S. of rectum.

Anus

It is a longitudinal slit situated on a raised papilla at the base of the *telson*. It is surrounded by circular and radial muscle fibres. The former forms the sphincter and the later acts as dilator.

Histology

The wall of midgut consists of 4 layers. Beginning from inside, these are (i) An epithelial layer of tall columnar cells resting on a basement membrane, having small basal or replacing cells. (ii) A thin layer of circular muscle fibres. (iii) A thick layer of longitudinal muscle fibres enclosing a plexus of blood lacunae. (iv) A connective tissue sheath.

The rectum also has the same layers except that its epithelial lining is covered by cuticle and there is an additional layer of longitudinal muscle fibres inside the circular muscle fibres.

Digestive gland

The *hepatopancreas* is a large and physiologically very important gland. It consists of a pair of orange-red coloured lobes lying close to each other surrounding the cardiac stomach and has a massive appearance. It is a racemose gland which arises as a pair of tubular outgrowths from the midgut. These outgrowths grow, divide and redivide, thus forming a compact network which is held together by connective tissue. The wall of each tubule forming the gland is made up of four layers : the inner most layer of *epithelial cells*, a thin *basement membrane* on which epithelial cells rest and the outer most layer or *tunica properia* formed of a network of connective tissue and *muscle fibres*. The epithelial layer has four kinds of cells (i) *Hepatic cells* which are columnar shape and with their cytoplasm filled with fat droplets (ii) *Granular cells* (iii) Irregularly scattered *ferment cells* and (iv) *Basal* or *replacing cells*.

The hepatopancreas performs following functions :

1. It secretes enzymes for digestion of carbohydrates, fats and proteins.
2. It serves to store glycogen, calcium and fat.
3. It helps in absorption of digested food material.

Food and feeding

Palaemon is omnivorous, feeding upon aquatic weeds, algae, mosses, insects and debris of bottom. It feeds at night. The chelate legs, aided by the third maxillipedes, capture and convey the food to the mouth. The coxae of the second maxillipedes hold the food, while the incisor processes of mandibles cut it into smaller pieces which are swallowed with the help of maxillipedes, maxillulae and maxillae. Inside the buccal cavity the molar

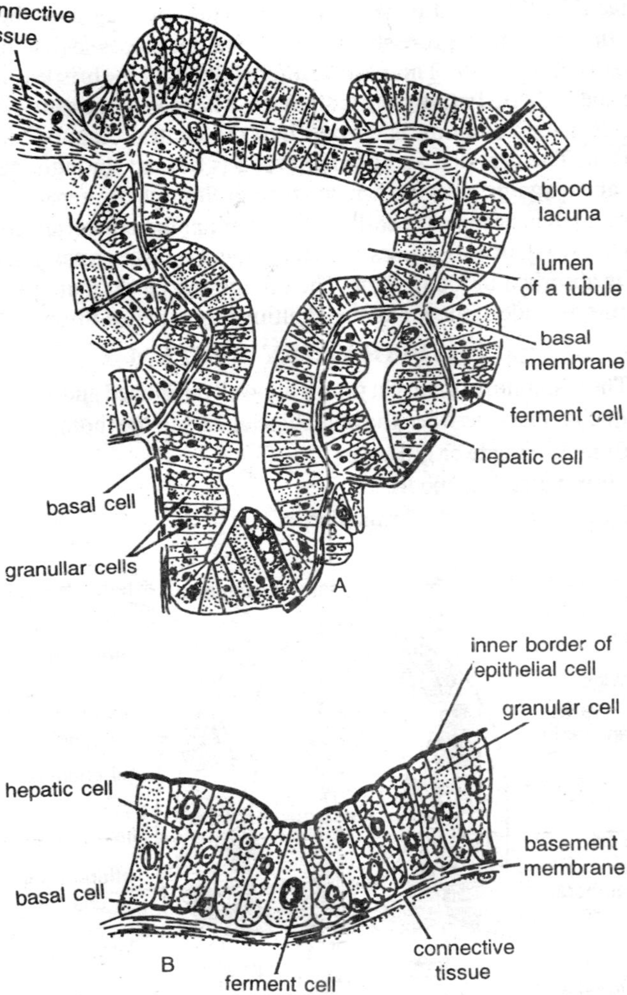

Fig. 2.24. *Palaemon*. A—V.S. of a part of hepatopancreas, B—A part of the wall of hepatopancreatic tubule.

processes of mandibles masticate the food, which is then conveyed to the cardiac stomach through the oesophagus.

Digestion

The digestion of food starts in the cardiac stomach where the hepatopancreatic juice from the pyloric stomach ascends up and mixes well with the food. Due to the alternate expansion and contraction of stomach wall the food is well churned. While passing over the hastate plate the food particles are still reduced to minor pieces so that the food is converted into a paste. This paste is filtered throught the bristles of combed plate and is carried to the pyloric stomach. The food in the semi-liquid and semi-digested condition is carried into the ventral chamber of pyloric stomach. The digested and liquified food is filtered through the pyloric filter and it again enters the hepatopancreas through the hepatopancreatic ducts and then becomes absorbed. The residual food material, consisting of undigested particles, passes on to the midgut and the remaining digested food is absorbed in the midgut. The undigested faecal matter pass on to the posterior side of the hindgut and ultimately expelled through the anus.

RESPIRATORY SYSTEM

The respiratory organs in prawn are well developed and are located in the gill chambers, between the branchiostegite and the thorax. These are :

1. Branchiostegite or gill cover.
2. Three pairs of epipodites.
3. Eight pairs of gills or branchiae.

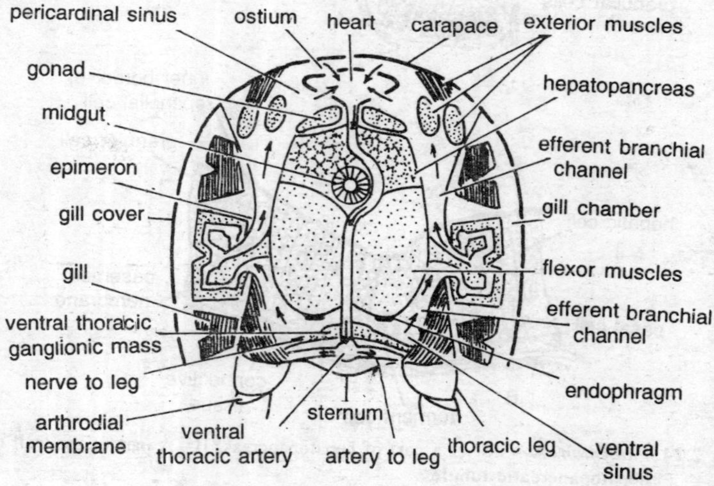

Fig. 2.25. *Palaemon.* Section through cephalothorax showing gill chamber.

1. Branchiostegite

The inner lining of branchiostegites is thin, membranous and highly vascular containing minute blood lacunae. These are constantly bathed in fresh water, thus forming the respiratory surface through which exchange of gases takes place between blood and water.

2. Epipodites

These are thin, leaf-like processes, one attached to each maxillipede. These epipodites are full of blood lacunae and, infact, constitute primitive type of gills. The epipodites lie in the anterior part of the gill-chamber below the scaphognathites. The epipodites of first pair are bilobed and larger than the others. Exchange of gases between blood and lacunae takes place between water and blood in epipodites.

3. Gills

There are 8 pairs of gills. These arise from the modification of epipodites and according to their position, they can be distinguished into three types :

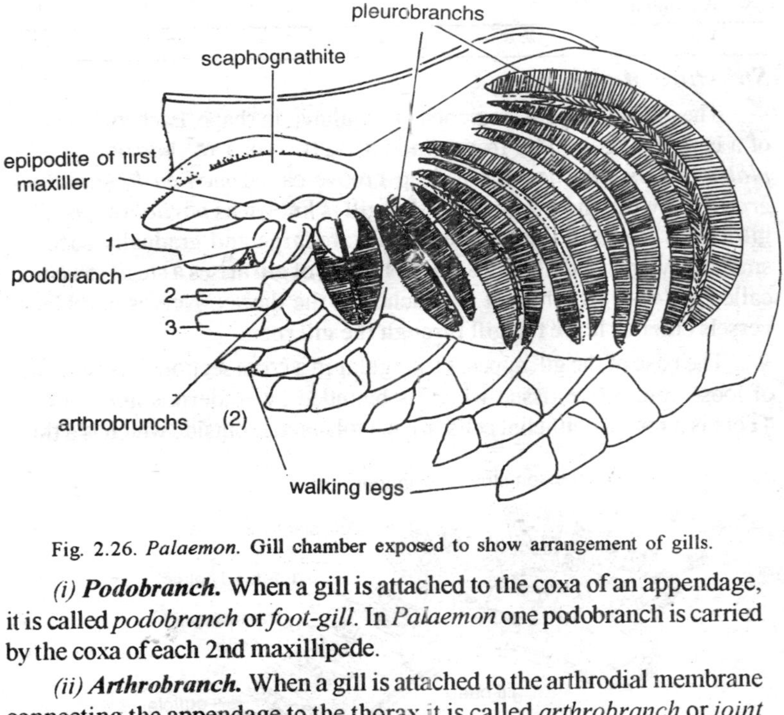

Fig. 2.26. *Palaemon.* **Gill chamber exposed to show arrangement of gills.**

(i) ***Podobranch.*** When a gill is attached to the coxa of an appendage, it is called *podobranch* or *foot-gill*. In *Palaemon* one podobranch is carried by the coxa of each 2nd maxillipede.

(ii) ***Arthrobranch.*** When a gill is attached to the arthrodial membrane connecting the appendage to the thorax it is called *arthrobranch* or *joint gill*. Each 3rd maxillipede bears two arthrobranches.

(iii) ***Pleurobranch.*** When a gill is attached to the lateral wall of the segment bearing limb, it is called *pleurobranch* or *side gill*. The last five gills on each side are pleurobranch attached to the lateral wall of the thoracic segments bearing five walking legs.

Branchial formula. The number and position of epipodites and gills can be illustrated by the branchial formula which is given in the following table.

	Appendage	*Epipodite*	*Podobranch*	*Arthrobranch*	*Pleurobranch*	*Total*
I	Maxillipede	1	-	-	-	1
II	Maxillipede	1	1	-	-	2
III	Maxillipede	1	-	2	-	3
I	Walking leg	-	-	-	1	1
II	Walking leg	-	-	-	1	1
III	Walking leg	-	-	-	1	1
IV	Walking leg	-	-	-	1	1
V	Walking leg	-	-	-	1	1
	Total	3	1	2	5	11

Structure of gill

The gills are almost crescentic or semilunar in shape. Each gill consists of a central axis or base from which arise two rows of rhomboidal *gill-plates* or *lamellae*, leaving a deep groove called *median longitudinal groove* between the two rows, such a gill is known as *phyllobranch*. The gill plates found in the middle of gills are large and gradually become smaller towards both the ends. In the middle of gill arises a small structure called *gill-root* by which it is attached to the thorax. Nerves and blood vessels enter or leave the gill through the gill roots.

The base of the gill appears triangular in a cross section. It is made up of loose conncetive tissue which is bounded by epidermis and cuticle. There is a row of epithelial cells in each gill-plates, outside which is a thin

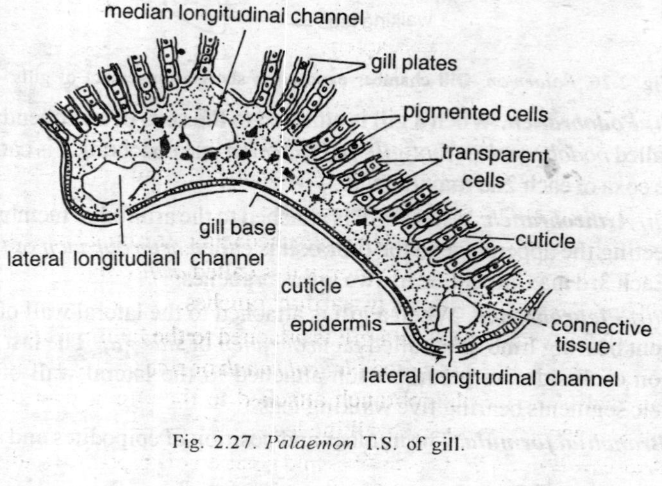

Fig. 2.27. *Palaemon* T.S. of gill.

layer of cuticle. The cells are of two types viz. *pigmented* and *transparent* alternating with each other.

Circulation of blood in gills

The gills are profusely supplied with blood. Three blood vessels run vertically in each axis - two *lateral longitudinal channels* along the two lateral margins and a *median longitudinal channel* in the middle. The lateral longitudinal channels are connected with each other by *transverse connective*, one of which, near the gill-root, communicates with the *afferent branchial channel* and receives blood from it. Each lateral channel gives a delicate *marginal channel* into each gill-plate which runs along the outer margin of the gill-plate and finally opens into the median longitudinal channel. Blood from the afferent branchial channel passes into the transverse connective from where it goes into the lateral longitudinal channel, then through the marginal channel into the median longitudinal channel which finally pours the blood, at the gill-root, into the *efferent branchial channel* by a short connection.

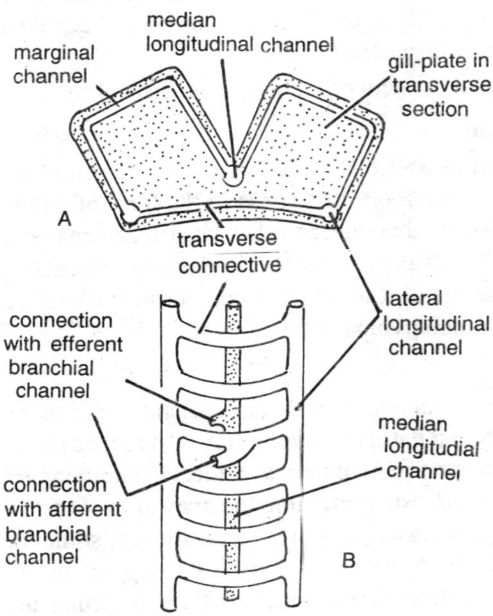

Fig. 2.28. *Palaemon*. A—Blood supply in gill plates, B—Blood channel in gill.

Mechanism of respiration

The exopodite of maxilla, called scaphognathite, is so situated that its anterior end penetrates into gill chamber and the posterior end lies outside. The gill chamber itself is covered dorsally and laterally but is open ventrally. The vibrating action of the scaphognathite bales out water from the anterior end of the chamber and simultaneously pushed a water current movement of the baler resuts in continuous entry of fresh water from behind and its subsequent exit from front. The incoming water current not only passes over the gills, epipodites but also washes the branchiostegites. The oxygen dissolved in water is taken into the blood and carbon dioxide passes out. The respiratory pigment, *haemocyanin* remains dissolved in blood. However, the haemocyanin content in the blood is low, therefore, the major part of oxygen is taken in as simple solution in plasma.

CIRCULATORY SYSTEM

In prawn the circulatory system is *open* or *lacunar type*, in which the veins and capillaries are totally absent. The blood vessels (arteries) open into spaces. These spaces are without a proper epithelial lining and are called *lacunae* or *sinuses.* The sinuses together form the body-cavity which looks like a coelom, but filled with blood and devoid of epithelial lining, is known as *haemocoel.*

The circulatory system of *Palaemon* consists of :

1. Pericardium

It is a wide thin-walled chamber around the heart and remains filled with *haemo-coelomic fluid.* Its floor is in the form of a thin horizontal septum which lies just above the gonad and hepatopancreas and is attached to the dorsal body wall along its anterior and posterior aspects. The septum is attached to the lateral walls of the thorax along the lateral aspect. The pericardial cavity pumps blood into the heart.

2. Heart

The heart is a triangular, muscular structure with its broader base facing posteriorly and the apex in front. It is found in the pericardial sinus along the median dorsal part of thorax. A median longitudinal strand, the *cardiopyloric strand*, extending from the apex of the heart to the dorsal wall of the pyloric stomach, and two lateral strands, stratching from the lateral angles of the heart to the body wall, support the heart in the pericardium. The walls of the heart are very thick, muscular and extermely elastic. There are 5 pairs of valvular openings, the *ostia,* in the heart. The first pair of ostia is situated in the mid-dorsal region. The second pair is found mid-ventrally. The third pair lies on the posterior border. The fourth pair lie anterio-laterally and the fifth pair posterio-laterally. The cavity of heart is not continuous but is traversed by a large number of interlocking muscle fibres.

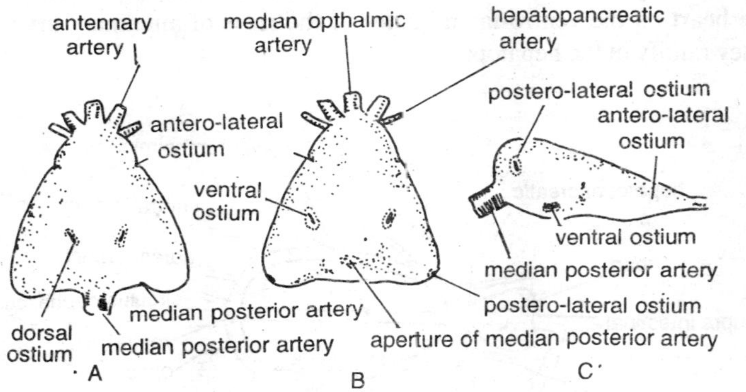

Fig. 2.29. *Palaemon.* Heart A—Dorsal view, B—Ventral view, C—Lateral view.

3. Arteries

The arteries are thick-walled muscular tubes, which arise from the heart and supply blood to different organs of the body. The main arteries are as follow :

(i) Median ophthalmic artery. It arises as a single median cephalic or ophthalmic artery from the apex of the heart. It runs forward to the head region in the mid-dorsal line above the renal sac. Here it joins the two antennary arteries.

(ii) Antennary arteries. A pair of these arteries arise from the apex of the heart one from either side of the median ophthalmic artery. Each runs obliquely forward along the outer border of the mandibular muscle. During its course it gives out the following branches :

(a) *Pericardial branch* to the pericardium,

(b) *Gastric branch* to the cardiac stomach, and

(c) *Mandibular branch* to the mandibular muscles.

Each antennary artery passes forwards and divides into two sub-branches.

(a) *Dorsal branch.* It sends an *optic artery* to the eye and then bends inward to meet its fellow of the opposite side forming a circular loop-like artery the *circular cephalicus,* which gives off a pair of *rostral arteries* to the rostrum.

(b) *Ventral branch.* It also divides into sub-branches, one of which supplies the antennule called *antennulary*. Second redivides into two : an *antennal artery* to antenna, and other *renal artery* to the renal organ.

(iii) ***Hepatic arteries.*** These are also paired arteries, which arise from the heart on the ventro-lateral sides of the roots of antennary arteries. They ramify in the hepatopancreas.

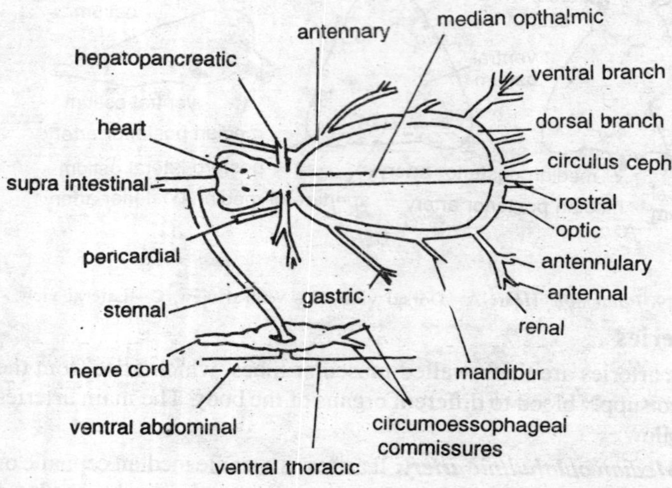

Fig. 2.30. *Palaemon.* Heart, main arteries, sinuses and veins.

(iv) ***Median posterior artery.*** It is small but stout artery which arises from the posterio-ventral surface of the heart and runs backwards. It immediately bifurcates into two branches.

(a) *Supra-intestinal artery.* The supra-intestinal artery passes straight backwards along the dorsal surface of the intestine extending up to the hind-gut, while it divides into two branches supplying blood to the intestine and the dorsal abdominal muscles.

(b) *Sternal artery.* The sternal artery is the stoutest of all others. It runs downwards and forwards through the hepatopancreas and after passing through an aperture in the middle of the ventral thoracic ganglionic mass to reach the ventral side, it gets divided into two branches. The ventral *thoracic branch* runs anteriorly up to the mouth and supplies the sternal region, first three pair of walking legs, maxillae, and maxillulae, etc. The *ventral abdominal branch* runs posteriorly upto the anus and supplies blood to the ventral abdominal region, last two pairs of legs, pleopods and hind- gut, etc.

4. Blood Sinuses

The various arteries, on reaching the organs, branch repeatedly into minute branches which do not form capillaries but end into wide spaces, *blood sinuses* or *blood lacunae*. All the blood lacunae of the body collectively form a pair of elongated but ill-defined *ventral sinuses*. These lie below the hepatopancreas and above the floor of thorax. The ventral sinuses communicate with each other at various places.

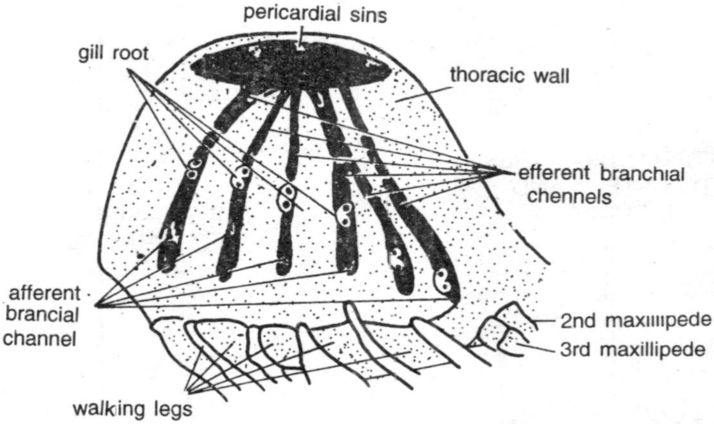

Fig. 2.31. *Palaemon*. Branchial channels.

5. Blood Channels

In *Palaemon*, there are 6 pairs of *afferent branchial channels* and 6 pairs of *efferent branchial channels*. These channels are lacunar in nature. The afferent branchial channels run along the inner surface of the thoracic wall and reach the gills through the gill-roots. Here each channel opens into the tranverse connective. The first afferent channel takes the blood to the podobranch and two arthrobranches. Remaining five channels supply the blood to the five pleurobranches. The blood is oxygenated while passing through the gills and this blood is carried to the pericardium by efferent branchial channels.

Course of blood circulation. It can be represented diagrammatically as shown below.

6. Blood

The blood is a thin, watery, almost colourless liquid containing floating

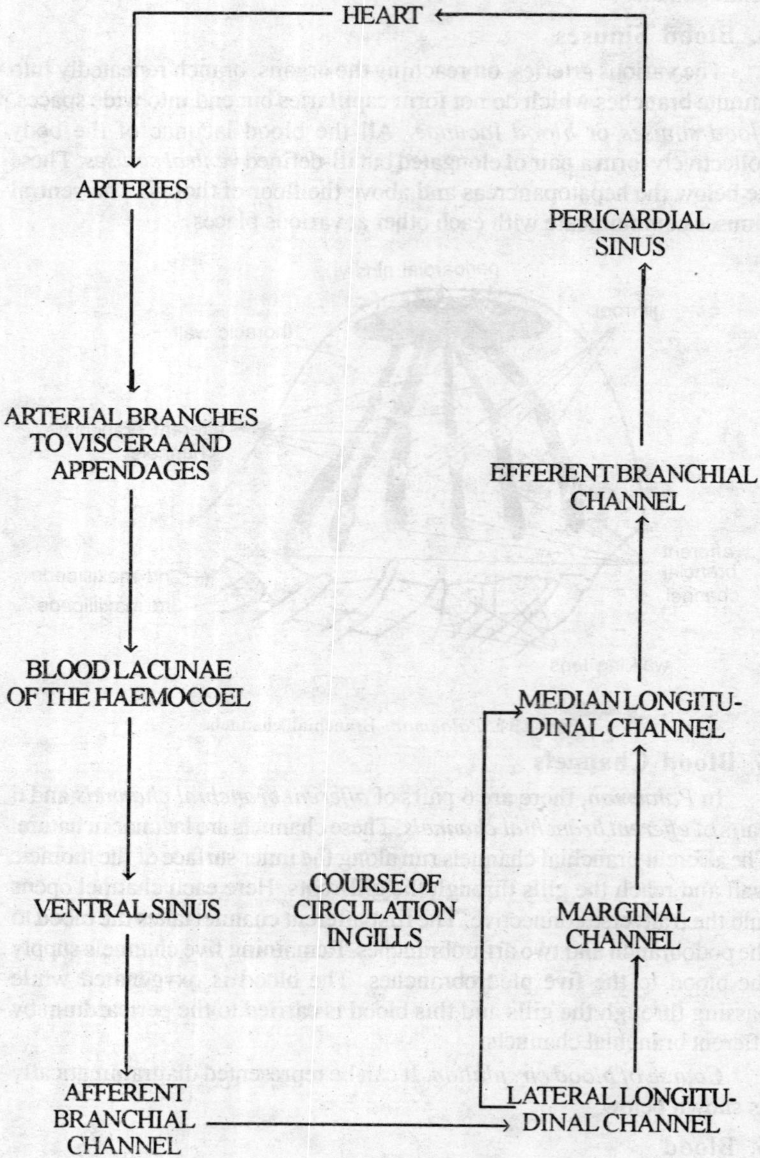

Chart showing course of circulation in *Palaemon*

colourless *amoebocytes*. The colour of the blood is faintly bluish, when it is oxygenated. The colour is due to the presence of a respiratory pigment i.e. *haemocyanin*. It is a compound of copper and protein. There is also a lipochrome pigment called *zoanerythin* in the blood. The blood has the power of coagulation when it comes in contact with air.

EXCRETORY SYSTEM

The excretory system of adult *Palaemon* consists of a pair of *antennary* or *green glands* and a median sac termed as *renal sac*.

1. Antennary or green glands

The antennary glands are opaque white organs of the size of a pea-seed, are enclosed inside the coxae of each antenna. Each gland consists of 3 parts :

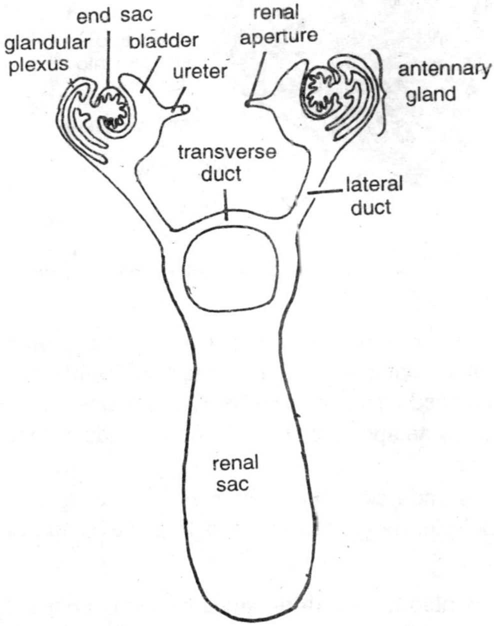

Fig. 2.32. *Palaemon.* Antennary gland.

(i) End sac

It is a small bean-shaped structure lying in between the labyrinth and bladder. Its wall is produced internally into folds and consists of two layers, outer thick layer of connective tissue with a number of small blood lacunae, and an inner thin layer of excretory epithelium. The cells are large with finely granular cytoplasm and rounded nuclei. The cavity of end-sac contains a large blood lacuna and communicates with the labyrinth by a single aperture.

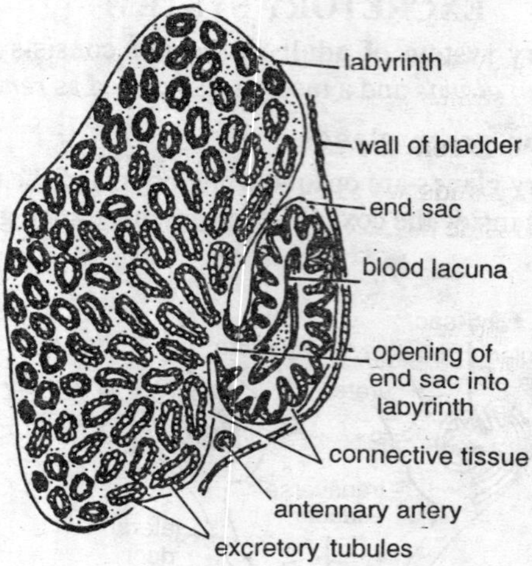

Fig. 2.33. *Palaemon.* Section of antennary gland.

(ii) Labyrinth

The *labyrinth* or *glandular plexus* is quite larger in size than the end sac and lies on its outer side. It is a mass of highly convoluted and extensively branched excretory tubules. The tubules open, on one side, into the end-sac by an aperture and on the other side, into the bladder by several apertures.

The wall of each excretory tubule is formed of a single-layered excretory epithelium whose cells have spherical nuclei and striated border.

(iii) Bladder

The urinary bladder is a thin-walled sac which consists of a single layer of excretory epithelial cells. It is situated on the inner side of the end-sac, Internally, its inner wall is continued as a short, excretory duct or *ureter*, which opens to the outside by a small renal or excretory aperture located on a papilla on the inner side of the coxal podomere of the antenna and in front of the labrum.

From the bladder, a narrow *lateral duct* runs posteriorly along the oesophagus. The lateral ducts of both the sides are joined by a transverse connective. These lateral ducts open into an elongated renal or *nephroperitoneal sac.*

2. Renal sac or nephroperitoneal sac

It is a large median sac lying just beneath the dorsal shield. It covers the entire cardiac stomach and reaches the gonads posteriorly. Anteriorly it communicates with bladder of each side by the lateral duct. The wall of renal sac is made up of a single layer of flattened excretory epithelial cells.

Physiology of excretion

The antennary glands perform two functions, *excretion* and *osmoregulation*. These glands are richly supplied with blood so that by ultrafilteration, the water and dissolved substances are passed into the end sac. The filterate passes into the labyrinths. Where selective reabsorption takes place. The useful substances are pass back into the blood. The remaining fluid, better to say the urine, flows into the bladder and the renal apertures. The urine contains mainly ammonium compounds, some *urea* and *amino acids*.

Osmoregulation

Palaemon is a fresh water crustacean so that due to higher salt concentration of body the surrounding water continually diffuses into the blood through highly permeable gills. Excessive water is eleminated along with nitrogenous wastes that is why the urine of prawn is hypotonic.

Integument

The integument is also regarded as an organ of excretion in young stages. The nitrogenous wastes are deposited over the integument, which is cast off now and then during ecdysis.

• NERVOUS SYSTEM

The nervous system of prawn is similar to that of earthworm in many respects. It consists of :

1. Central Nervous System
2. Peripheral Nervous System
3. Autonomic Nervous System.

1. Central Nervous System

The central nervous system includes a pair of cerebral ganglia, a pair of circum-oesophageal commissures and a double ventral ganglionic nerve cord.

*(i) **Cerebral or supra-oesophageal ganglia.*** The cerebral gauglia is a bilobed structure situated at the base of rostrum in front of the

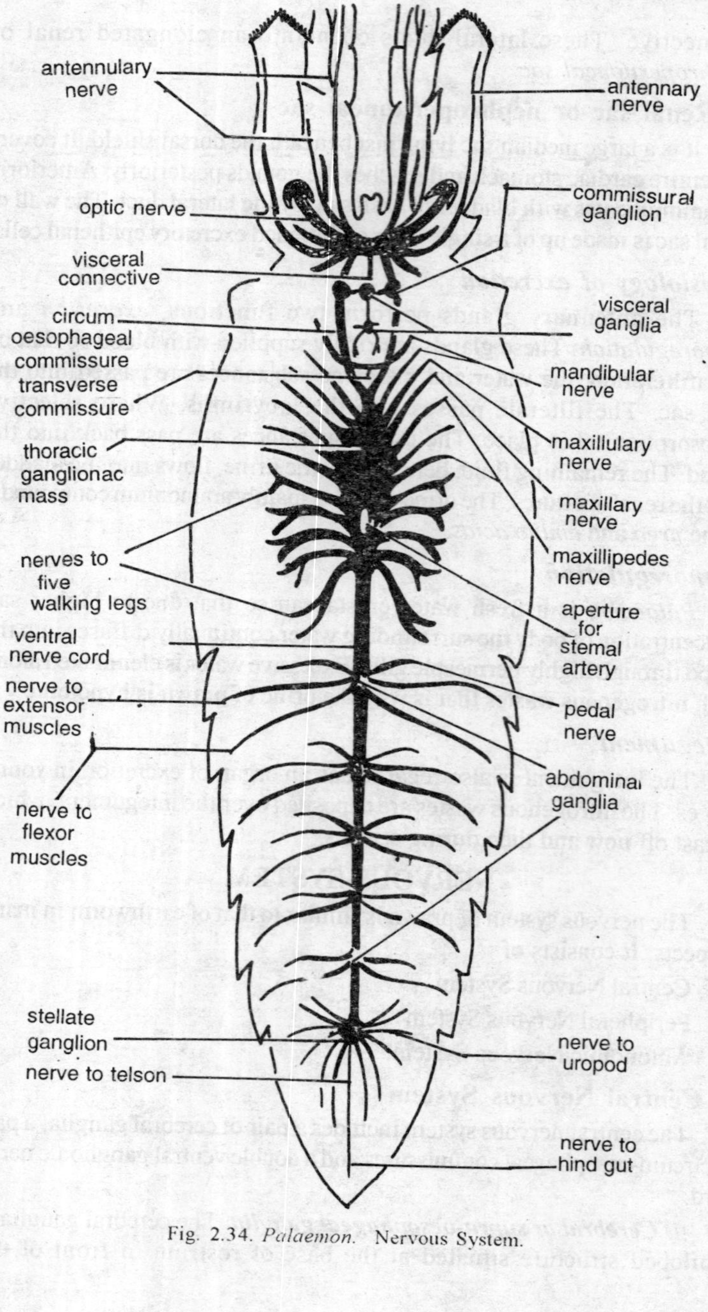

Fig. 2.34. *Palaemon.* Nervous System.

oesophagus. It consists of three pairs of fused ganglia, as it appears from the three pairs of nerves originating from it.

(ii) Circum-oesophageal commissure. There are two stout nerves, which arise from the posterior end of brain and run backwards and downwards around the oesophagus. These unite ventrally with the sub-oesophageal ganglia which form the indistinguishable anterior part of the ventral thoracic ganglionic mass. Each commissure bears a small *commissural ganglion* near its anterior end, and gives off small nerve to the mandible of its side. The two circum-oesophageal commissures are connected with each other by a slender *transverse connective* near the posterior end.

(iii) Ventral nerve cord. It lies in the mid-ventral line of the body just above the sternal plates. The ventral nerve cord is actually double but due to fusion it has lost its double nature. It bears 17 pairs of ganglia. First eleven pairs of ganglia belong to *cephalothoracic region*. These 11 pairs of ganglia are fused to from a single, large, oval mass called the *ventral thoracic ganglionc mass.* It is perforated by the sternal artery near its hind end. The posterior 6 pairs of ganglia of the nerve cord belong to the abdomen. The last abdominal ganglion or *stellate ganglion* is larger than others due to having several post-abdominal ganglia fused with it.

2. Peripheral Nervous System

It is formed by the paired nerves which arise from the various parts of central nervous system.

(A) Five pairs of nerves arise from brain or cerebral ganglia are :

(i) Antennulary nerves. A pair of antennulary nerves arise from below the origin of optic nerves. Each nerve enters the antennule of its side into which it sends a *statocystic branch* to statocyst.

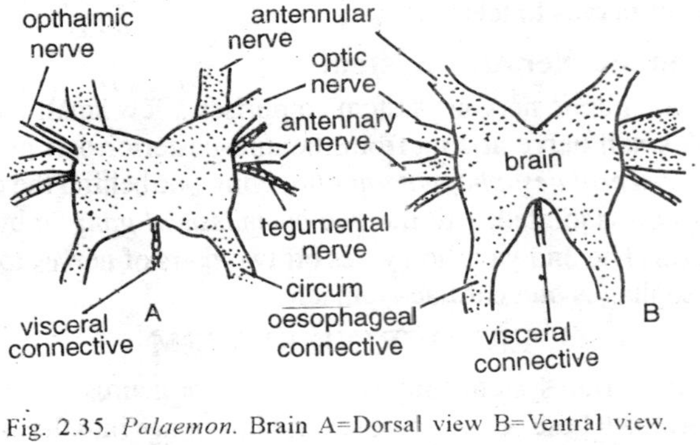

Fig. 2.35. *Palaemon.* Brain A=Dorsal view B=Ventral view.

(ii) ***Optic nerves.*** A pair of stout optic nerves arise from the dorsal surface of the brain, one on each side, and supply the eye-stalks.

(iii) ***Opthalmic nerves.*** A pair of opthalmic nerves arise from the brain, one on each side, and supply the occular muscles in the eye-stalks.

(iv) ***Antennary nerves.*** A pair of stout antennary nerves originate from the ventral surface of the brain. Each nerve divides into two branches, the outer innervating the squama and the inner innervating the feeler of antenna.

(v) ***Tegumental nerves.*** A pair of slender tegumental nerves arise just behind the origin of antennary nerves. These nerves innervate the labrum.

(B) ***Mandibular nerve.*** It arises from the outer side of each circum-oesophageal connective and supplies the mandible of its side.

(C) ***Nerves from cephalothoracic ganglionic mass.*** 11 pairs of nerves are given out from the mass.

(a) *Mandibular nerves* for mandibles.

(b) *Maxillulry nerves* for maxillulae.

(c) *Maxillipede nerves*. There are three pairs of maxillipede nerves which supply to maxillipedes.

(d) *Peraeopod nerves.* These are 5 pairs in number and supplying to the walking legs.

(D) ***Nerves from abdominal ganglia.*** Each of the first four abdominal ganglia gives off following pairs of nerves in each segment.

(a) *Pedal nerves.* A pair of pedal nerves to pleopods

(*b*) A pair of nerves to the extensor muscles.

(*c*) A pair of nerves to the flexor muscles of the succeeding segment.

(*d*) The Vth abdominal ganglion sends two pairs of nerves to the flexor muscles of sixth segment, two pairs of nerves to the uropods and two pairs of nerves to telson.

3. Autonomic Nervous System

The autonomic nervous system comprises a few small ganglia and nerves. A small nerve arising from the mid-posterior part of the brain, bears two *visceral oesophageal ganglia*, lying one behind the other. The first ganglion is jointed with the two *commissural ganglia* by a pair of connectives. The 2nd ganglion gives off two pairs of nerves to the walls of the oesophagus and cardiac-stomach.

ENDOCRINE SYSTEM

The endocrine system comprises a group of neurosecretory cells in each eye-stalk. These cells secrete several hormones into the *sinus gland* lying adjacent to them. From sinus gland, hormones are poured in the blood. These hormones are believed to regulate-colouration of body and compound eyes; deposition of lime salts in the exoskeleton and moulting. Moulting is controlled by two hormones *moulting inhibiting hormone* is secreted by X-organs in the eye-stalks and *moulting accelerating hormone* by the Y-organ presont beneath the adductor muscle of mandible.

SENSE ORGANS

The sense organs of *Palaemon* are :

(1) Statocyst, (2) Olfactory setae, (3) Tectile setae, (4) Compound eyes.

1. Statocysts

A statocyst lies attached to the dorsal wall of the precoxa of each antennule and opens through an aperture on the dorsal concave surface. Each aperture is covered by a fold of skin. A statocyst is a subspherical cuticular sac about 1-1.5 mm wide. Each statocyst is supplied by the statocystic branch of antennulary nerve.

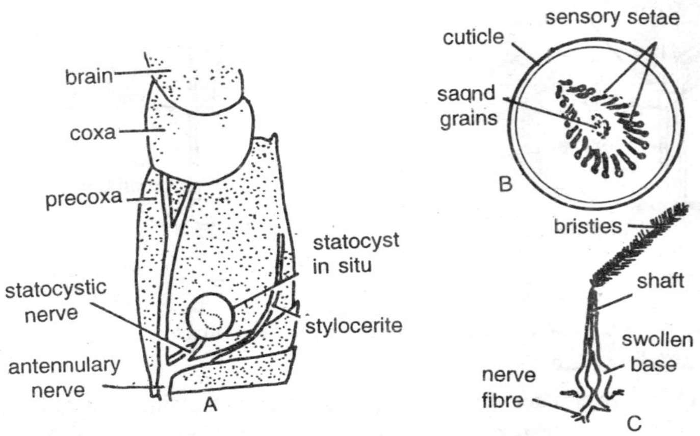

Fig. 2.36. *Palaemon.* A—Statocyst *in situ*, B—T.S., C—A single receptor.

The cavity of the capsule possesses a heap of sand particles lined by a number of delicate elongated statocystic setae. Each seta consists of a swollen *base* and a long *shaft* which is half bent upon itself and covered with fine bristles. The base movably articulates with the wall of the sac by means of an arthrodial membrane and receives a fine branch from the statocystic nerve.

The statocysts are the organs of orientation and equilibrium. In the normal swimming position. The sand grains lie on the floor of the statocyst under the influence of the force of gravity. When the body of swimming prawn gets tilted to one side it causes the displacement of sand particles in the same direction. This presses the sensory setae and stimulates them. The stimulated setae transmit this feeling to the brain and the prawn knows about its position and balances itself accordingly.

2. Olfactory setae

These are also known as *aesthetases*, occur on the middle small feeler of each antennule. Each seta consists of a basal segment or *shaft* and is attached segment of *blade*. Shaft is hollow and is attached to the integument while the later (blade) is free and bluntly rounded and is covered with a thin membrane. Small nerve fibres from the antennulary nerve supply these setae.

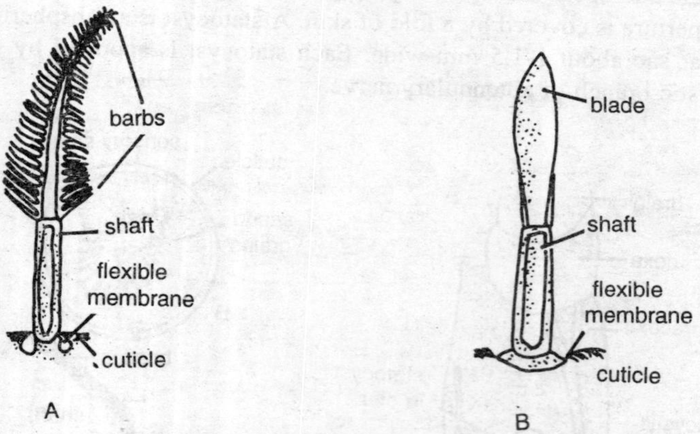

Fig. 2.37. *Palaemon.* A—Tactile seta, B—Olfactory seta.

3. Tactile setae

These are situated on the elongated feeler of the antennae and other parts of the body, especially on the appendages. Each seta consists of a swollen base or *shaft* and a distal segment or *blade* with its central axis. The blade is tapering and slightly inclind distally and has two rows of barbs. The tactile setae are sensitive both to the surface as well as water current but get stimulate only when moved.

4. Compound eyes

There is a pair of compound eyes, which are black and hemi-spherical. Each eye is situated at the base of short, movable and two jointed stalk. Each eye is contained in an *orbital notch* of the rostrum. Each eye is actually a composite structure being composed of a large number of visual units called as *ommatidia.* There are about 2500 such ommatidia in a compound eye. The surface of the eye is roughly hemi-spherical and is divided into a large number of small square facets. Each facet corresponds to a single ommatidium. All the ommatidia are similar, displayed radially, lying side by side and separated by a dark pigment cells. Each ommatidium consists of following parts.

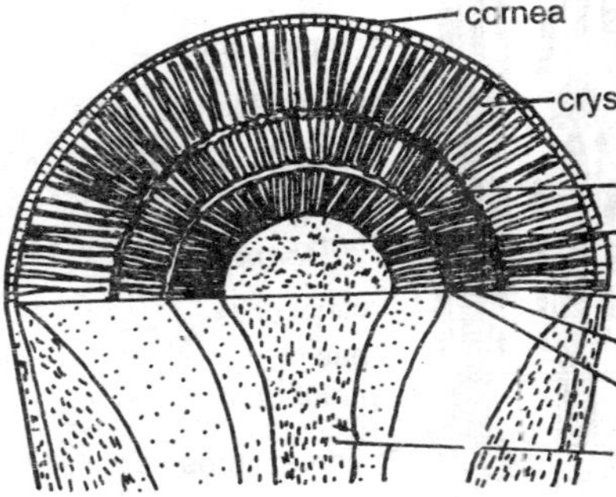

Fig. 2.38. *Palaemon.* L.S. of compound eye (diagrammatic).

(i) ***Cornea.*** The eye is protected externally by a transparent covering of cuticle known as *cornea.* It is distinguished into facets like those of graph paper.

(ii) ***Corneagen cells.*** These are flat, one pair, and of epidermal in origin. They secrete a new corneal layer at the time of ecdysis.

(iii) ***Cone cells.*** Beneath the corneagen cells lie four elongated cone cells or *vitrellae*, which constitute a transparent, homogenous *crystalline cone*. The inner ends of cone cells are long and tapering.

This part of the eye from the cornea to the posterior end of the cone cells is called *dioptrical region* as it is concerned with the focusing of the light on the underlying region known as *receptor region.*

(iv) ***Rhabdome.*** The inner ends of cone cells lie upon an elongated, spindle-shaped, transversely striated structure known as *rhabdome*. It is secreted and surrounded by a group of seven elongated *retinal cells*. The inner end of the rhabdome is supported by a basal membrane beyond which it is continuous with the nerve fibres. The rhabdome and retinal cells form the *receptor region.*

(v) ***Nerve fibres.*** The ommatidia are innervated by nerve fibres of the optic ganglion.

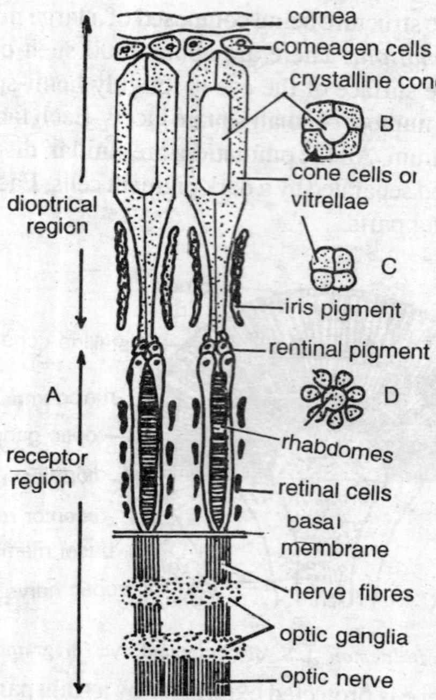

Fig. 2.39. *Palaemon*. (Two ommatidia in L.S.).

Each ommatidium is ensheathed by densely pigmented cells which optically separate from the adjacent ommatidia. There are two sets of these pigmented cells: the *iris pigmented cells* forming the *iris sheath*, round the dioptric region, and the *retinal pigmented cells* that form the *retinal sheath* round the receptive region.

Working of the eyes

The eye of prawn is so constricted that while it can swiftly spot out moving objects and is an efficient photoreceptive organ, it can neither focus properly nor can form a good clear image. Mounted on a movable stalk, it can move on the head and gives the animal almost 360^0 vision. Each ommatidium of the compound eye produces a separate image of an object present just infront of it. Therefore, the image formed of an object by this eye consists of several pieces contributed by many adjacent ommatidia just like a mosaic-art net. That is why it is called a *mosaic vision.* The nature of the image depends upon the intensity of light.

(a) ***Apposition Image.*** In bright light the pigment cells spread in such a way that they completely separate optically one ommatidium from adjacent. No light can pass through the chamber walls from one ommatidium to the other. In this condition the rays of light, which strike the cornea obliquely are absorbed by the pigment cells without producing a visual effect. Only those rays of light, which pass directly through the centre of cornea, can traval through the ommatidium and reach the rhabdome to form an image of a part of object. As a result the complete image formed is a mosaic of several components, places in juxtaposition. Such an image is called *apposition image*.

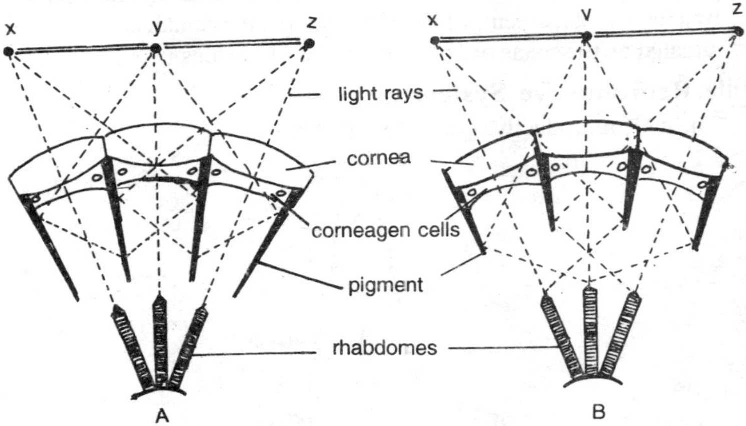

Fig. 2.40. *Palaemon* A=Apposition or mosaic vision, B=Super-position image.

(b) *Super-position Image.* In the weak light, the pigment cells migrate partly towards the outer and partly towards the basal ends of the ommatidia. When this is the condition, even the oblique rays of the source of light are capable of forming a point of image, after passing through a number of ommatidia. Thus, a combined image is thrown on the retinular layer. In this way, an erect superposition image is formed as the rays refracted by several crystalline cones are superimposed on the focus of retina. The superposition image is less distinct in detail than the apposition image, but it is more sensitive to weak intensities of light.

REPRODUCTIVE SYSTEM

The sexes are separate and there is a well marked sexual dimorphism. The males and females differ in several respects.

1. The males are larger than females.
2. Males have thoracic legs more closely arranged than in the females.
3. Males have narrower abdomens than those of the females.

4. In males the 2nd pair of chelate legs are longer and have a thicker covering of setae and spines.
5. Males have *appendix masculina* arising from the appendix interna of each pleopod of the second pair.
6. In males, the epimeres of abdominal segments are smaller in size than in the females, where they are bigger in size for carrying eggs.
7. In males, the paired genital and thoracic openings are situated on the arthrodial membrane between the 5th pair of the walking legs covered by a flap of integument, while in the females, the genital apertures are situated on the coxae of 3rd pair of the walking legs.

Male Reproductive System

The male reproductive system comprises:

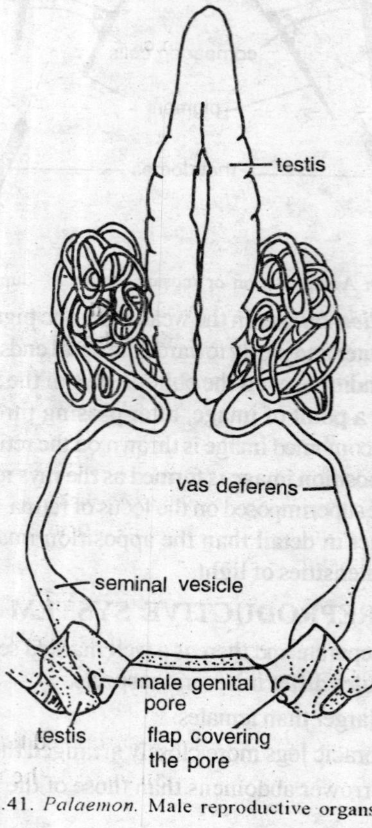

Fig. 2.41. *Palaemon.* Male reproductive organs.

1. Testes.
2. Vasa deferentia
3. Vesicula seminalis or seminal vesicles.

1. Testes

The testes are median dorsal structures situated in the posterior part of the thorax above the hepatopancreas and below the pericardium. They extend anteriorly upto the renal sac and posteriorly upto the first abdominal segment. These are soft white structures which are fused anteriorly to from a single lobe, while at the posterior end these remain in close contact. The two testes enclose a central space or gap which gives way to the *cardiopyloric strand* connecting the heart to the cardiac stomach.

Each testis consists of a large number of coiled, hollow thin walled *seminiferous tubules* buried in connective tissue. The cavity of each tubule is lined by a single layer of *epithelium*, the cells of which undergo *spermatogenesis* to form the *spermatozoa*. A mature sperm consists of a rounded cytoplasmic body, containing a large, dark crescentic nucleus, and a tail-like blunt process.

2. Vasa deferentia

From the outer surface of the posterior end of each testis arises a very long and narrow tube, the *vas deferens*. It is differentiated into two parts : the proximal highly convoluted part which emerges out of the testis and the distal straight part which runs vertically downwards between the

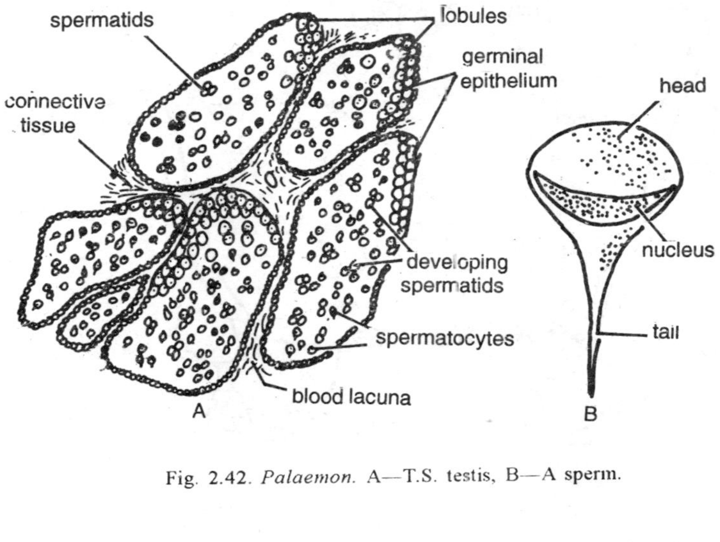

Fig. 2.42. *Palaemon*. A—T.S. testis, B—A sperm.

abdominal flexor muscles and the thoracic wall. Each runs inwards to open into the vasicula seminalis.

3. Seminal vesicles

The distal part of each vas deferens enlarges to form a *seminal vesicle* or *vesicula seminalis* near the coxa of 5th walking leg. It is club-shaped structure which opens to the outside through the male genital pore on the inner surface or coxa of 5th leg. The male genital pore is covered by a small flap-like piece of skin. The vesicula seminalis store the sperms.

Female Reproductive System

The female reproductive system comprises following organs :

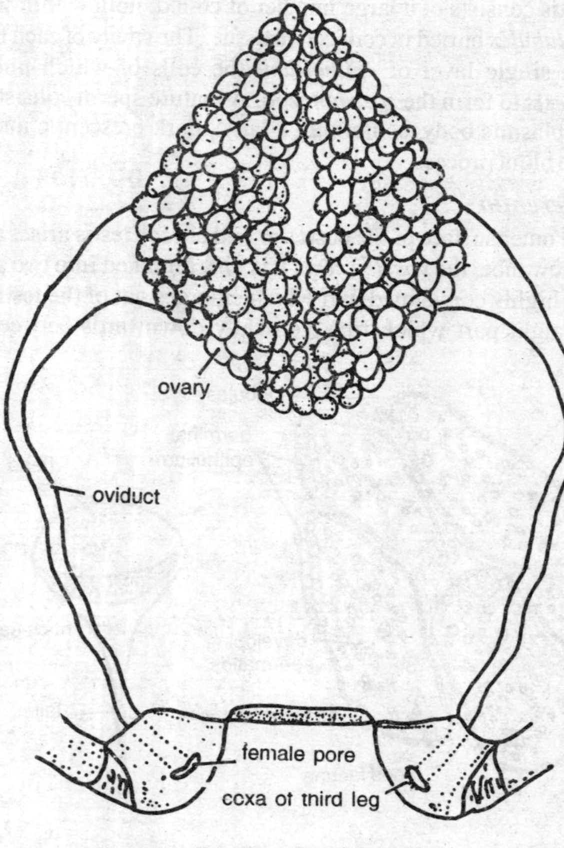

Fig. 2.43. *Palaemon.* Female reproductive organs.

1. Ovaries
2. Oviducts.

1. Ovaries

A pair of ovaries which are compact and sickle-shaped bodies touching each other at both the ends but leaving a gap in the middle for the passage of the *cardiopyloric strand*. The shape and size of the ovaries vary with age and the season of the year. Each ovary is enclosed within a membranous capsule and is made of numerous radial rows of ova at various stages of development. The immature ova lie towards the centre while the mature ova towards the surface of the ovary. The latter are large nucleated cells with plenty of yolk material.

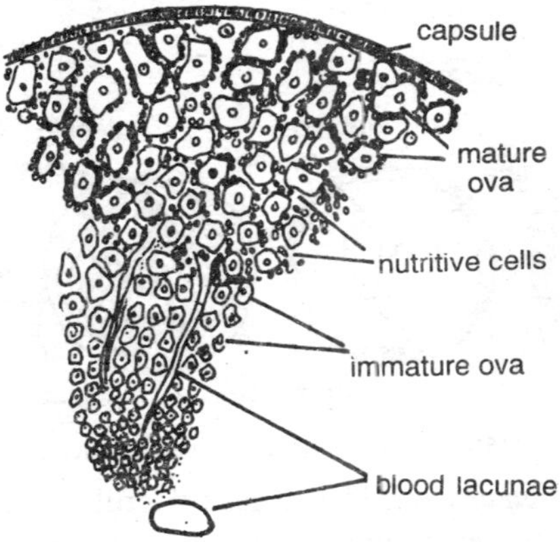

Fig. 2.44. *Palaemon*. T.S. of ovary.

2. Oviducts

From the middle of the outer surface of each ovary arises a short thin-walled tube known as *oviduct*. Its anterior end is broad and forms the oviducal funnel. It lies opposite to the surface of ovary. Each oviduct runs vertically downwards to open to the exterior by female genital pore. The pore is situated on the inner side of coxa of 3rd walking leg.

Fertilization

Breeding season in *Palaemon* is from May to July. During the season, the females carry eggs attached to their pleopods. Except in breeding season, the females are found without eggs. Fertilization is external. When ova are laid, they are attached to the setae on the pleopods with the help of a sticky secretion of glands situated both on the appendages and on the body segments. The male, later on, deposits spermatophores near the genital openings, where fertilization takes place. It is not known yet that how male deposits sepermatophores.

Development

In *Palaemon lamarrei, P. malcolmsonii* etc. the development is direct i.e. without larval stage. However, in other prawns (*Penaeus, Lacifer*) the development is indirect i.e. with larval stages.

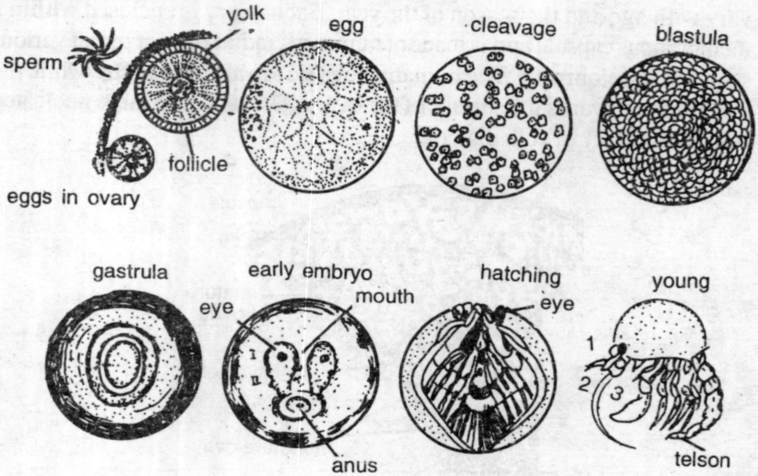

Fig. 2.45. *Palaemon*. Showing development.

The cleavage is *superficial*. The nucleus of fertilized egg, lying in the centre, divides and redivide forming a number of nuclei. These nuclei are later on enclosed by masses of protoplasm, but areas of the separate cells do not become marked off by furrows cutting right through the egg. As a result of segmentation, a blastoderm is formed, which appears at first on the ventral side of the egg. Its formation begins at one point on the surface and proceeds gradually from this point, which always represents the future ventral side of the egg. Gastrulation starts from the posterior end of the ventral side which is marked by the first formation of the blastoderm.

Gastrulation

A small invagination develops at a time when the blastopore is not completely formed. This invagination marks the first step towards gastrulation and small archenteron is formed with its opening, the *blastopore*. When the blastopore closes, the cells of the endodermal vesicles lose their epithelial nature. From the lateral walls of the endoderm vesicle arise elements which form the *mesoderm* at a later stage. *Endoderm* is formed by the cells, which arise from the floor of endodermal vesicle and pass on to the yolk and multiply therein. Each of these endodermal cells swallows the surrounding food yolk, with the result the yolk mass breaks into smaller spheres. Afterwards the midgut is formed by the inter-penetration of the food yolk by the cells of gastrula. Fore-gut and hind-gut also arise by ectodermal invagination forming *stomodaeum* and *proctodaeum*. Both reach the midgut and finally fuse.

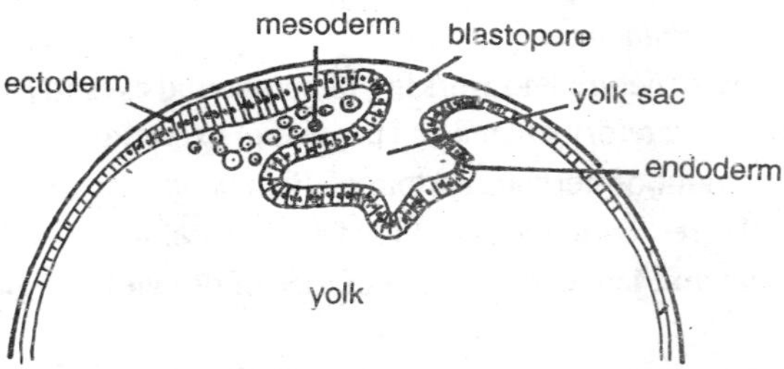

Fig. 2.46. *Palaemon.* V.S. of early gastrula.

When the mesoderm appears for the first time, it consists of a mass of cells near the point of origin. These cells multiply rapidly and spread out apparently in an irregular manner between the ectoderm and the food yolk.

The ectoderm gives rise to the external integument and an internal skeleton. Nervous system develops from ectodermal thickening. Vascular system is developed from the mesenchyme. Growth of the main parts of the embryo in size, during which the entire yolk is consumed, results in a small individual which is similar to adult except size. After several moults the adult is formed.

Power of Regeneration

Prawns and certain other crustaceans like lobsters, crabs, cryfishes etc., have a power of regeneration which is greater in yound stages but very slow in adults. For example, if the tip of eye is removed, normal regeneration takes place. If the entire eye stalk along with optic nerve is removed, an antenna like structure is formed. Such type of regeneration in which normal organ is not developed, is called *heteromorphosis*.

Crustaceans have the power of *autotomy*. If a thoracic leg is broken in some mishap, its distal segments are spontaneously thrown off to feed the enemy.

Revision Questions

1. Give an account of respiratory organs of Prawn.
2. Describe the cephalic and thoracic appendages of *Palaemon.*
3. Give an account of digestive system of Prawn and its mechanism of feeding.
4. Give an account of Arthropod eye and explain the theory of mosaic vision.
5. Describe in detail the circulatory system and the course of blood circulation in *Palaemon.*
6. Describe the structure and working of compound eyes in Prawn.
7. Describe the excretory organs and process of excretion in *Palaemon.*
8. Give an account of nervous system of *Palaemon.*
9. Describe the reproductive systems of *Palaemon.*
10. Draw a neat and labelled diagram of T.S. of prawn through the gill region.
11. Write short notes on : (a) statocyst, (b) Hastate plate, (c) Green gland, (d) Ommatidium, (e) Antennary gland.

3

Prawn Fishery

Prawn, a crustacean, commonly known as shrimps, is considered by most people a fish. They are a rich source of proteins and vitamins A and D. Their muscles contain a considerable amount of glycogen and free amino acids, making their flesh tastey. Due to its nice taste, it is supposed to be a cherished delicacy to serve as food through out the world.

Prawn fishery means capturing of prawns from natural resources as well as their culture. The prawn fishery of India is next to that of U.S.A. in the world. In certain areas, particularly on West Coast, prawn catches are exceeding. Prawn production in Kerala and Mumbai Coasts is 92.4% of the total of all India prawn catch. The earning from the export of prawns and their products has been increased from 17 million rupees in 1960 to about 950 million rupees in 1975 and, it is increasing day by day.

Kinds of Prawn Fishery

Depending upon the areas and nature of waters, from where prawns are captured in India, there are following types of prawn fishery :

1. Shallow water prawn fishery

It is located on the West Coast where fishing is restricted to shallow water near the coast. During rainy season the shoals come near the Malabar coast and the fishermen catch them with their cast-nests.

2. Estuaries and back water

The principal areas of production are the back water extending along the southern half of Kerala Coast and the mouths of numerous hill-streams in the Malabar area. Chilka lake is much rich in prawn production.

3. Fresh-water prawn fishery

Fresh-water prawns are caught from rivers, lakes and ponds throughout the country.

4. Marine prawn fishery

It is confined to shallows coastal areas upto 40 meters deep. Important prawn productive areas along the Indian coastlines include the Kanya-kumari, Cochin, Parur, Cranganore (West Coast). Ennore, Pulicat, Collair (along East Coast).

Indian Prawns of Commercial Value

A number of species of prawns are found distributed in water are of different sizes. About 73 species and 20 genera belonging to the families Penaeidae, Pandalidae, Hippolytidae, Sergestidae and Palaemonidae. Some species are good for culture while several others are caught in very small quantities. The following species of prawns are important from culture point of view.

1. *Penaeus carinatus.* It is the largest species about 25 cms. in length but seldom fished in large number.
2. *P. indicus.* It is slightly smaller species measuring about 20 cms. in length. This is the commonest commercial species found along both the coasts.
3. *P. monodon.* It is the largest species measuring up to 30 cms. in length. It occurs along East and West Coasts.
4. *Metapenaeus monoceros.* It is common in estuaries and brackish water lakes like Chilka, Pulicat and Collair. It measures upto 15 cm. in length.
5. *M. dobsoni.* It is common in the brackish waters of Kerala. It measures 12.5 cms. in length.
6. *M. brevicornis.* It occurs in low laying paddy fields of Bengal. It measures 12.5 cms. in length.
7. *Parapenaeopsis sculptilis.* It occurs on both the coasts but good catches are obtained in Mumbai and West Bengal. It measures 14 cm. in length. This is also found in Chilka lake.
8. *P. styliferus.* This is found on the West Coast and fished maximum in Malabar coast. It measures 11 cms. in length.
9. *Macrobrachium carcinus.* This is a freshwater prawn and migrates to brackish water for breeding. It measures 30 cm. in length.
10. *M. malcolmsonii.* This is a migratory species generally found in Chilka lake towards the close of monsoon. It measures 15 cms. in length.
11. *Palaeomon fluminicola.* It prefers to inhibit fresh or brackish water and reported to ascend the river Ganges.
12. *P. tenuipes.* This is found on the West Coast.
13. *Acetes indica.* It is caught in large number in certain river mouths and coastal waters. It measures only 2-4 cms. in length.

Food of Prawn

The prawns feed on vegetable matters, living organisms and also the dead organic materials. Thus they are *omnivorous*. They mainly feed on algae, moss and other aquatic weeds. The food of *Macrobrachium malcolmsonii* comprises of small insects, sea weeds and algae. *Penaeus indicus* feeds on varying quality of food material viz. vegetable matters, crustaceans, detritus and other organisms present in surrounding.

Culture of Fresh Water Prawn

Macrobrachium carcinus is big sized prawn found in fresh water rivers, fields and low salinity areas of estuaries. The prawns collected from fresh water rivers or fields are transferred into the tanks which are well aerated. The transport of these collected prawns should be performed in plastic bags or in bamboo-baskets with aquatic plants for aeration.

Fig. 3.1. Indian coastline showing prominent *prawn productive* areas.

For fertilization purpose one male and female prawns are kept in an aquarium with a capacity of 60 litres of water. After 24 hrs. of mating the female starts laying eggs. For commercial exploitation the spawning has to be planned. Spawning tanks of different size should be prepared. For 10 pairs, a tank of 200 × 100 × 40 cms. is suitable, while for 24 pairs, a tank of 300 × 150 × 40 cms. would suitable. The proper aeration and cleanliness of tanks is most important.

Mixing of sea water with fresh water is well suited for the proper development of eggs. During development, when the colour of eggs is changed from orange to light gray, some more sea water should be again added to make a 5% concentration. As the larvae are hatched from eggs, they should be immediately transferred to other cement tanks from the spawning tanks. The size of these cement tanks is about 200–300 × 50 – 70 × 25 cms. These tanks should be filled with the water of spawning tanks. The pH of these tanks should be between 7–8 and the temperature between 24 and 30°C. Moulting of larvae takes place. The artificial food like steamed eggs custurd, fish ball and fish eggs should be given to larvae. This food supply is continued for 4–5 times per day till the larvae are 2–3 days old up to the metamorphosis. Now they should be transferred to fresh water tanks for acclimatization and to be fed on artificial diet.

The young ones of 5 cms. length i.e. about 60 days old, can be reared in fresh or slightly brackish water at 22–32°C in well aerated ponds or rice fields. The artificial pond of size 50 × 20 × 1.5 meters is best for operation and can be managed economically.

For normal growth of prawns in their natural habitat. The natural food can be made by providing 200 kg. cow dung mixed with 10 kg. lime per hectare per month is sufficient. For the supply of food in the form of living organisms small pieces of waste fish, insects, earthworms and plant food to be managed for each pond.

Within 6 months, under proper natural conditions the prawns can develop up to 20 cms. in length and 100 gms. in weight. Two crops of prawns can be harvested per year by this method.

Culture of Marine Prawn

Several factors determine the success of marine prawn farming. These are as follows :

1. Selection of Site

The most suitable site is along the tidal flats and marsh lands adjacent to brackish water estuaries. The site of irregular shape and uneven bottom are not fit for prawn farming as they reduce the effeciency of harvesting.

2. Water Quality

There should be a regular supply of clean and well aerated water to fulfil the water lost through evaporation, sewage and drainage during the course of management operation. The temperature and salinity conditions of water must be suitable to the species to be reared.

3. *Quality of Soil*

The soil having clay composition with a pH 6.5–7.5 is usually suitable as it tends to hold water. The soil having more than 50cms. of siit on the bottom is not suitable.

4. *Availability of Seed*

The naturally occurring species in a given locality would most likely dictate the species to be suitable in the area.

5. *Other factors*

The surface area for commercial and economic farming should not be less than 5 hectares. The site should be accessible by road at all the times.

Preparation of Farm

For the preparation of ponds, it is essential to drain off the water and sundry the bottom followed by light tilling if possible. During this period 400 kg. per hectare agricultural lime should be applied. This operation is necessary as lime absorbs the excess of carbon dioxide. This also supply calcium to the prawns during moulting. Commercial fertilizers such as rice barn, poultry and cattle dung, night soil and sewage wastes are used to increase the fertility of soil. In this medium algal growth is accelerated. It is desirable to drain the pond completely once in a year to increase the rate of microbs production.

Methods of Prawn Fishing or Collection of Prawns

The Indian coasted region of 4500 kms. with a huge area of continental shelf has great scope of prawn fishing throughout the year. Fishing normally requires the *crafts* and *gears*.

1. *Fishing crafts*

Several types of crafts are used for the prawn fishing in India. Locally made traditional crafts such as wooden catamarans, boats and canoes have been used since long time. They are known by different names according to regions. In Kerala, *Vallom* is used which is operated by 5–6 persons. It is 8.3–10.3 meters in length. From Kakinada to Masulipatam, *Masula* is used. It is operated by 8–12 persons and 8.5–12 meters in length. The boat used is Maharashtra is known as *Machwa* and is operated by 7–9 persons. It is 9–14 meters in length. *Dinghi* is used in Orissa. It is operated by 8–10 persons. It is 10.4–13.3 meters. *Nauk* is used in West Bengal. It is operated by 8–10 persons and 10.4–13.2 meters in length.

Motorised boats were first of all introduced in Maharashtra. This has enabled fishing in a shorter time and up to greater depths. Steel trawlers have come under use recently.

2. Fishing gears

Several types of gears are used to catch prawns and lobsters like nets, traps, hooks and hands.

(i) *Nets.* Generally nets are made up of cotton, nylon, hemp or in combination of these. The spaces of nets should be specially kept in view so that the prawns may not be able to pass through them.

The simplest type of net used is *cast net*. It is operated by single person by throwing the net over the water surface and later on drawn out. Some nets are bag like in appearance and called *bag-nets.* The nets with bag or without bag when tied with stakes are called *stake nets.* The bag nets are towed by the crafts which are under movement and prawns are caught themselves in bags are called *drag nets.* When the upper margin of the mouth of the drag nets is fixed to the beam and the lower margin is free hanging, the net is known as *beam trawls.* Movable nets which are kept hanging by attaching floats on the upper margins and sinkers on the lower margins are termed *seines. Boat seines, shore seines* and *drift nets* are conventionally used to catch prawns in the coastal waters. Smaller types of nets, *scoop nets* and *hand nets* are used.

(ii) *Traps.* The traps, made up of bamboo mats, are the earliest devices to catch prawns. They are left over night in shallow waters against the water current to collect prawns. The traps used are Ottal, Gutta guli, Uduka, Muchu, Uted, Thapa, Chhapa, Polo and Poluha at different places. Along the canals of Malabar *chain-dragging* method for prawn collection is employed. A heavy chain of iron is dragged over the bottom by two cones connected by short bamboo poles in front and rear. The moving chain disturb the prawns which jump in agony and fall into the canoes.

Preservation and Processing of Prawns

A part of the catch is marketed fresh for consumption, but a major portion is processed and preserved.

1. Spoilage

After fishing, prawns are subjccted to *spoilage* or *putrefaction*. The spoilage starts in the boat where prawns are stored for six hours without ice because they deteriorate upto 4 hrs. At 28°C prawns keep up their original quality, but after 8 hrs. they may become unfit for eating. The spoilage is mainly of two types. In *chemical spoilage*, tissue contents, such as free amino acids are leached out leading to loss of taste, nutritive value and weight. *Bacterial spoilage* is due to bacterial contamination from outside. Free fatty acids are released from muscles by enzymatic hydrolysis of triglycerides and phospholipid, essential constituents of lipids, which later on undergo auto-oxidation producing carbonyl and thereby affecting the quality of prawn by introducing rancidity. Prevention of spoilage is known as *preservation*.

2. Chilling and Freezing

Most effective methods of controlling spoilage are chilling and freezing. After fishing the prawns should be stored in ice till they are processed. The quantity of ice used should be kept in the proportion of 1:1, ice to prawn by weight. The storage time in which prawns remain in good condition is known as *shelf time*. Refregerated sea water serves as a good preservative medium. Peeled prawns are store in a better condition than the whole prawns.

In storage the prawns begins to blacken within a few days. It can be prevented by *glazing*. Glazing means covering iced prawns with water.

3. Beach drying

The most widely practised method is beach drying along the coastal areas. The prawns are simply spread over the sand for drying by the solar heat. These dry prawns are marketed as such or after shelling.

4. Pulp making

The prawns are boiled in weak brine, dried in sun, paked into jute bags and are beaten with clubs. The shells separate and a pulp or paste like mass is left which is marketed in the Far East.

5. Pickling

The prawn pulp after boiling and shelling is pickled with oil, spices and vinegar in Malabar coast.

6. Blanching

Before freezing and packing, the fresh prawns are dipped in boiled brine for one to three minutes. It is followed by immersed in cooled water (2°C), the process called glazing.

7. Smoking

Prawns are first boiled in salt water followed by drying in kilns by burning moist bark. Smoking in practised in Andhra Pradesh and Orissa.

8. Semi-drying

Cleaned prawns are blanched in brine, shelled by hands, treated by

25% brine, and then half-dried in solar heat. They can be stored for two months.

9. Freezing and Canning

Headless prawns are frozen directly, without boiling. They are chilled down and placed in freeze storage at 10°F until transportation.

Canning is costly and the prawns can be kept indefinitly without use of any preservative. For canning, the prawns are cleaned, cooked at high temperature under pressure and packed in cans with the addition of oil. The cans are then sealed airtight and sterilized by heat to kill bacteria.

Export of Prawns

The prawns have become an important source of earning foreign currency because of their export to various parts of world. The prawns are exported in the form of frozen prawns, canned prawns, pickles, prawn powder etc. They are exported to Australia, Belgium, Canada, Denmark, France, Greece, Germany, Japan, Muscat, Oman, Newzealand, Sweden, Switzerland, Singapore, U.S.A., U.K., Spain etc.

Prawn Meal and Manure

The chitinous shells, if dried and powdered, provide the *prawn meal.* It is a valuable protein rich poultry feed. Mixture of prawn meal and shark meal can be used as a food for pigs.

Pollution and Prawn Fishery

Industrialization leads to pollution which is a great problem in many places. The waste products of tannery, textile and paper mills, sugar and distillery mills etc. are extremly poisonous. Concentration of one ppb of DDT in water can kill or paralyse prawns and other aquatic animals. Oil and its components can also affect prawn in many ways. Thus prawn fishery is adversely affected by pesticides, effluents, metals and oils.

4

Larval Forms in Crustacea

The embryonic development is followed by metamorphosis which may be either very well marked or supressed. In most of the curstaceans the egg hatches to produce *Nauplius larva* which is free swimming but in some cases the nauplius stage is passed within the egg so that there is no free swimming nauplius larva produced after hatching. In crabs, fresh water prawns and cryfishes the development is direct i.e. without larval stage. In others several larval forms are met within Crsutacea and special terms are applied to each one of them. Following larval forms are met within Crustacea :

Nauplius larva

The nauplius larva is oval in shape behind any sign of external segmentation. It is characterised by the presence of 3-pairs of appendages. The first pair later on becomes antennules and the 2nd and 3rd pairs becomes antennae and mandibles. The antennules are uniramous where as the antennae and mandibles are biramous. The antennae are provided with jaws whereas the mandibles are without jaws. Thus the antennae have masticatory function. The median eye is the only organ of vision. Besides eye there is a pair of antennal glands. The mouth is guarded by a large labrum and the alimentary canal is straight and made up of foregut, midgut and hindgut. The anus may or may not present at the time of hatching. A heart is still not formed. At the posterior end there are two pairs of setae. The nauplius swimms in water by the stock of all the appendages. In Branchiopoda and Ostracoda the nauplius larva undergoes a series of moultings or ecdysis to become the adult. During gradual process the adult characters are gradually attained.

In order Cladocera the development is direct so that the nauplius stage is passed in the egg and when the egg hatches an adult type of crustacean is produced.

In subclass Ostracoda a bivalved nauplius is produced without biramous appendages i.e. all the appendages are uniramous.

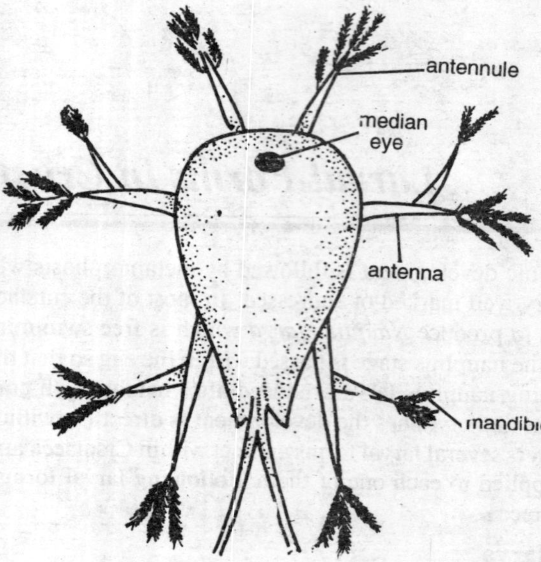

Fig. 4.1. Nauplius larva.

In subclass Copepoda immediately after hatching a nauplius is produced and after moult it is convested into metanauplius larva.

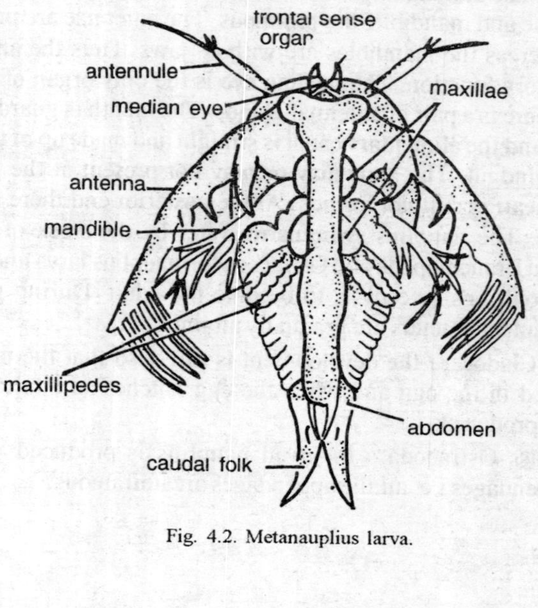

Fig. 4.2. Metanauplius larva.

Metanauplius larva

It is characterised by the presence of several pairs of appendages. Besides 3-pairs of each antennules, antennae and mandibles, it develops another four pairs of appendages. These appendages become the maxillulae, maxillae and first two pairs of maxillipedes of the adult. Each mandibles acquires a masticatory process.

The body consists of an oval cephalothorax, an elongated trunk-region and abdomen, which terminating in a caudal fork provided with setae.

Cypris larva

In subclass Cirrpedia, the free swimming stage is nauplius. The larva undergoes a moult which gives rise the cypris larva. The cypris larva is characterized by the presence of bivalved shall, like that present in Ostracoda. In cypris larva the bases of antennules are modified into attachment discs which give attachment to the larva aided by the secretion of cement glands. There is a pair of compound eyes near the median eye to detained from nauplius larva. Antennae and mandibles are absent, sometimes the mandibles are rudimentary. 6-pairs of thoracic appendages are present in thoracic region. The abdomen is very short and ends into a pair of caudal fork or style. After swimming for some time, it attaches itself to some object with the help of attachment discs and secretion of cement glands and get transformed into pupa. Later the pupa becomes adult. In the parasitic crustacean like *Sacculina* free swimming nauplius and cypris are produced but the cypris larva attach itself to a crab and gradually becomes a tumour like adult.

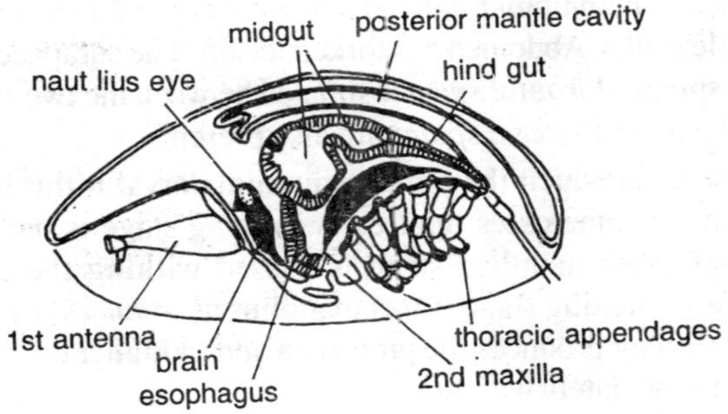

Fig. 4.3. Cypris larva.

Thus in the case of Brachipods, Ostracodes, Copipodes and Cirripedes, there is a gradual metamorphosis. The adult is formed by a series of moults. The larval forms gradually give more and more of adult characters and this is just like the incomplete metamorphosis of insect. But in the case of Malcostraca there is abrupt several transitional changes from one larval form to another and all these forms are very much different from their adults. The metamorphosis may be compared with complete metamorphosis of insects, for example the silk moth.

Protozoea

In the case of Decapod crustaceans (*Penaeus*) the egg hatches to produce a nauplius larva which moults to form another larval stage, the *protozoea.* The *carapace* becomes larger and covers the dorsal surface anteriorly. The seven pairs of appendages, which were already formed in the metamorphosis, become well-developed and are capable of functioning. The antennae are still locomotory organs. Antennule is 4-jointed, antenna possesses a 3-jointed endopodite and a four-joint exopodite. The mandibles are palpless, toothed and are *masticatory* in function. Abdomen is still unsegmented and without appendages. However, rudiments of the remaining posterior six thoracic segments are also formed. Later on, the third pair of thoracic limb also develops, which forms the third maxillipede of the adult. Anterior five abdominal segments are also distinct. Formation of paired eyes has also begun.

The protozoea again moults and gives rise to the zoea larva.

Zoea larva

The body of zoea has a distinct, highly developed cephalothorax and distinctly 6-segmented abdomen. The cephalic appendages are will developed. Abdominal appendages are lacking. Zoea is characterized by the absence of mandibular palps. Eyes are large and compound situated on movable stalks. Abdomen has forked telson. The carapace is produced into long spines. *A frontal spine* is situated between the two eyes, a *dorsal spine* and a pair of *lateral spines* are also present.

The stage at which these free swimming larval forms begun varies vary much. In some cases the free swimming stage is nauplius but in some other cases nauplius stage is passed withing the egg, so that immediately hatching stage is metanauplius as in *Lucifer.* In the case of *Sergestes* the egg produces the protozoea and in higher decapods zoea is produced immediately hatiching.

Mysis or schizopode

In *Penaeus*, zoaea larva instead of changing into megalopa changes into mysis which is very similar to the adult *Mysis.* It possesses five pairs of posterior thoracic legs which are biramous with flagellar expopdite

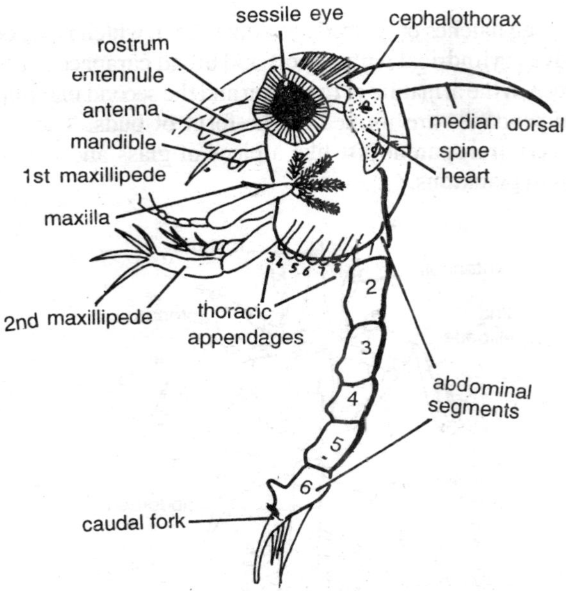

Fig. 4.4. Zoea larva.

and help in locomation. The abdomen is elongated with well developed biramous pleopods. The mysis larva metamorphoses into the adult prawn.

In lobsters the egg hatches into the mysis larva.

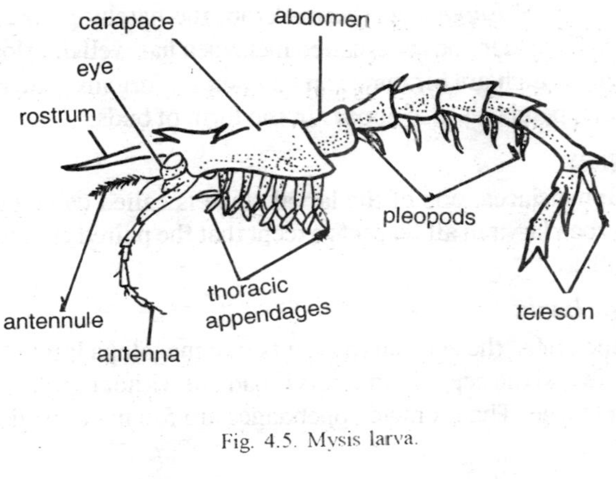

Fig. 4.5. Mysis larva.

Alima larva

In *Squilla* egg hatches out directly as alima larva, which is a modified zoea. The body is cylindrical with a short and broad carapace. The larva differs from zoea in the armature of the telson and the second maxillipedes. Abdominal appendages are formed in the form of buds. The larva is pelagic. It is very transparent just like a piece of glass and is found in large numbers in planktons.

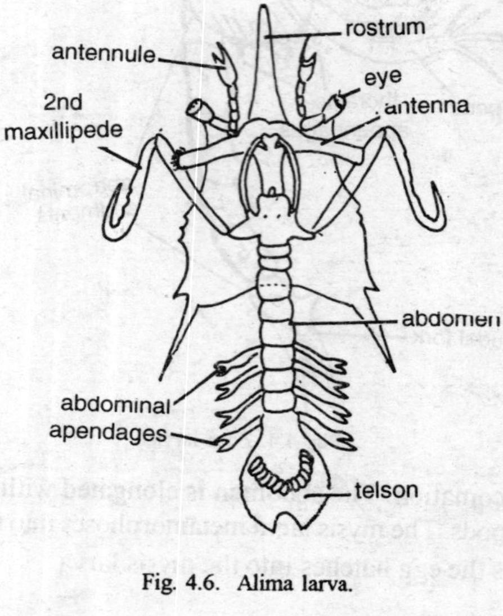

Fig. 4.6. Alima larva.

Metazoea larva

In the case of *Eupagurus* (hermit crab) the hatching takes place at zoea stage. The older zoea also called metazoea has well developed third maxillipedes which are biramous and swimming organs. The 6-pairs of abdominal appendages also appear in the form of buds.

Calyptopsis

In Euphausiacea, one of the larval stage is called calyptopsis. It is similar to zoea larva in all respects except that the paired eyes are sessile i.e. not stalked.

Megalopa larva

In some crabs, the zoea larva moults into megalopa larva. It is crab-like creature and the cephalothorax is broad but without spines. Eyes are stalked and large. The thoracic appendages are 5-pairs of walking legs.

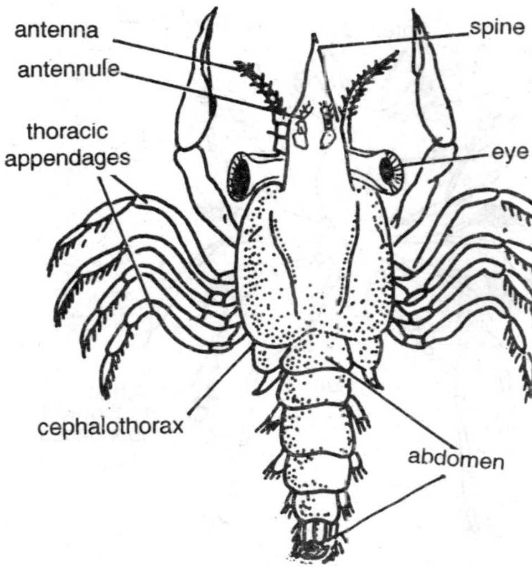

Fig. 4.7. Megalopa larva.

The abdomen is also well formed and bears biramous pleopods. The megalopa swims on the surface of the water leading a pelagic life but, later on, sinks down to the bottom and moults into a typical crab.

In hermit crab, the metazoea leads to glaucothoe stage. It corresponds to the megalopa stage of Brauchyura. It has a large symmaterial abdomen and adult appendages.

Phyllosoma larva

In rock lobster, *Palinurus,* and its allies hatching takes place as phyllosoma larva or *glass crab* which is a greatly modified mysis state. It is strange looking, leaf-like, transparent, with eyes at the ends of long stalks. The broad depressed cephalic and thoracic shields have the transparency of glass while limbless abdomen is very small. Only 6-pairs of thoracic appendages are present in newly hatched larva. The maxillipedes are rudimentary or absent, the second are uniramous followed by 4-pairs of very long biramous legs. Phyllosoma undergoes several moults before reaching the adult stage.

Significance of larval forms

Life history of higher crustaceans show that as the later stages are added, the earlier stages are supressed. The nauplius occurs as a free swimming larva in lower crustaceans but in higher forms hatching takes place either mysis or zoea stage and the nauplius stage passes into egg. In

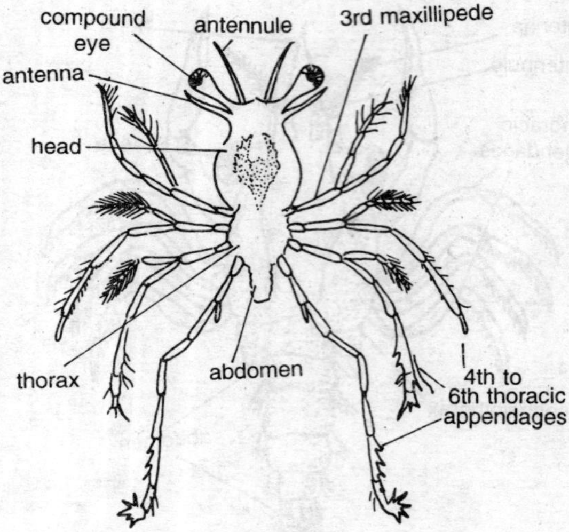

Fig. 4.8. Phyllosoma larva.

certain respect the nauplius features are as follows :

1. The preoral position of antennae.
2. Presence of masticatory endite at the base of antenna.
3. Presence of biramous mandible.
4. Presence of median eye.
5. Frontal sense organs.

The biramous character is also found in some copeodes and ostracods. It appears probable that the nauplius was establishing as a larval form at a period of development before the divergence of the ancestral groups of crustacean and probably that is the reason why the nauplius stage appear in some form of life-history of some of the crustaceans. None of the earlier adult crustacean shows a nauplius like form but rather a multisegmented form of the annelidian type. Similarly none of the primitive crustacean that have none of the zoea larvae but zoea-like adults. As mysis occur in the development of decapod and resembles each other inspite of the fact that their adults are entirely different. This shows the structure of crustacean ancestors. if we accept that "Ontogeny recapitulates phylogeny" i.e. the various larval forms give ancestral history during development but this is considered only partially correct. The explaination that has been given to divert these forms is that they have adapted to special environmental conditions at different time. however. some phylogenetic significance attached only to nauplius larva because of primitive features of crustacean stock. The other larval forms have no phylogenetic significance and have developed in accordance with influence of environment which were different from time to time.

Revision Questions

1. Write an essay on larval forms in Crustacea.
2. Write short notes on (i) Zoea larva, (ii) Alima larva, (iii) Cypris larva, (iv) Nauplius larva.

5

Parasitism in Crustacea

Parasitism is such an association between two organisms of the same or different species in which one lives on or inside (parasite) the body of the other (host), receiving nourishment and shelter without any compensation for the host. Therefore, in this association the parasite is benefitted and the host in harmed.

Parasitism is seen in several groups of Crustacea, such as Copepoda, Branchiura, Cirripedia, Isopoda and Amphipoda. However, it is not seen in Branchiopoda and Ostracoda.

Effect of parasitism upon parasite

Changes from a free-living condition to a parasitic life involves changes in the body of the organism. Due to parasitic mode of life certain new characters are developed due to new environmental requirements, similarly certain characters, which are of no use for new environment, are lost. The peculiarities of the crustacean parasites in relation to parasitism are as follows :

1. ***Degeneration.*** The parasites have lost organs no longer used in the adult form. They look sac-like in shape due to parasitic degeneration.
2. ***Adhesive organs.*** Suckers and hooks are formed specially for the attachment to the host.
3. **Feeding.** The feeding occurs by suction or absorption through organs. The alimentary canal is reduced or absent.
4. ***Reproduction.*** The reproduction is vast and wasteful to ensure chances of fertilization. In some cases hermaphroditism is induced due to parasitism. The life-cycle is not much complicated. There is no alternation of hosts and no alternation of generations.
5. ***Effect on host.*** In some cases there is no noticeable effect on the host. However, in some cases, the hosts may show remarkable effects on metabolism due to parasite.

Parasitic Copepoda

The Copepoda contain many of the parasitic forms. There are a few parasitic cyclopods and harpacticoids and the orders Notodelphyoida, Monstrilloida, Caligoida and Lernaeopodoida are exclusively parasitic. There are over 1000 species in these four orders and they display a great diversity of structural modifications and variations in life-cycles.

Fishes are the common hosts of caligoid and lernaeopodoid copepods. These parasites are commonly called *fish lice*. The fish lice are ectoparasitic and attach to the gill-filaments or fins or the general integument. Other copepods are commensal or endoparasitic within polychaete worms, in the intestine of echinoderms, and in the alimentary canal of tunicates and bivalves. Some copepods are parasitic on other crustaceans.

All degrees of modification from the free-living copepod forms are exhibited by these parasites. This is illustrated by the general survey of some parasitic copepods.

1. *Ergasilus*. It is an ectoparasitic organism. It lives attached to the gills of several fishes. The antennae are modified into hooks for holding on to the host. The eyes are absent and the trunk appendages are highly reduced.

2. *Salmincola*. A special attachment button and thread are produced by a frontal gland on the head. The button attaches to the gill-filament of

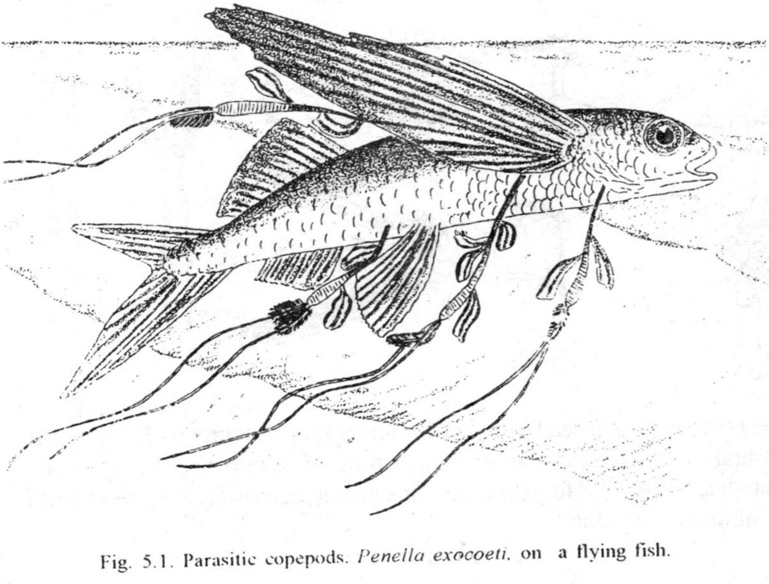

Fig. 5.1. Parasitic copepods. *Penella exocoeti*, on a flying fish.

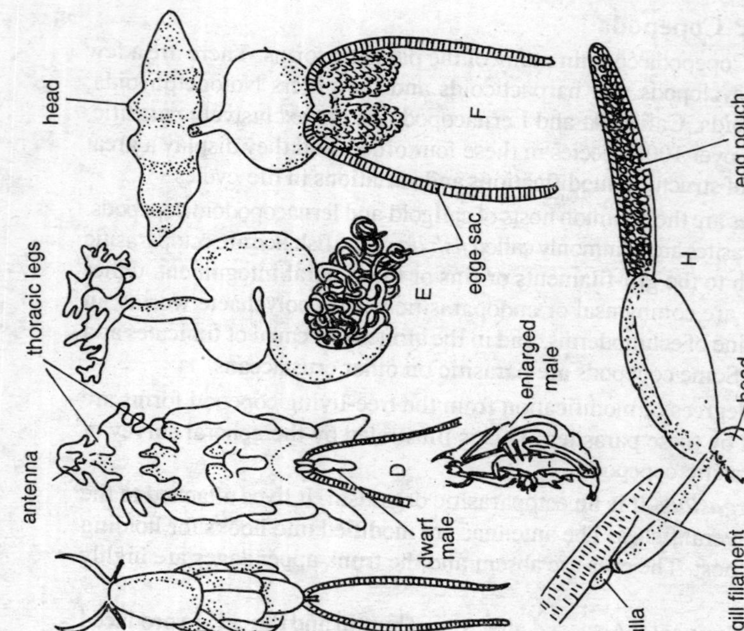

Fig. 5.2. Parasitic copepods. A. *Ergasilus*, B-*Caligus*, C-*Anthosoma*, D-*Chondracanthus*, male and female, E-*Lernaea*, F-*Leisteria*, G-*Mytilicola*, H-*Salmincola*.

the host, and the thread provides for temporary attachment. Later on the frontal gland produces another button or bulla into which the second maxillae are inserted for permanent attachment. External segment is reduced or altogether absent.

3. *Monstrilla.* The mouthparts and antennae are absent in adult but two dorso-lateral and one median eyes are present. The larva is endoparasitic within the blood vessels of polychaetes, possesses 2 or 4 large absorptive processes. The adult after emerging from the host, throws off antennae to assume a brief pelagic existence as its digestive tract is vestigial.

4. *Lepeophtheirus.* The mouthparts are adapted for piercing and sucking. The labrum and labium together form a tube containing the stylet-like mandibles. The antennae, maxillae and maxillipedes are also used to lacerate the surface of host.

5. *Caligus.* The sexual dimorphism is not well marked. Both the sexes live on the skin or inside gill chambers of fishes as ectoparasites. They can swim rapidly.

6. *Spyrion.* Morphologically degradation is more marked. Female is about 70 mm. long with a swollen head embedded in the skin of fish, while remainder of body hangs freely in water.

7. *Chondracanthus.* It is parasitic on the gills of various marine fishes. The sexual dimorphism is exhibited. Female is large, depressed, unsegmented and bearing paired irregular lobes. Appendages are degenerate. Mouth is provided with 3 pairs of sickle-shaped jaws. The male is much smaller and attached permanently to the female near genital aperture with the help of prehensile antennae.

Larval parasitism is exhibited by the Monstrilloida. On hatching, the free-swimming nauplius locates a polychaete or prosobranch gastropod host and burrows through the body wall. Inside the body of host the larva so atrophies that all of its typical crustacean characters are lost. The emergent stage immediately moults into an adult, which is free-swimming but lacks mouthparts or gut.

Parasitic Branchiura

Branchiura contains 75 species of common blood-sucking ectoparasites on the skin or in the gill-chamber of freshwater and marine fishes and on some amphibians. These are commonly known as *fish-lice* or *carp-lice*. The most common genus of carp-lice is *Argulus.* The most striking differences between branchiurans and copepods are the presence of a pair of sessile compound eyes and a large shield-like carapace covering the head and thorax. The abdomen is small, bilobed and unsegmented, and ending into a minute caudal furca.

The head bears a pair of sessile lateral compound eyes. The antennae (two pairs) are very small and the first pair is provided with a large claw for attachment to the host. The mouth is suctorial. The labrum and labium form a sucking tube, the *mouth cone.* In *Argulus,* there is also a large sheathed hollow spine located in front of the mouth cone. The spine is used for puncturing the skin of the host and is supplied by a gland that is supposedly poisonous. The second maxillae are heavy and uniramous terminating in claws. The maxillipeds are absent. The four pairs of thoracic appendages are large and biramous with swimming setae and with a proximal extension of exopodite. The branchiurans can detach and swim from one host to another.

The sexes are separate and both the sexes crawl freely over the surface of host and leave regularly at the breeding seasons. The breeding seasons

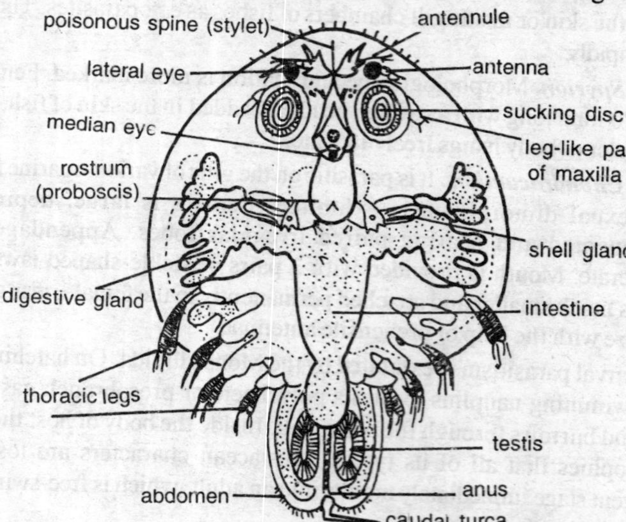

Fig. 5.3. *Argulus.*

are spring and autumn. Fertilization is internal and the eggs are deposited on the bottom. A post nauplius stage hatches from the egg and is parasitic like the adult. The branchiurans continue to moult after reaching sexual maturity.

Parasitic Cirripedia

Cirripedia includes the familiar marine animals known as *barnacles.* Cirripedes are the only sessile group of crustaceans, aside from the parasitic forms, and as a result they are one of the most aberrant group of Crustacea. Many cirripedes are parasitic and exhibit various degrees of degeneration according to nature of parasitism.

1. *Laura.* It is an Ascothoracica. It is an endoparasite in Octocorallian corals. The mouthparts are modified for piercing and sucking. The thoracic appendages are reduced and less than 6 pairs in number. The intestine has extensive diverticula into the mantle. The females are larger than the males.

2. *Trypetesa.* It is also a member of Ascothoracica. It bores into the shells of certain molluscs. Four pairs of thoracic limbs are present and the abdominal segments are absent.

3. *Proteolepas.* It was found by *Darwin* in the mantle cavity of the stalked barnacle (*Alepas*). It is a small, maggot-like animal. The mantle, trunk appendages and anus are absent. The body is divided into rings.

4. *Sacculina and Peltogasterella.* Both belong to Rhizocephala. These are exclusively parasitic on decapod crustaceans. They have worldwide distribution *Peltogasterella* is parasitic on hermit crab (*Eupagurus*), where as *Sacculina* is parasitic on other crabs. The adult body has no segmentation, appendages, digestive system and sense organs. The body is a tumour-like structure attached to the abdomen of its host by a short peduncle or stalk. A thin sac-like mantle encloses the visceral mass which is made up of hermaphrodite gonad. A root-like system of numerous delicate rhizoids arises from the stalk to penetrate to all parts of the body of host to absorb nutrients. A typical *nauplius* larva hatches from each fertilized egg into the brood pouch. The larvae escape from the brood pouch into the sea by an opening. The larvae develop into *cypris* larvae. The larva attaches to the bristle of a crab by means of its antennae and becomes *kentrogen* larva. The kentrogen penetrates the body of host to become an endoparasite cell mass. In a few weeks the endoparasite becomes external as a small sac having female gonad. A cypris larva, destained to become male attaches itself to the female and injects into its brood pouch a cell mass that later differentiation into a testis. The female is now converted into a hermaphrodite organism.

The effects of parasite are seen in the host. The generative organs of the host become reduced or completely degenerated. This phenomenon is known as *parasitic castration*. The infested male host assumes female secondary sexual characters. The female host shows merely parasitic castration and no male secondary sexual characters are developed.

Parasitic Isopoda

Isopoda is one of the largest group of Crustacea. There are about 4000 species, most of them are free living and several are parasitic in nature.

1. *Cancricepon.* It is a parasite in the gill-chambers of certain crabs.

Piercing mouthparts are present. The female is larger and the male is smaller and remains attached to female.

2. *Bopyrus.* It is a parasite in the gill-chambers of prawns. The mouthparts are suctorial and the mandibles are piercing. Maxillae are absent. The female is large, shows greater degeneration and asymmetry. The male is very small and permanantly attached to female.

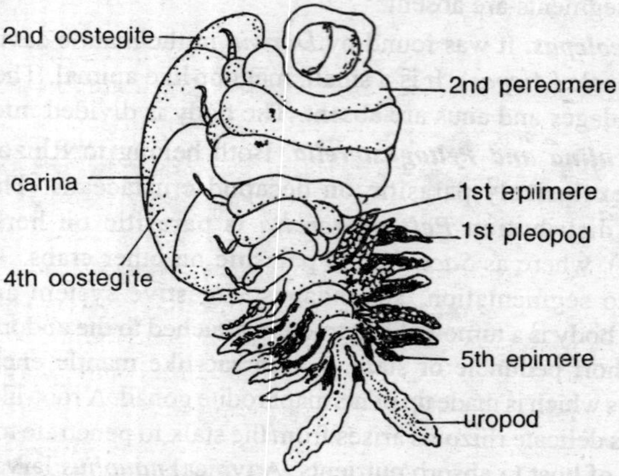

Fig. 5.4. Parasitic epicaridean isopods. The bopyrid *Cancricepon elegans*, parasitic in the gill chambers of certain crabs.

3. *Cryptoniscus.* It is a parasite on rhizocephalan cirripedes. It is protandrous hermaphrodite. The adult female is extremely degenerate.

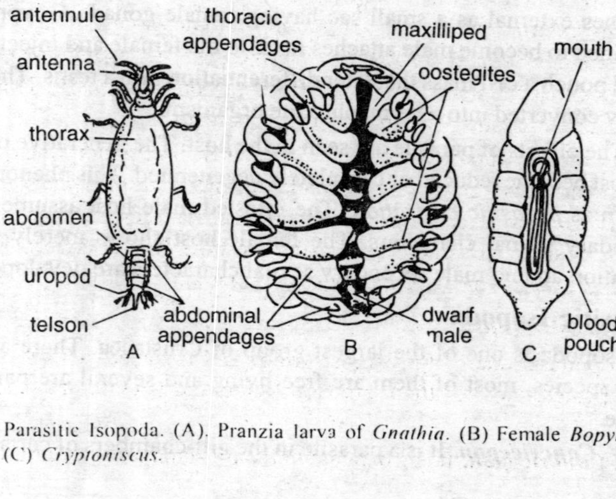

Fig. 5.5. Parasitic Isopoda. (A). Pranzia larva of *Gnathia*. (B) Female *Bopyrus*, (C) *Cryptoniscus*.

Parasitic Amphipoda

There are more than 5500 species of group Amphipoda. Most of them are free-living marine organisms. However, few are ectoparasites on fishes (*Lafystius* and *Trischinostoma*). *Cyamus* is an amphipod commensal on whales. It is commonly known as *whale-louse*. The mouthparts are suctorial. The legs are clawed, adapted for clinging to the skin of whale. The abdomen is vestigial and the thorax is dorsoventrally flattened.

6

Periplaneta americana

The cockroach is a familiar type of the Class Insecta belonging to Order Blattaria. These are very ancient insects and are found practically all over the world except the polar regions. There are about 2600 species of cockroaches, of which, the common species in India are *Periplaneta americana, Periplaneta australasiae*, *Blatta orientalis* and *Blatta germanica*. *Periplaneta americana* was originally designated as *Blatta orientalis* by *Linnaeus*, later changed to *Blatta kakeralac* by *De Geer* and finally the present name was given by *Burmeister*.

SYSTEMATIC POSITION

Phylum	-	Arthropoda
Class	-	Insecta
Sub-class	-	Pterygota
Division	-	Exopterygota
Order	-	Blattaria
Family	-	Blattidae
Genus	-	*Periplaneta*
Species	-	*americana*

Habits and Habitat

They are generally found in warm damp places particularly in kitchens, cup-boards, hotels, drains, ships, trains and even sewers. Cockroaches eat every thing fit for human consumption, besides they will eat leather, paper, cloth and wool, thus they are *omnivorous*. Sometimes when the food is not available they turn *cannibal* and eat each other. Cockroaches are *nocturnal* i.e. they come out to feed at night. Cockroaches are *cursorial* i.e. run very fast though they can fly also. The eggs are laid in oothecae. The young ones hatching from oothecae resemble the adults in mode of life except that they are smaller in size.

EXTERNAL MORPHOLOGY

Shape and Size

The body of cockroach is narrow elongated, bilaterally symmetrical and dorso-ventrally flattened. It is about 4 cm. in length and 2 cm. in width.

Colouration

The colour of body is reddish brown. The tergum of the first thoracic segment bears two dark patches surrounded by a light brown margin.

Exoskeleton

The entire body is covered by a hard brown-coloured chitinous exoskeleton. Each segment of the body has hardened plates called *sclerites* which are joined to each other by thin flexible *articular membranes.*

Division of body

The body is divisible into three distinct regions :

(1) Head, (2) Thorax and (3) Abdomen

1. Head

The head comprises the anterior body region of cockroach. It is sclerotized capsule formed by the fusion of several sclerites, often distinctly seen in the embryo. It lies at right angle to the longitudinal axis of body and is highly movable in all directions due to a flexible neck.

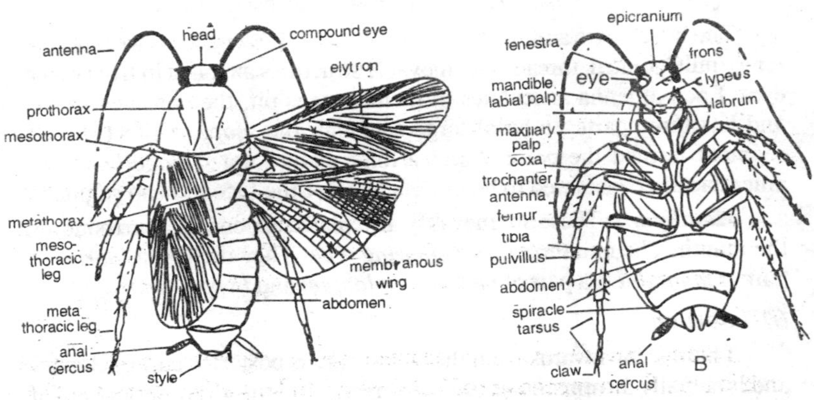

Fig. 6.1. *Periplaneta.* A—Dorsal view, B—Ventral view.

The head bears large compound eyes which are broad dorsally. A dark sclerite covers the top of the head and extends downwards between the eyes. It is called *vertex.* In the nymph the vertex is divided into two *epicranial plates* by an inverted 'Y' shaped *epicranial suture.* During moulting of the nymph, the head capsule splits at the epicranial suture, and the epicranial suture disappears in adult. The unpaired triangular plate, lying between the arms of the epicranial suture is the *frons*. Below the frons is *clypeus* and *labrum*. Covering the sides of the head and lying below the compound eyes are cheek sclerites or *genae*. In the angle between the eyes and antennae are two small pale coloured areas called *fenestrae*, which are undeveloped *ocelli*. At the back of the head is a large rectangular *occpital foramen* bordered by an arched sclerite, the *occiput*.

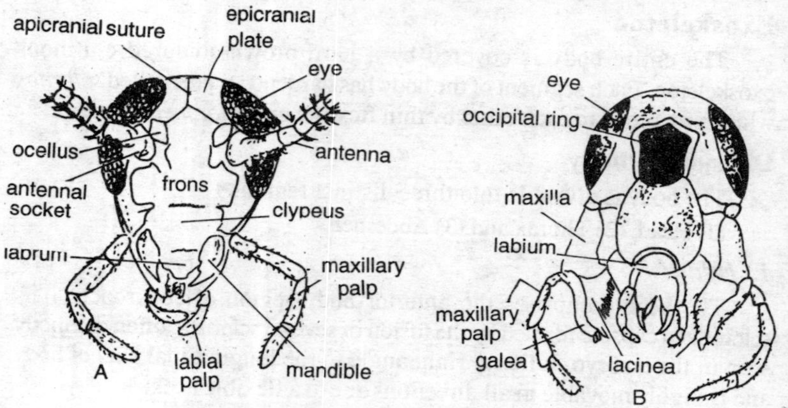

Fig. 6.2. *Periplaneta*. Head A—Front view, B—Hinder view.

The head bears a number of appendages. The *antennae* are a pair of long, multi-jointed, thread-like, movable structures attached in front of the eyes. Each antenna articulates in membranous pit, the *antennal socket* and has three parts : a basal large segment, the *scape*, a short second segment termed the *pedical* and a long many jointed *flagellum*. The antennae are considered to be modification of the second head segment.

Mouthparts. These are movably-articulated appendages surrounding the mouth. The mouthparts are *chewing type*. They include a *labrum*, a pair of *mandibles*, a pair of *maxillae*, a *labium* and *hypopharynx*.

(i) Labrum

It hangs downwards from the membranous posterior part of clypeus and is actually an appendage of 3rd segment. It forms the anterior wall of the preoral cavity and is called the *upper lip*. It bears on each side of its inner anterior surface, a sclerite-*torma*, and two rows of *gustatory setae*.

(ii) Mandibles

The mandibles articulate with the genae and lie on the sides of mouth just behind the labrum. These are the appendages of 4th segment and have ball and socket attachment with the head. These are heavily sclerotized and each bears 3-teeth like structures and a small serrated molar area. The upper inner margin bears a membranous lobe, the *prostheca* from which sensory hairs arise. Each mandible is worked by four muscles.

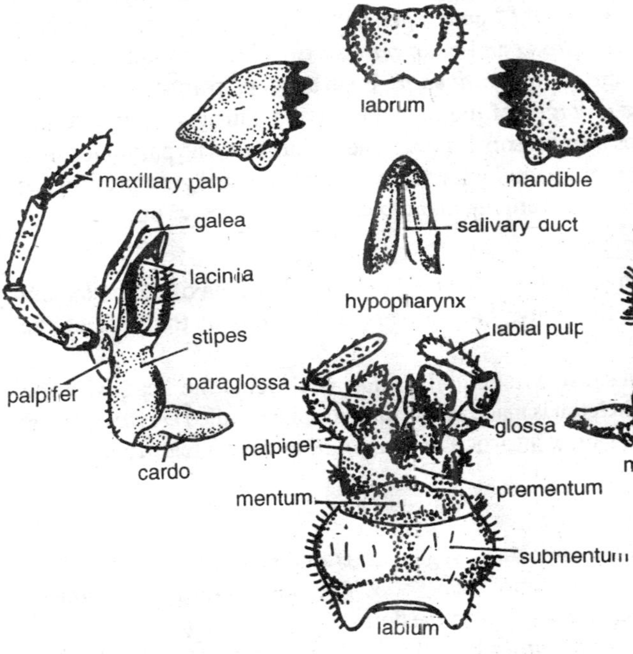

Fig. 6.3. *Periplaneta.* Mouthparts.

(iii) Maxillae

Two maxillae lie below the mandibles. They are regarded the appendages of 5th segment. Each maxilla is biramous appendage consists of three parts: The basal *protopodite*, inner *endopodite* and outer *exopodite*. The protopodite is made of two joints, proximal *cardo* and distal *stipes*. The exopodite is a long structure arising from a small projection, the *palpifer*, on the outer side of the distal end of stipes. It is formed of 5 joints and known as *maxillary palp*. The endopodite is attached to the distal end of the stipes internal to maxillary palp. It is made up of two pieces, outer *galea* and inner *lacinia*. The lacinia bears bristles along the inner margin, and ends in the two sharp claws. The maxillae hold food by the claws of lacinia and bring it to the mandibles for mastication. They are also used for cleaning of antennae, palps and front legs.

(iv) Labium

The labium lies behind the first pair of maxillae. It is formed by the fusion of second pair of maxillae along the middle line. The basal joint of the fused protopodites is the *submentum*, the second is *mentum* and the third is *prementum*. The endopodites of labium are almost separate, and are together known as the *ligula*. Each half of ligula comprises two pieces the inner smaller *glossa* and outer *paraglossa*. The exopodites are also separate. They are called *labial palps*. Each labial palp arises from a short projection, the *palpiger,* of the premantum and consists of three joints. The labial palps bear sensory bristles. The labium takes no part in feeding, but the glossae and paraglossae together termed the *ligula* prevent the loss of food particles from the mandibles thus act as lower lip.

(v) Hypopharynx

It is the tongue or *lingua* which lies between the maxillae and labium inside the mouth cavity. The salivary duct opens at the base of the hypopharynx.

Neck. A narrow, elastic and a short neck joins the head with the thorax. Its exoskeleton is made up of chitinous plates called *cervical plates.* It is nearly extension of an articular membrane and not a segmental region of the body.

2. Thorax

The thorax consists of three segments (i) anterior *prothorax*, (ii) middle *mesothorax* and (iii) posterior *metathorax*. The exoskeleton of each segment consists of a *notum* or *tergum* on the dorsal side, a *sternum* on the ventral side and a *pleuron* on either lateral side. The tergum (pronotum) of the prothorax is the largest and projects forward to conceal the neck and a part of head also. The mesotergum and metatergum are smaller and rectangular. The sclerites of each segment and also those of the adjacent segments are joined by thin, soft, flexible *arthrodial membranes* to permit movement.

Legs. Thorax bears three pairs of legs, one pair per segment. All the legs are similar in structure. They are called *prolegs, mesolegs* and *metalegs,* on the basis of the segment which bears them. Each leg bears five segments:

(i) *Coxa* which is attached with the thorax between pleuron and sternum, forms a movable joint.

(ii) *Trochanter*, which is smaller than coxa, can move freely on it.

(iii) *Femur*, which is long and broad.

(iv) *Tibia*, which is long and of uniform thickness.

(v) *Tarsus*, which is the last segment of the leg. It is made up of 5 movable podomeres. The last podomere, ending into two small-hooked claws, is often called *pretarsus*. Between the claws, there is a delicate, hair-covered porous pad, the *pulvillus* or *arolium*. The tarsal segment bears many fine hairs, while at its lower edges are soft adhesive pads, the *plantulae*. The pads afford grip on the smooth and slippery surface where claws fail to hold on.

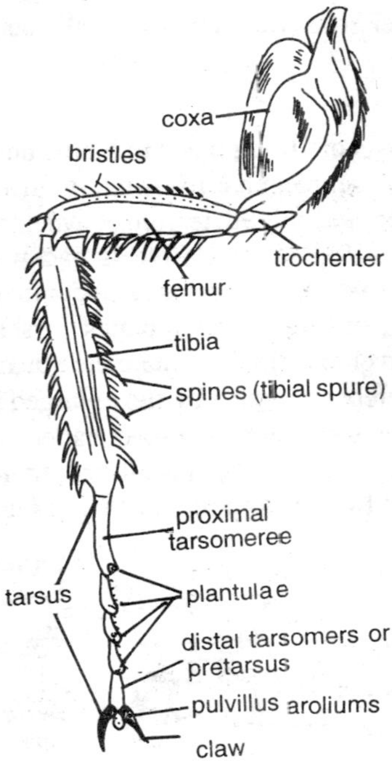

Fig. 6.4. *Periplaneta.* A leg.

Wings. Besides the legs, the thorax also bears two pairs of wings. According to their position, the wings are called the *mesothoracic* and *metathoracic*. The wings are formed as lateral extensions of the integument between the tergum and the pleuron near the anterior end of the segment. The first pair of wings are heavily sclerotized and are known as *elytra* or *tegmina*, they are protective and cover the hind-wings in a folded state, the left tegmina partly overlaps the right. The second pair of wings are membranous and larger, they lie folded below the tegmina when at rest. Each wing has two thin membranous layers of cuticle enclosing tubular and chitinous prolongations of the haemocoel between them known as *veins* or *nervures*. They are highly sclerotized and enclose small trachea and even nerve fibres.

The movements of each wing, during flight, are effected by two sets of muscles, the *direct* and the *indirect* muscles. The direct muscles are so called because they are inserted directly on the sclerites at the base of the wings. The indirect muscles are not inserted on the wings. One group of indirect muscles runs dorsally and longitudinally between the mesothorax and metathorax, the other group runs dorso-ventrally between the tergum and sternum.

3. Abdomen

The abdomen consists of 10 segments in adults but 11 segments in embryoes. Typically, each segment has its own tergum and sternum. The 8th and 9th terga of female, and only 9th tergum of male, are largely covered by the 7th tergum. The 10th tergum projects backwards beyond the abdomen and is cleft into two lobes whereas all abdominal segments have their own terga, there are only nine sternal plates. First sternum is small but all others are well developed. In the female the 8th and 9th sterna are invaginated and completely concealed by the enlarged boat shaped 7th sternum. This large sternum encloses a big space called *oothecal chamber* or *gynovalvular chamber*, where ootheca is formed. In males the 8th and 9th sterna are concealed by 7th sternum but there is no such chamber,

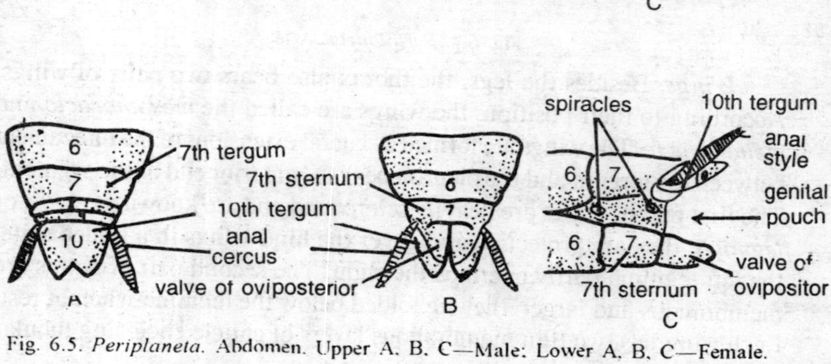

Fig. 6.5. *Periplaneta*. Abdomen. Upper A. B. C—Male: Lower A. B. C—Female.

however, there are a pair of small, thin and unjointed *anal styles* projecting back from the 7th sternum. Anal styles are absent in female. Other paired outgrowths are called *anal cerci* arise from the 10th tergum. These cerci are 15 jointed, longer and thicker structures. These are present in both the sexes.

In male there are a group of chitinous hooks and plates near the end of the abdomen, these are the *gonapophyses* which help in copulation. Some chitinuous structures in the female have a similar position and form an *ovipositor*.

In the 10th segment just below the tergum is *anus* supported by a pair of *podial plates*. The *gonopore* in both the sexes occurs below the anus and is surrounded by gonopophyses.

Spiracles. There are 10 pairs of small slit-like apertures on the lateral sides of the body, called *spiracles*. Two pairs are larger and situated on the thorax, one between the pro- and meso-thorax and other between meso- and meta-thorax. Rest of the 8 pairs are called *abdominal spiracles* present on the sides of abdominal segments 2-8, each being located between the two adjacent terga and two adjacent sterna.

Stink glands. The intersegmental membrane between the 5th and 6th abdominal terga is depressed on either side as *stink glands*. These glands produce a secretion with a characteristic odour.

BODY WALL

The body wall has three distinct layers :

(i) Cuticle, (ii) Epidermis, (iii) Basement membrane.

(i) Cuticle

It is the outermost layer and is differentiated into three regions : an outer *epicuticle*, middle *exocuticle* and inner *endocuticle*. The epicuticle has an outer layer of wax- like lipoid and an inner layer of hard protein, but the chitin is absent. It bears movable and fixed bristles. The exocuticle is a thick, laminated, flexible layer of chitin. It bears pigments and is sclerotized. It gives both rigidity and elasticity. The endocuticle is made up of protein and chitin arranged in horizontal lamellae. A few *pore canals* are present in exo-and endocuticle.

(ii) Epidermis or hypodermis

It comprises a single layer of columnar epithelial cells resting on the basement membrane. It secretes cuticle. The gland cells, which secrete movable hair-like setae that project from the surface of cuticle, are present in epidermis.

In adult cockroach large oval cells are present below the epidermal cells called *oenocytes*. The oenocytes probably secrete wax.

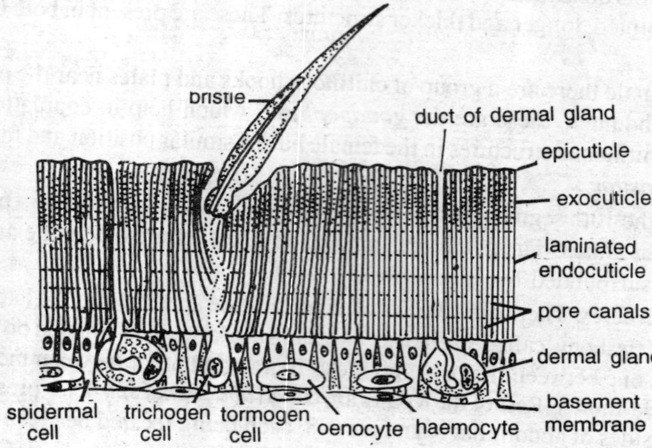

Fig. 6.6. *Periplaneta.* T.S. of body wall.

(iii) Basement membrane

The basement membrane is generally 0.5 micrometer or less in thickness and in electron microscope appears as a continuous, amorphous granular layer. It is composed of mucopolysaccharides.

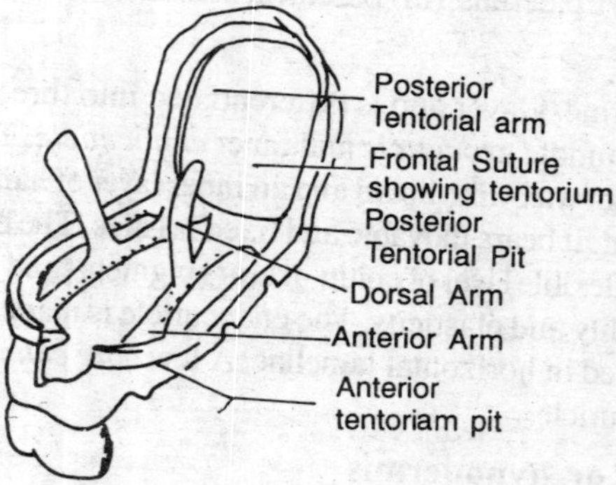

Fig. 6.7. *Periplaneta.* Tentorium.

Endoskeleton

The endoskeleton in cockroach is made up of chitin. The endoskeleton of head is known as *tentorium*. It lodges in the posterior part of the head.

It consists of a chitinous plate with a small aperture in it. Three pairs of processes arise from this plate; *anterior arms* attached with clypeus and frons, *posterior arms* attached with the posterior part of the head and *dorsal arms* bent towards the antennae.

Body cavity

The true coelom is reduced and is represented by the cavities of gonads. The body cavity is filled with blood and is called the *haemocoel*. Loose tissues known as *fat bodies* or *corpora adiposa* are found in the body cavity.

DIGESTIVE SYSTEM

The digestive system comprises the *alimentary canal* and *digestive glands* are associated with it.

1. Alimentary Canal

The alimentary canal comprises three regions.

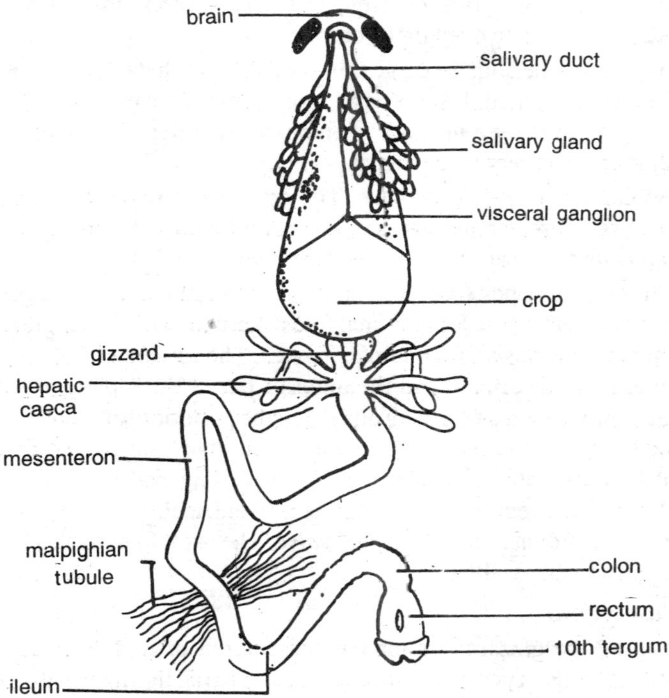

Fig. 6.8. *Periplaneta*. Alimentary canal.

1. Foregut or Stomodaeum
2. Midgut or Mesenteron
3. Hindgut or Proctodaeum.

1. Stomodaeum

Its inner wall is made up of epithelial cells and lined with cuticle. The stomodaeum comprises: preoral cavity, pharynx, oesophagus, crop and gizzard.

(i) *Preoral cavity.* It is bounded anteriorly by the labrum and clypeus, posteriorly by labium and laterally by mandibles and maxillae. Inside the cavity the hypopharynx is present. It bears the opening of salivary duct. The mouth lies on the roof of the preoral cavity. The mouth leads into a tubular pharynx.

(ii) *Pharynx.* It is a vertical canal that rises up in the head capsule and passes over the tentorium, and then bends to open into the oesophagus near the occipital foramen.

(iii) *Oesophagus.* It is a narrow tube that is present in the neck and extends into thorax to open in the crop.

(iv) *Crop.* The crop is a large thin walled, pear-shaped sac extending well into the abdomen. It serves as a reservoir for storing food. Its outer surface is covered by a network of tracheae. Its internal epithelial and cuticular lining is very much folded.

(v) *Gizzard or proventriculus.* The crop leads behind into a small, thick walled, muscular and cone shaped proventriculus. Anteriorly it bears an *armarium* and posteriorly a *stomodael valve.*

The armarium possesses internally six longitudinal folds that greatly reduce its lumen. These longitudinal folds alternate with six longitudinal folds alternate with six longitudinal grooves. The cuticular lining of each longitudinal fold forms (a) in the anterior part a thick plate produced centrally into strong, sharp teeth and (b) in the posterior part, a *thin plate.* Behind each longitudinal fold the cuticular lining of the gizzard forms a soft cushion-like lobe, the *pad* or *pulvillus,* with long, backwardly directed hair. Behind the region of the pads the proventriculus extends into the lumen of the midgut as the *stomodael valve*. The latter folds back on itself, and is thus double-walled.

2. Mesenteron

Behind the gizzard is a tubular mesenteron or midgut lined with endodermal cells, it performs the tasks of completing the digestive process and absorption of food. The endodermal cells of the midgut break down food particles, while secretory cells and their contents are discharged into

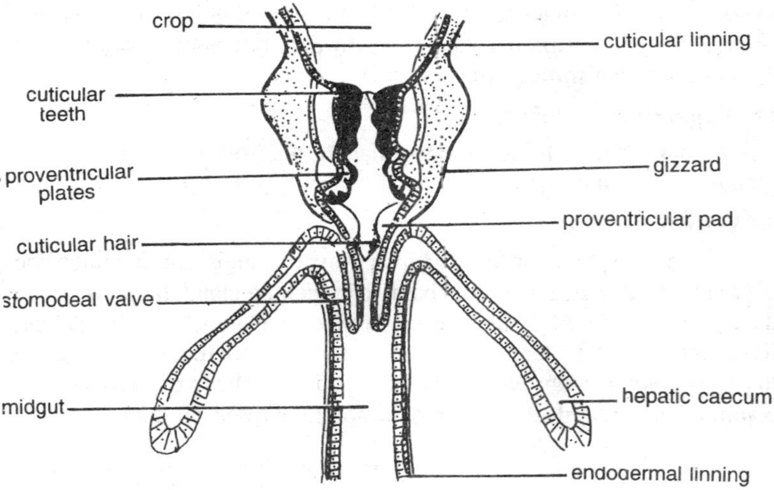

Fig. 6.9. *Periplaneta*. Section through gizzard and mesenteron.

the lumen and fresh cells are formed. From the anterior end of the midgut arise eight tubular *hepatic caeca*. From the posterior end 70-80 very fine, thread-like yellow *Malpighian tubules* are given out.

3. Proctodaeum

Its wall is also made up of ectodermal epithelium and lined internally with cuticle. It consists of the three parts *ileum*, *colon* and *rectum*. At the junction of mesentercn and ileum are present Malpighian tubules arranged

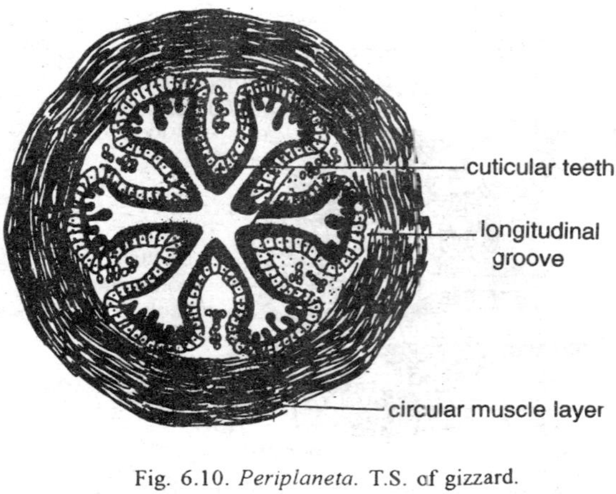

Fig. 6.10. *Periplaneta*. T.S. of gizzard.

in six groups. The inner surface of rectum is raised into six longitudinal ridges, the *rectal papillae* which help in the absorption of water. The rectum opens out through the anus.

II. Digestive glands

Associated with the alimentary canal, the digestive glands are salivary glands and hepatic caeca.

1. Salivary glands

There are a pair of diffused whitish salivary glands, one on either side of crop in the thorax. Each gland has two branching glandular parts whose ducts form a common glandular duct. Between two branches of each gland lies a thin walled sac-like *reservoir*. The ducts of both the reservoirs unite to form *common reservoir duct* which opens into the hypopharynx. The common glandular duct opens into the common reservoir duct.

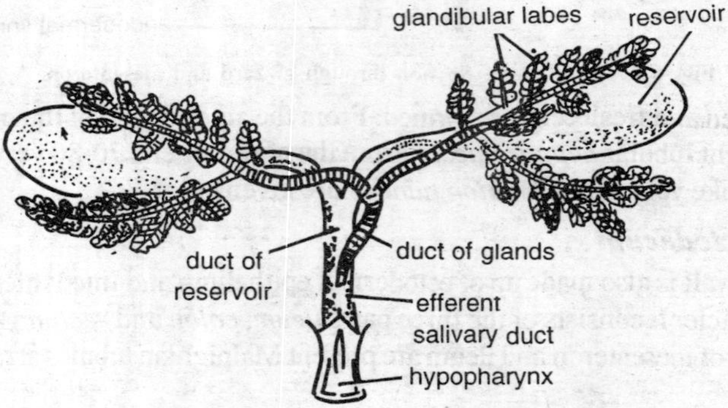

Fig. 6.11. *Periplaneta*. Salivary glands.

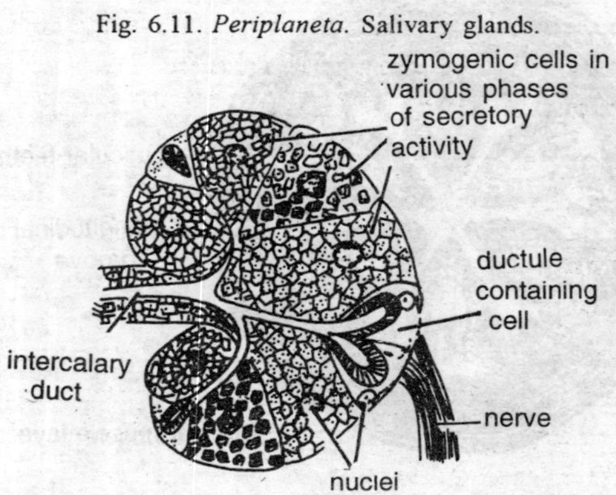

Fig. 6.12. *Periplaneta*. Section of salivary gland.

Each gland consists of several secreting *acini* or *lobules* in grape-like clusters and connected together by fine tubules. Two types of cells are found in acini, *zymogenic cells* and *ductule containing cells.* Both types of cells secrete saliva which contains enzyme *zymase* and mucoid substance. *Intercalary ducts* from the various acini join each other and are connected to their reservoir where saliva is stored.

2. Hepatic caecae

These are 7-8 finger-like, hollow tubes, ending blindly and open into the midgut. They secrete digestive enzymes.

Food and Feeding

The cockroach is an omnivorous animal and feeds on all sorts of food except metals, bricks, stones etc. The food is located by the sense of smell, whose receptors lie on the antennae and palps. The food is caught and masticated with its mandibles and maxillae.The broken food is pushed into the pre-oral cavity with the help of labrum and other mouthparts. During chewing the food is mixed with saliva brought into the preoral cavity by the common salivary duct. *Amylase* of saliva converts starch into sugar. Digestion of food started in pre-oral cavity by saliva continues in the crop. Enzymes from midgut also come into the crop through the gizzard for the digestion of some fats and proteins. From the crop the food reaches gizzard where it is ground by the contraction and expansion of muscles and the chitinous teeth. The food is than filtered by the bristles. In the mesenteron the food is acted upon by the enzymes secreted by hepatic caecae and the lining of midgut. The secretion contains many enzymes such as *peptase*, *trypase*, *invertase* and *lipase*. The digested food is absorbed by the lining of midgut and hepatic caecae.

In the midgut, *peritrophic membrane* is formed round the food for the protection of the lining of it from the abrassion by hard indigestible particles in the food. The membrane is a thin chitinous tube formed of the secretion from the funnel-like extension of the gizzard round the food. It is permeable to both, the enzymes and digested food.

The undigested food reaches the rectum through hindgut. The absorption of water takes place in the rectum and the undigested food is converted into almost solid faeces. The faeces is temporarily stored in the rectum from where it is passed out at interval through the anus.

CIRCULATORY SYSTEM

The circulatory system is poorly developed. It is of *open type* i.e. the blood vessels open into spaces and not into capillaries. Thus the blood comes in direct contact with tissue. There are three main parts in the circulatory system.

1. Haemocoel

The haemocoel is devoid of true epithelial lining. It is filled with haemolymph. It is partitioned by a dorsal and a ventral diaphragms into three sinuses - (i) the dorsal *pericardial sinus*, containing the *drosal blood vessel* or *heart*, (ii) the middle *perivisceral sinus*, containing the gut, and (iii) the ventral *perineural* or *sternal sinus*, containing the nerve cord.The diaphragms or septa are provided with pores or *fenestrae*.The ventral diaphragm also extends as septa into the legs. In each leg the septum divides its cavity into two sinues, one for the outward and the other for inward flow of blood.

The pericardial septum is slightly bulged upwards.There are a series of paired triangular muscles, the *alary muscles* attached to the pericardial septum by their broad bases and to the terga by their pointed ends.There are 12 pairs of alary muscles in *Periplaneta.*

2. Heart

The heart is a long, muscular, contractile tube extending along the mid-dorsal line just beneath the terga of thorax and abdomen. It is enclosed in the pericardial sinus. It is composed of 13 chambers, each chamber

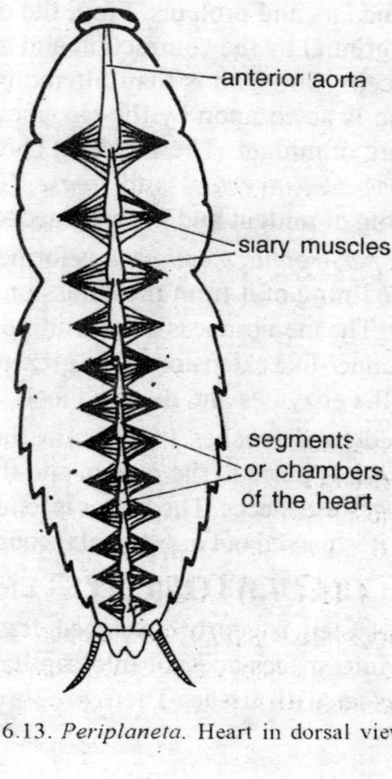

Fig. 6.13. *Periplaneta.* Heart in dorsal view.

communicating by a valvular opening (auricular valve) with the one infront of it. The chambers communicate to the pericardial sinus by paired lateral openings called *incurrent ostia.* Ostia are also valvular allowing the entry of blood from the pericardial sinus into the heart. The heart is closed behind and is continued infront as a short tube, the *aorta* which opens into the haemocoel of the head. It is provided with *aortic valves.*

Nutting (1951) has described that in roach, *Blaberus,* there are 6 paired lateral segmental blood vessels of which 2 thoracic and 4 abdominal pairs. Through these blood vessels the blood leave the heart.

3. Blood

The *blood* or *haemolymph* is contained in the haemocoel. It is only the extracellular fluid and makes 15-75% volume of the insect.

It consists of liquid plasma and numerous colourless blood cells or haemocytes. There are about seven types of haemocytes (*Wigglesworth,* 1965).The principle function of the haemocytes is *phagocytosis* i.e.the ingestion of small solid particles. Some haemocytes fulfil the function of food storage since they contain inclusion of glycogen and fat. In insect the respiratory pigment is absent in blood and thus plays no role in respiration. Instead it helps in transportion of digested food materials, collects nitrogenous waste products from the tissues and brings to the extractory organs for their removal, healing of wounds, transportation of the hormones etc.

Circulation of blood

The heart is the principal pulsatory organ and undergoes rhythmical contractions and relaxations which are brought about by the muscle fibres of its walls.The rhythm of the heart appears to be purely myogenic (*Miller,*

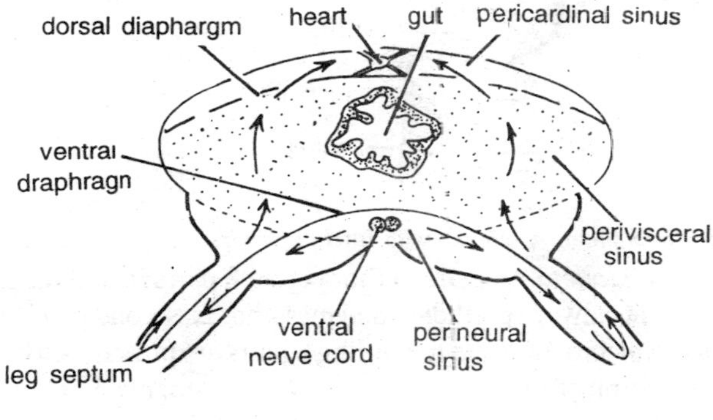

Fig. 6.14. *Periplaneta.* Course of blood circulation in T.S. of a thoracic segment.

1974) though it is influenced by nervous system. Contraction of the heart takes the form of a wave of peristalsis which runs forward from the posterior end. The blood is poured into the head sinus through the dorsal aorta. There is an *accessory heart* at the base of each antenna which sends haemolymph into the antenna. From the head sinus, the haemolymph flows into the perivisceral and sternal sinuses. Now the alary muscles contract, so that the pericardial septum becomes flat. This increases the capacity of pericardial sinus. The haemolymph, therefore, flows from periviscreal sinus to the pericardial sinus through the perforations of diaphragm. On relaxation of the alary muscles, the pericardial diaphragm bulges upwards, this presses blood which enters the heart through ostia. The rate of heart beat in *Periplaneta* is 49 beats per minute.

RESPIRATORY SYSTEM

In the cockroach respiration takes place by means of internal air tubes known as *tracheae*. These ramify through the organs of the body and its appendages, the finest branches being called *trachioles.* The air generally enters the tracheae through *spiracles,* which are lateral in position in the thoracic and abdominal regions.

1. Spiracles

The spiracles are derived from the mouths of the ectodermal invaginations which give rise to the *tracheal system*. Ten pairs of spiracles

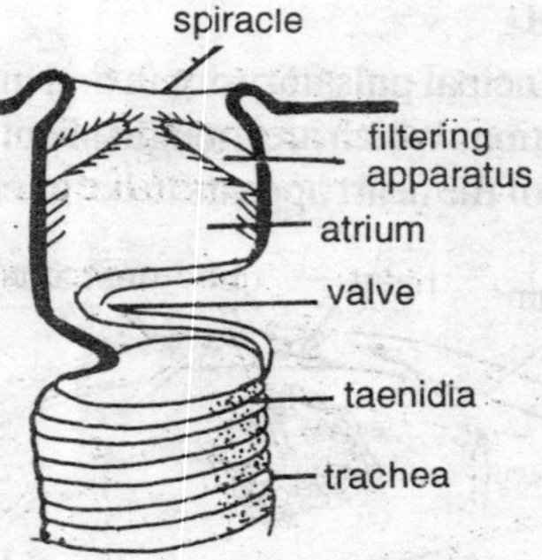

Fig. 6.15. *Periplaneta.* A spiracle with filtering apparatus, valve and atrium.

are found in *Periplaneta* : 2-pairs in thorax and 8 pairs in abdomen. The first pair is present between prothorax and mesothorax, second pair between mesothorax and metathorax and remaining 8 pairs are present between the sternum and pleuron of first to eighth abdominal segments. Each spiracle is surrounded by an annular sclerite called *peritreme* and leads into a chamber called *atrium* from which the trachea arises. The spiracle is

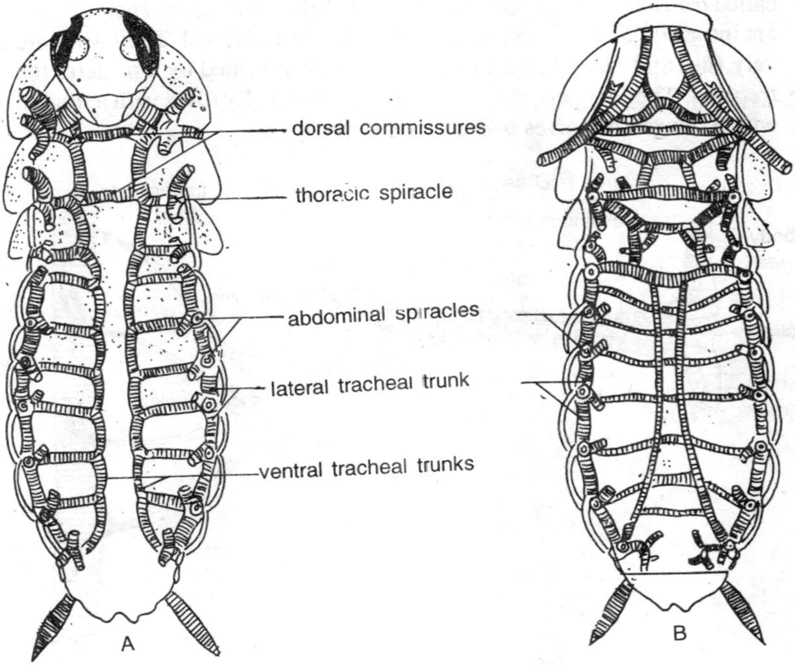

Fig. 6.16. *Periplaneta.* The tracheal system.

provided with the *closing apparatus*. The apparatus consists of one or more muscles which are associated with cuticular parts and by closing the spiracular aperture, prevents excessive loss of water vapour. The atrium is a specialised region, lacking *taenidia* or *intima*, provided with hairs which prevent the entry of sand particles with air. The first pair of spiracles is the largest. They are in the form of oblique slits. The second pair is smaller and simpler than first pair.

2. Tracheae

The tracheae are elastic, cuticular tubes, which, when filled with air, have a salivary gland appearance in dissections. The cuticular lining or *intima* is continuous with that of external body surface and secreted by the tracheal matrix cells derived from epidermis. A delicate basement membrane forms the outermost coat of the tracheae. There are 3 pairs of large, parallel, longitudinal *tracheal trunks*, one dorsal, one ventral and one lateral in position. These are connected together by *transverse commissures*.

The tracheae branch and rebranch to from ultimate finer branches called *tracheoles*. They have a diameter of only one micron. Their cavities are intracellular, i.e. each tracheole is made of a single cell. Their walls are very thin and devoid of intima, instead they are lined by a protein, the *trachein.* They are permeable to water and are usually filled with a fluid in which oxygen dissolves and diffuses to the tissue.

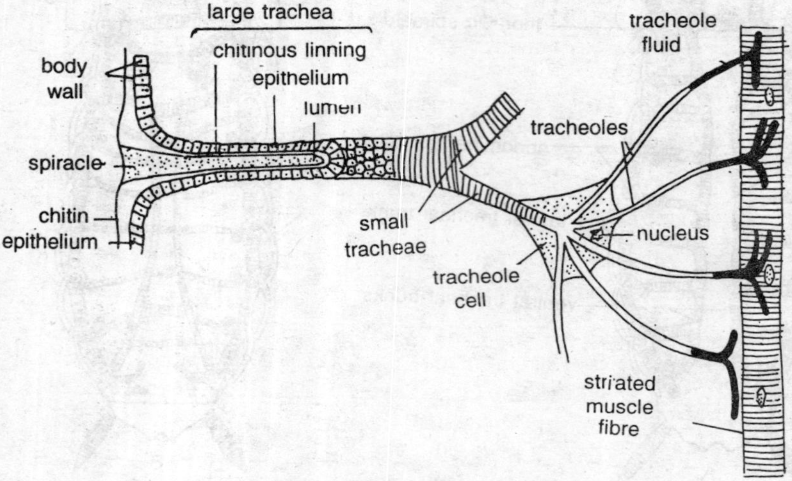

Fig. 6.17. *Periplaneta.* A trachea.

Mechanism of Respiration

By the contraction and expansion of *tergosternal muscles* the terga and sterna contract and expand as a result of which *inspiration* and *expiration* takes place. Expiration is an active process where as inspiration is passive. The first pairs of thoracic and abdominal spiracles, always remain open but the remaining open during inspiration but close during expiration. Air enters the spiracles during inspiration and comes to the tracheae, then it comes to the tracheoles which contain fluid, the oxygen dissolves in the fluid and reaches the cells of the tissues. Opening of the spiracles and subsequent diffusion of air are due to the stimulation of spiracles by carbon dioxide. In expiration some carbon dioxide may pass out through the spiracles but the major part of it diffuses out through the cuticular covering of the body. Carbon dioxide also dissolves in the plasma and reaches the body surface which is permeable to gases and allows carbon dioxide to pass out.

When the cockroach is running, the oxygen requirement increases, the special respiratory movements of the abdomen start. The fluid present in the tracheoles is withdrawn into the body cells and the air rushes forward to take its place, the oxygen reaches itself directly to the cells of the body tissues. It is not supplied through the medium of blood or *haemolymph*.

EXCRETORY SYSTEM

The excretory system is simple in cockroach. The principal excretory organs are the *Malpighian tubules*, working in conjunctions with the hindgut. Some haemocoelomic structures, such as the *nephrocytes, fat bodies* etc. have also been regarded as excretory organs in the past.

Malpighian tubules

They are long, slender blind tubes lying in the haemocoel where they are freely bathed by the haemolymph. They open at their proximal ends into the intestine, near the junction of mesenteron and hindgut. They are regarded as derivatives of ectodermal proctodaeum. The Malpighian tubules are richly supplied with a reticulum of fine tracheae whose larger branches serve to maintain these organs in position. Their number is between 60 to 150 and are arranged in 6-8 bundles. Each tubule is about 16 mm. long and 0.5 mm. in diameter and is lined by glandular epithelium with a characteristic *brush border*. The glandular cells extract the uric acid, water and inorganic salts from the haemolymph and secrete them into the lumen from where they flow into the hindgut to be eliminated with the faeces. The reabsorption of water takes place in the rectum. The Malpighian tubules have been observed to show peristaltic movements that bring more and more haemolymph in their contact.

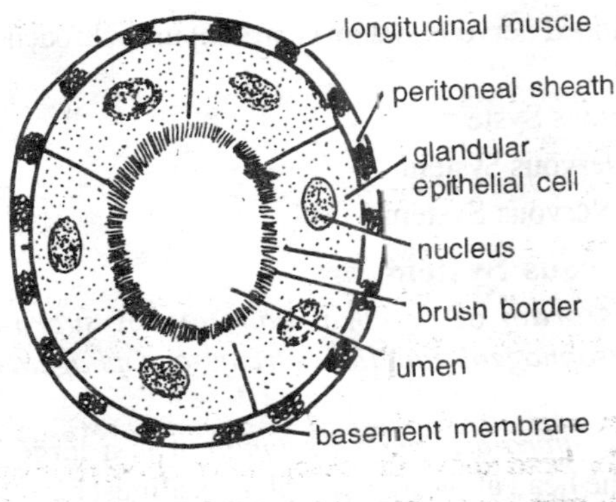

Fig. 6.18. *Periplaneta*. T.S. of Malpighian Tubule.

The role of Malpighian tubules in regulating water balance and ionic concentration presents complex biophysical problems still not fully elucidated.

Fat body cells

The fat-body is composed of irregular masses or lobes of rounded or polyhedral cells. These are usually vacuolated and contain inclusions of various kinds.The fat bodies are derived from mesoderm by differentiation. The nitrogenous waste products (uric acid and urates) are present in the form of excretory granules in specialized *urate cells* of the fat body. Throughout the life these products are deposited in these cells.

Besides urate cells the other types of cells present in the fat body are: (i) *Trophocytes* which store fat and glycogen as reserve food materials. (ii) *Oenocytes* which are believed to secrete cuticular lipids and it has been suggested that they may be the site of ecdysone production and (iii) *Mycetocytes* which are filled with symbiont micro-organisms that help in amino acid synthesis.

Uricose glands

The mushroom glands of male cockroach possesses long, blind tubules at its periphery called *uricose glands* or *utriculi majores.*

These glands store uric acid and discharge it over the spermatophore during copulation.

Cuticle

Some excretory substances probably accumulate in cuticle of the body wall shed along it during *ecdysis* or *moulting.*

NERVOUS SYSTEM

A well developed nervous system is present in cockroach. It comprises three main parts:

I. Central Nervous System

II. Peripheral Nervous System

III. Autonomic Nervous System

I. Central Nervous System

It comprises *brain* or *cerebral ganglion* or *supra-esophageal ganglion, sub-oesophageal ganglion, circum-oesophageal connectives* and *nerve cord.*

The *brain* or *supra-oesophageal ganglion* is a large and bilobed mass located in the head above the oesophagus almost between the bases of the antennae. It represents three fused ganglia of the head region. It is divided into three parts - the *protocerebrum*, *deutocerebrum* and *tritocerebrum.* The remaining three ganglia of the head fuse to form the *sub-oesophageal ganglion* which lies below the oesophagus, posterior to the brain. The brain and the sub-oesophageal ganglion are connected together, on either side, by a *circum-oesophageal connective*.

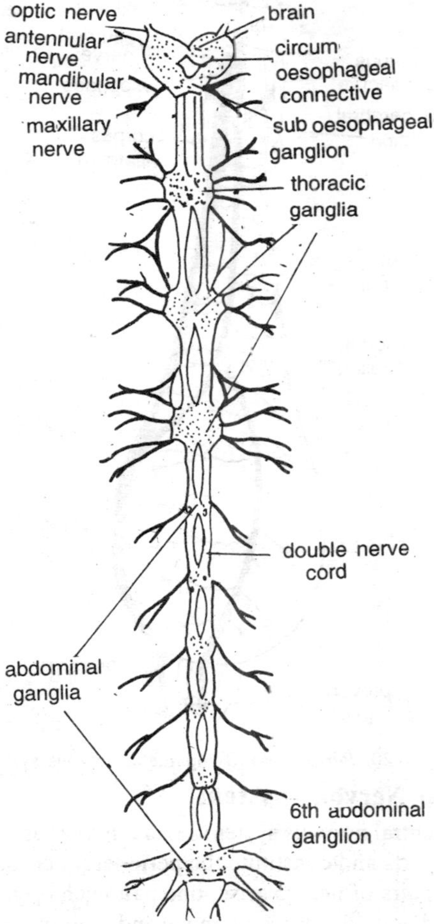

Fig. 6.19. *Periplaneta.* Nervous system.

From the sub-oesophageal ganglion, a double *ventral nerve cord* runs posteriorly along the mid-ventral line of the thorax and abdomen. The nerve cord bears nine ganglia, three in the thorax and six in the abdomen. The last abdominal ganglion is the largest, representing many fused ganglia of the posterior abdominal segments.

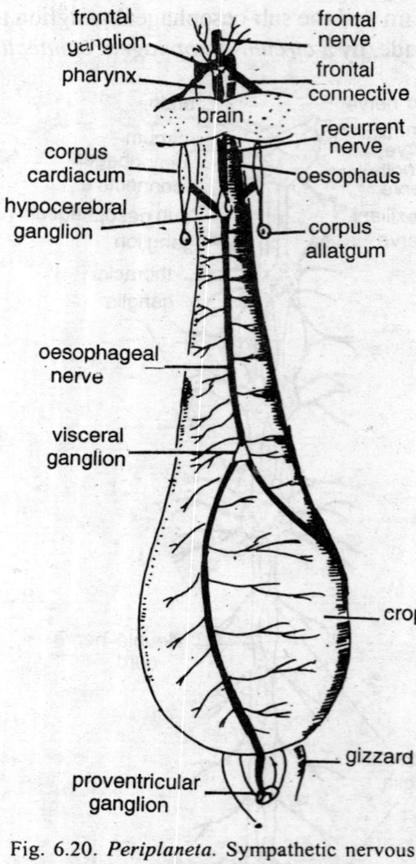

Fig. 6.20. *Periplaneta.* Sympathetic nervous system.

II. Peripheral Nervous System

From the central nervous system several nerves are given off, which go to different parts and constitute the peripheral nervous system. From the brain three pairs of nerves arise, one pair each *optic*, *antennary* and *labrofrontal* and go to the eyes, antennae and labrum respectively. From the sub-oesophageal ganglion also three pairs of nerves arise and go to the mandibles, maxillae, and labium as *mandibular*, *maxillary* and *labial nerves*. Several pairs of nerves arise from each ganglion of the nerve cord and they innervate the various parts of their segment; but from the last abdominal ganglion five pairs of nerves arise and go to the last five segments of the abdomen, one pair to each segment, reproductive organs, copulatory appendages and anal cerci.

III. Autonomic Nervous System

It comprises a few ganglia and connectives. A *frontal ganglion* is situated on oesophagus just in front of the brain. It is connected with the brain by a *frontal connective*. From the frontal ganglion a median *recurrent nerve* passes backwards and connects frontal ganglion to the *hypocerebral ganglion* present on oesophagus. From the hypocerebral ganglion three nerves arise, two-lateral and one median. The lateral nerves meet the corpora cardiaca. The median nerve runs backwards over the oesophagus and joins the *ingluvial ganglion* which is situated on the crop. From the ingluvial ganglion, a pair of nerves proceed backward over the alimentary canal and connect with *proventricular ganglion*.

SENSE ORGANS

Cockroach has a variety of sense organs meant to catch the stimuli of touch, taste, smell, sound and vision. Basically all these organs, except the visual organs, are build on the same pattern and are in the form of minute projections called *sensillae*.The sensillae may be present in enormous numbers, and each is supplied with a sensory nerve fibre.

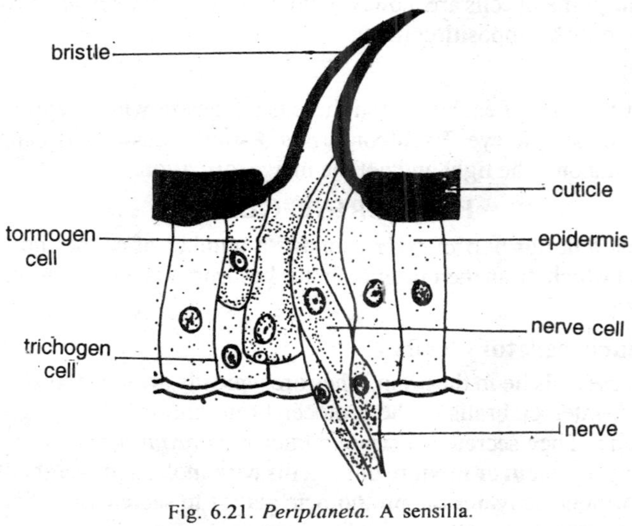

Fig. 6.21. *Periplaneta.* A sensilla.

Sensillae

A typical sensilla is derived from a hair-like process of the cuticle or seta which is set on a small circular membrane. Below this membrane is a large *sense cell* connected on one side, with the nerve fibre and on the other with the setal process. Usually two more cells are associated with seta. A large *trichogen cell*, which is responsible for secreting seta and a *tormogen cell* which gives rise to the membrane. The typical sensilla is modified into two types - the *tactile sensilla* in which there is a single sense cell and the *chemical sensilla* in which there is a group of such cells. In the latter, the seta itself is short and permeable to the fluids or vapours. The *acoustic sensillae* are so modified as to be present in groups and associated with certain membranes which vibrate by sound waves.

In *Periplaneta,* tactile sensillae are distributed over the general body surface but specially on antennae, palpi, anal cerci and distal segments of the legs; olfactory sensillae are present on the antennae; gustatory sensillae on the palpi and other mouthparts and the legs, and cell acoustic sensillae on the base of the abdomen.

Compound eyes

Of all the organs of special sense, the most important is a pair of large, sessile, *compound eyes* in the form of black, kidney-shaped patches on the lateral sides of the head capsule. Each consists of about 2,000 visual elements or units, called *ommatidia,* similar in structure to those already described for the prawn.

The pigment cells are, however, not retractile in *Periplaneta* so that it can see only by apposition image.

Ocelli

At the base of each antenna there is a fenestra which represents an ocellus or simple eye. Each comprises a single lens and is capable of preseiving only the light and not the image formation.

ENDOCRINE GLANDS

The anterior part of body contains a number of endocrine glands, many of which, from their position in the head, are known as *retrocerebral glands*.

1. Neurosecretory cells

These cells lie in the brain. They are typically arranged in clusters in the pars-intercerebralis of the protocerebrum and in a lateral group on each side. They secrete a hormone known as *brain hormone* which is probably a protein or mixture of proteins with molecular weight 40,000. The hormone activates the prothoracic glands to secrete their hormone. The neurosecretory cells are modified motor neurons.

2. Corpora cardiaca

Situated in close association with the aorta behind the brain is a pair of small bodies, the *corpora cardiaca*. Each is connected to the protocerebrum by a pair of nerves and to the hypocerebral ganglion by a single nerve. The hormones of corpora cardica are those produces by its *intrinsic cells* and by the neurosecretory cells of the brain. The intrinsic cells are known to secrete at least two metabolic hormones, the *hyperglycemic* and *adipokinetic hormones.*

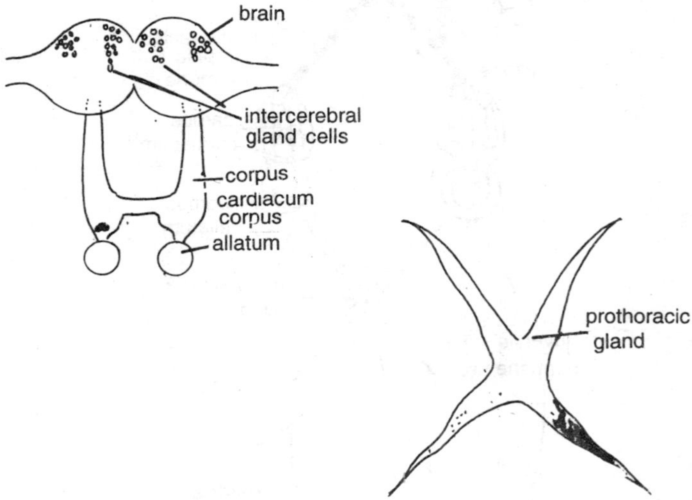

Fig. 6.22. *Periplaneta.* Endocrine glands.

3. Corpora allata

These endocrine glands are situated laterally behind the brain, usually near the corpora cardiaca. They are paired, spherical or oval bodies. The hormone secreted by these glands is known as *juvenile hormone* or *neotenin* whose most important function is to inhibit the realization of imaginal characters during postembryonic development. During last larval or nymphal instar the corpora allata become inactive and the resulting change in hormonal balance leads to metamorphosis. In adults it exerts a gonadotropic influence in the adult females. There is no gonadotropic effect in males but their secretion may be needed for full development of the accessory reproductive glands and for normal sexual behaviour (*Barth* and *Lester,* 1973).

4. Prothoracic glands

These are paired structure, also known as *thoracic glands, pericardial glands, ecdysial glands* or *ventral glands.* They are fairly large and irregular glands situated in prothorax and are well supplied with

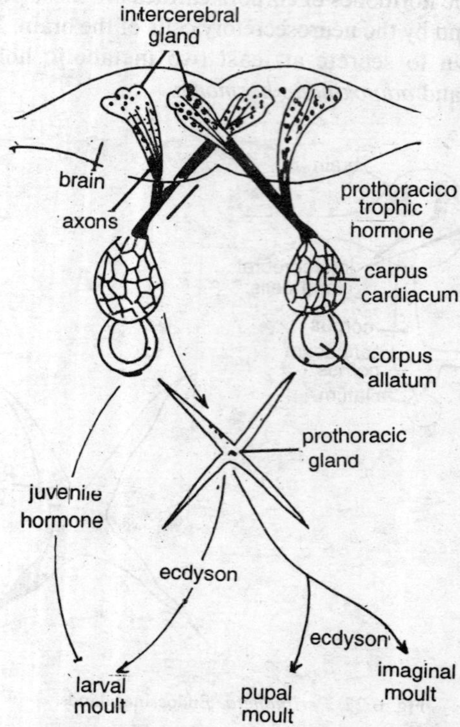

Fig. 6.23. *Periplaneta*. Interaction of hormones.

tracheae and in most cases they are provided with nerves. They secrete a hormone known as *ecdysone* which controls moulting of the nymph. These glands atrophied completely in adult insect.

REPRODUCTIVE SYSTEM

The cockroach is unisexual with a distinct sexual dimorphism.

Male Reproductive Organs

The male reproductive system comprises a pair of *testes*, *vasa deferentia*, *ejaculatory duct*, *seminal vesicles*, *mushroom gland* and *conglobate gland*.

Testes

A pair of testes, lying embedded in the fat body, on either dorso-lateral side of the 4th to 6th abdominal segments. Each testis is a trilobed structure and consists of numerous small whitish *follicles* arranged in longitudinal series. In the adult they become nonfunctional and reduced.

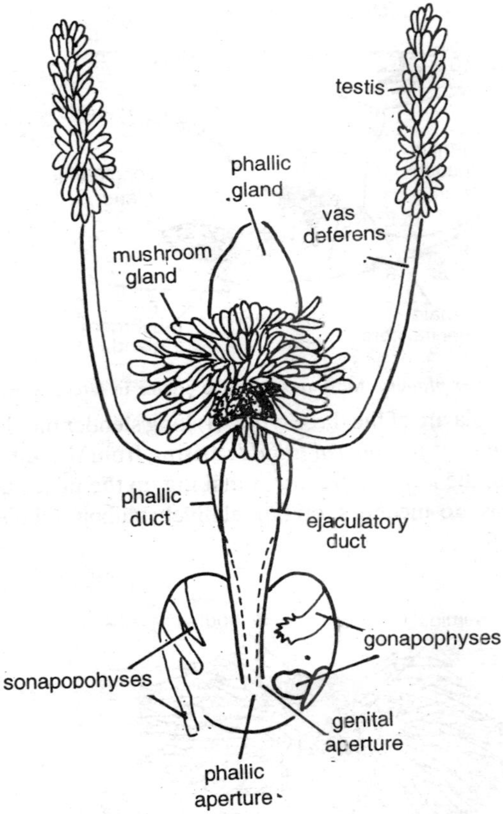

Fig. 6.24. *Periplaneta.* Male reproductive organs.

Vasa deferentia

From the hind end of each testis, there starts a fine duct, the *vas deferens*. The two vasa deferentia run backward and inward to open into a wide median duct, the *ejaculatory duct*.

Ejaculatory duct

The ejaculatory duct arises in the 7th segment runs backwards and opens out at the male genital aperture.

Mushroom or utricular gland

It is present at the junction of vasa deferentia and ejaculatory duct. The mushroom gland or utricular gland consists of a number of blind finger-like diverticula opening into the anterior end of the ejaculatory

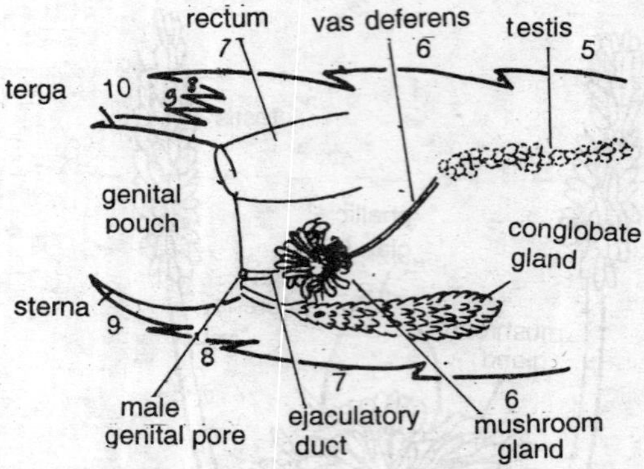

Fig. 6.25. *Periplaneta.* Male reproductive organs in lateral view.

duct. The diverticula are of the three types: (i) long slender tubules known as *utiriculi majores* or *peripheral tubules* arising from the anterior end, (ii) short tubules, the *utirculi breviores* making up the major portion of gland and (iii) postero-median short but relatively bulbous tubules known as *seminal vesicles.*

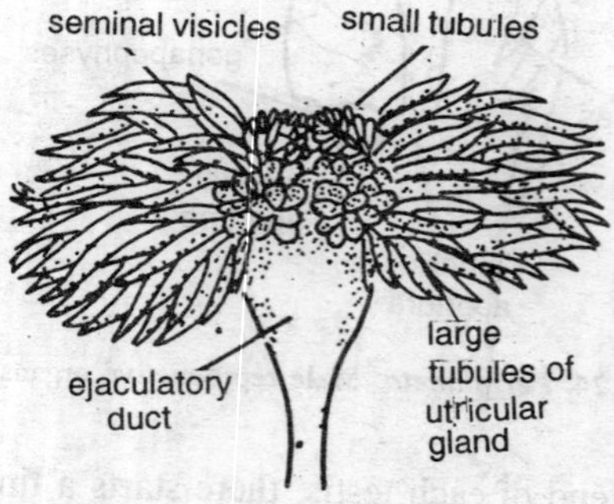

Fig. 6.26. *Periplaneta.* Uricose gland.

Phallic or conglobate gland

It is a large, white elongated, sac-like structure lies beneath the ejaculatory duct that opens out near the male genital aperture. Its function is unknown.

External genitalia

The genital pouch lies below the 9th and 10th abdominal terga. It contains the dorsal anus and ventral male genital opening. Associated with the latter are the *phallic organs* or *gonapophyses* which help in copulation.

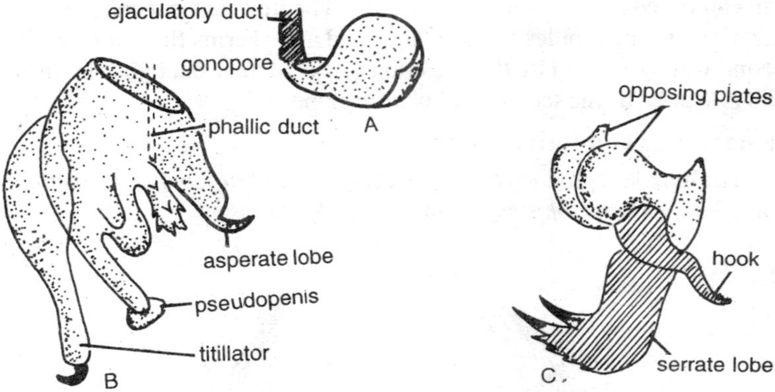

Fig. 6.27. *Periplaneta.* A—Ventral phallomere, B—Left phallomere, C—Right phallomere.

The gonapophyses consist of three irregular chitinous processes, the *left*, *right* and *ventral phallomeres*. (i) The right phallomere consists of two horizontal opposing plates, a sickle-shaped hook. (ii) The left phallomere possesses a broad base, from which arise several lobes or arms. A long slender lobe, the *titillator*, with a curved hook, lies on its extreme left. On its right is a short lobe, the *pseudopenis*, with a broad hammer-like head. Inner to it are three short, soft lobes, one of which, called the *asperate lobe*, bears a curves hook. (iii) The ventral phallomere is a brown broad plate without lobes. It lies close to the gonopore.

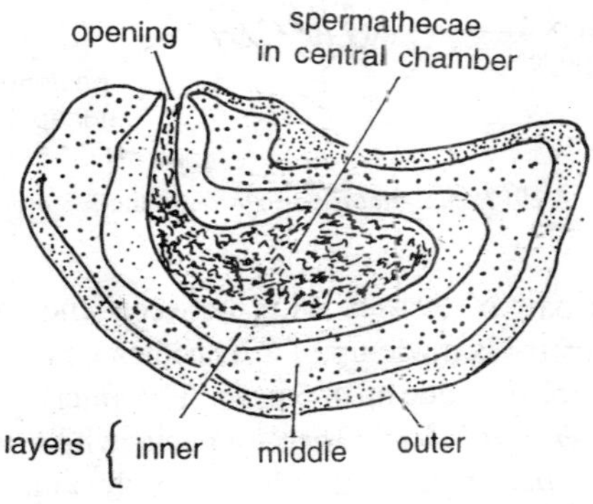

Fig. 6.28. *Periplaneta.* A spermatophore.

The sperms pass into the seminal vesicles and are glued together to form a *spermatophore* by the secretion of seminal vesicles. A spermatophore is pear-shaped body, about 1.33 mm. in diameter and its wall composed of three non-cellular coats. The first coat is secreted by the secretion of long tubules of utricular gland and sperms flow into it. The second wall is formed by the secretion of ejaculatory duct and the third wall is formed by the secretion of phallic gland.

Female Reproductive Organs

The female reproductive organs comprise, *ovaries*, *oviducts*, *vagina*, *gynatrium*, *collaterial glands* and *spermathecae*.

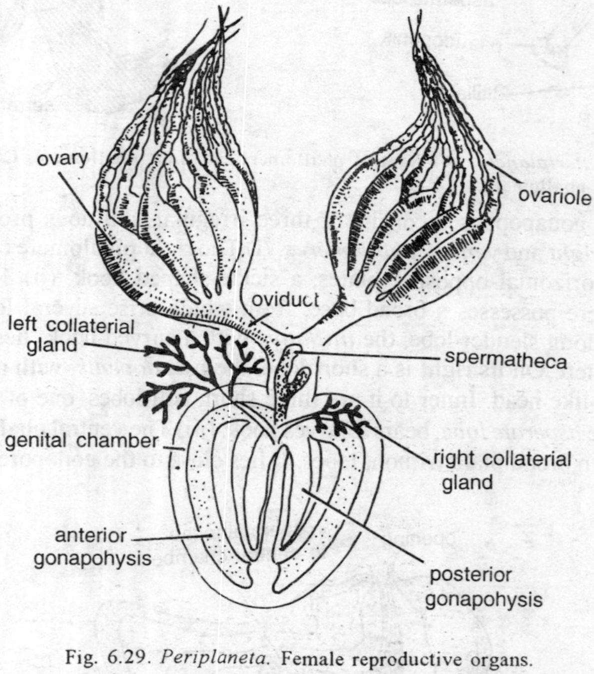

Fig. 6.29. *Periplaneta.* Female reproductive organs.

Ovaries

There are a pair of ovaries lying between 2nd to 6th abdominal segments. Each ovary is made up of 8 ovarioles which are yellowish in colour. Each ovariole is made up of 5 parts: The anterior most thread-like structure called *terminal suspensory ligament* which is syncytial. It is followed by *germarium* containing cells at an early stage of egg formation. It is followed by *vitellarium* containing mature ova and gives beaded appearence. It is followed by *egg chamber* which contains only one mature ovum. It is followed by *pedicel*. Posteriorly, all the eight ovarioles of each ovary join to form a wide oviduct.

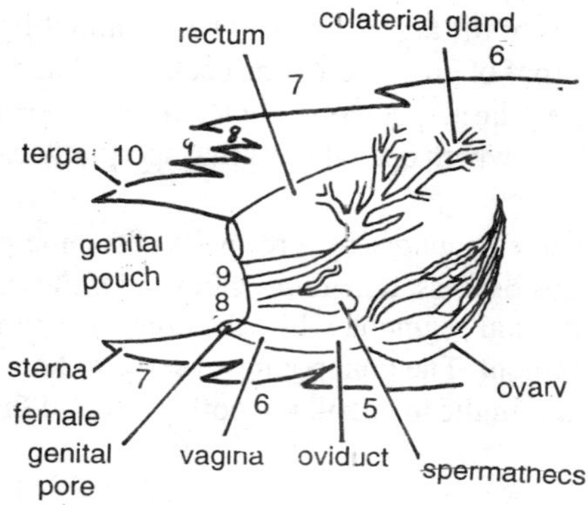

Fig. 6.30. *Periplaneta.* Female reproductive organs in lateral view.

Oviducts

Each oviduct is formed by the union of ovarioles. Both the oviducts unite to form a common oviduct or vagina.

Vagina

It is a formed by the union of both the oviducts. The vagina opens into the gynatrium by a slit-like female genital pore situated in the 8th segment.

Spermathecae

There is a pair spermathecae of unequal size, one spermatheca being larger than the other. The two unite to form a short common duct which opens into the genital chamber on a small spermathecal papillae. It is claimed by some workers that there is a single spermatheca and it has a lateral coiled caecum.

Collaterial glands

A pair of highly branched collaterial glands lie behind and above the ovaries. The left gland is more opaque and well developed. The right gland is transparent and less developed. The two open into gynatrium or *genital pouch* through their openings. The secretion of these glands provides material for the formation of *ootheca* or *egg case* round a group of eggs.

External genitalia

The gynatrium or genital pouch is formed by the 7th, 8th and 9th sterna. The sternum of 7th segment is boat-shaped and splits posteriorly into two parts, the *apical lobes* or *gynovalvular plates*. This forms the floor and side walls of the brood pouch. The 8th and 9th sterna form the anterior wall and roof of brood pouch respectively. The pouch is divided into two chambers a large posterior *oothecal chamber* and an anterior *genital chamber* into which open the vagina, spermatheca and collaterial glands.

The gonopore is surrounded by three pairs of symmetrically arranged, plate-like, chitinous process, the *gonapophyses*. Of these, one pair arises from the 8th abdominal segment, while the other two pairs belong to the 9th abdominal segment. The gonapophyses serve to hold the ova as they are laid and also aid in the formation of ootheca round them. They, thus, act as *ovipositors*.

Copulation

The breeding season of cockroach is from March to September. Copulation occurs in night. The male and female cockroaches come opposite to each other in such a way that the posterior ends of their abdomen touch each other. The male opens the genital chamber of the female with the help of its titillator and pushes is phallomeres into the genital chamber of the female. At the same time the dorsal phallomere opens the male genital aperture through which the spermatophore is deposited on the spermathecal papilla. The phallic secretion is now poured over the spermatophore which fixes the spermatophore firmly to the spermathecal papillae and harden to form the third wall of spermatophore. The copulation lasts about an hour and a quarter, after which the two cockroaches separate. The sperms pass from the spermatophore into the spermathecae slowly in the course of the next twenty hours after which the empty spermatophore is discarded.

Ootheca Formation

The mature eggs are released alternately by each ovary into the common oviduct. They slowly move backward and come into the genital chamber through the female genital pore where they are fertilized. Soon after fertilization, the collaterial glands pour their secretion over the fertilized eggs forming a hard covering over them called *ootheca*. The ootheca is shaped and moulded by the ovipositor and the walls of oothecal chamber. The ootheca is 12 mm. in length. On one side it has a straight crest with a serrated margin. It contains 16 fertilized eggs standing vertically in two rows. The female cockroach moves about with ootheca for several days and then keeps it in a dry, dark and safe place.

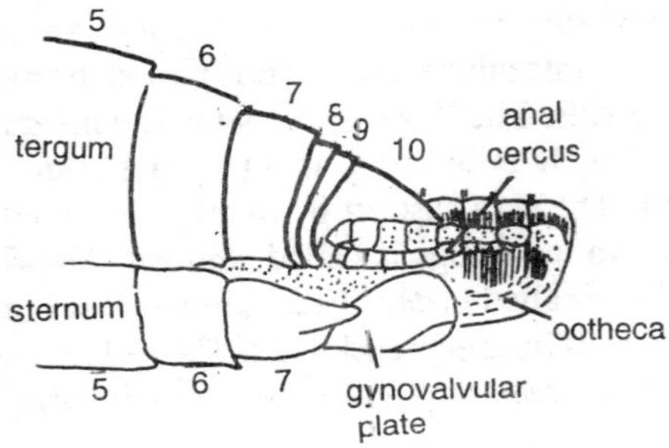

Fig. 6.31.*Periplaneta*. Ootheca protruding from abdomen of female.

Development

The eggs are enclosed in ootheca. Each egg is laterally compressed being concave on one side (future ventral side) and convex on the other.

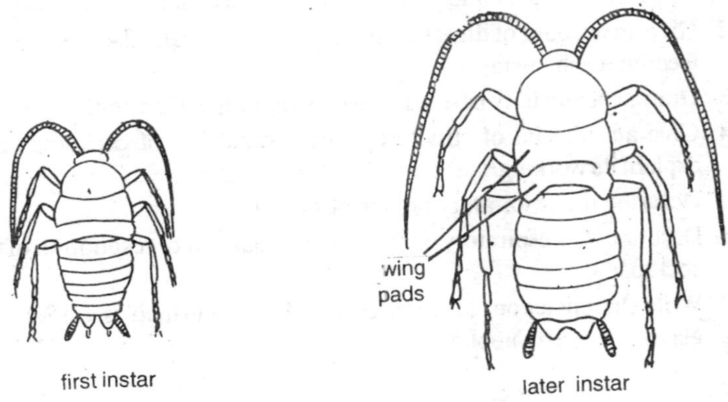

Fig. 6.32. *Periplaneta*. A—First instar nymph, B—Second instar nymph.

Cleavage is *superficial*. The nucleus divides and redivided and surrounded by a stellate mass of protoplasm forming the cleavage cells. When cleavage-cells are produced in large number, they migrate to the periphery of the egg forming *blastoderm*, surrounding the yolk. The egg becomes a blastula, gastulation takes place shortly after. A differentiation in the gastrula results in an early embryo which separates from the blastoderm by an amniotic cavity. Its three germ layers gradually form the various organs of nymph.

Hatching

A split of the dorsal keel of the ootheca releases the nymphs. At the time of hatching the cockroaches are white in colour and bear dark eyes, soon the integument becomes dark. The young ones are devoid of wings but in other respect they resemble the adult. These are called *nymphs*. They gradually grow into adults. During this period the nymphs shed the integument at intervals. The process of casting the integument is called *ecdysis* or *moulting*. It is brought about by a hormone secreted by prothoracic gland. The skin cast off by nymphs is called *exuviae*. The interval between two moultings is called *stadium*. The form of nymph betwen two moults is called *instar*. There are about six or seven moults and an equal number of instars in cockroach. The metamorphosis is called *paurometabolic development* or *gradual metamorphosis*.

Revision Questions

1. Describe the systematic position and external characters of cockroach.
2. Give an account of digestive system of *Periplaneta* also describe the feeding mechanism.
3. Draw neat and fully labelled sketeches of the mouthparts of cockroach.
4. Give an account of structure of compound eye of cockroach and explain its working.
5. Describe the respiratory organs of *Periplaneta*.
6. Describe the respiratory organs and mechanism of respiration of the cockroach.
7. Write short notes on : (a) Ootheca, (b) Malpighian tubules, (c) Salivary gland, (d) Spermatophore.

7

Grasshopper or Locust

Grasshopper or locust belongs to order Orthoptera of class Insecta. It is easily available and is harmless to handle. It has a relatively large size that is why convinent for dissection. Being specialized, it exhibits essential features of insects better than other forms.

Habits and Habitat

Grasshoppers are *voracious* herbivorous having a worldwide distribution. They are found in greatest abundance in places with open grasslands and abundant leafy vegetation, where there is plenty of food and place to breed.

They are diurnal insects, i.e., they are active during the day time. They feed on the leaves of ak plant. There is a tripple mode of locomotion-crawling, leaping and flying. It is unisexual. Males and females differ externally in size, wing length and genitalia. It is oviparous. The eggs are laid in holes in the ground. The nymphs resemble the adults in structure and mode of life, except that they lack wings and are more brightly coloured. There is no parental care.

EXTERNAL FEATURES

Shape and size. Grasshopper is relatively a large insect reaching upto 8 cm. in length. Body is narrow, elongated, cylindrical bilaterally symmetrical.

Colouration. The usual body colour is yellowish and brownish with different markings and colour spots. The pigment in the chitin provides the protective colouration to the body matching the enviornment. The ak plant (*Calotropis*) grasshopper (*Poecilocerus pictus*) is blue and yellow in colour.

EXOSKELETON

The entire body is covered by a hard cuticular exoskeleton secreted by the underlying *epidermis* also called *hypodermis*. The exoskeleton is composed of thick plates, the *sclerites,* which are fused at some places and movably articulate to others by thin, soft, flexible membranes, the *arthrodial membranes.* The exoskeleton serves many functions. It protects the body against mechanical injury, checks desiccation, provides hard surface for muscle attachment and forms sensory bristles.

Pigments in and under cuticle provided a protective colouration by which grasshoppers resemble their surroundings.

Division of body

In grasshoppers, the body is divided into three typical regions, viz., the *head, thorax* and *abdomen.*

Head. The anterior movable region or head appears as a single structure, but it is made up of 6 embryonic segments fused together, as indicated by the presence of paired appendages of *hypognathous* type, i.e., lying vertically and its broader mouth-bearing side directed downwards.

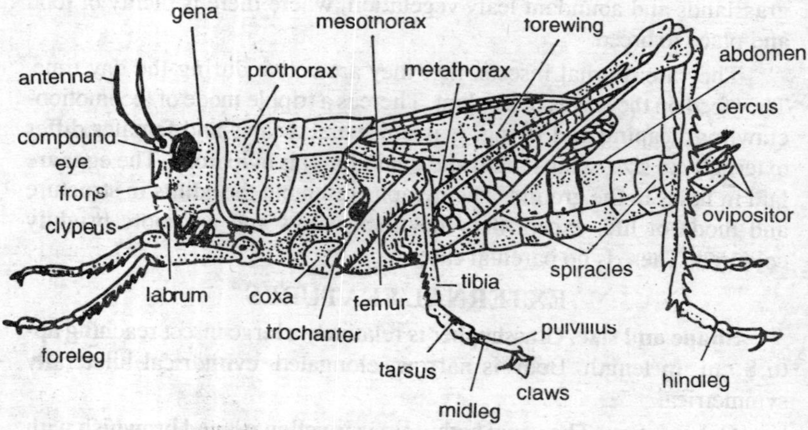

Fig. 7.1. *Grasshopper.* (Lateral view).

Exoskeleton of head (head capsule or epicranium) is made of firmly-jointed chitinous plates or sclerites. Its dorsal region is known as *vertex,* lateral sides as *cheeks* or *genae* and the front portion as *frons.* A broad rectangular plate below frons is the *clypeus.*

Head bears one pair of dorso-lateral *compund eyes*, three frontal simple eyes called *ocelli,* one pair of segmented *antennae* and a set of chewing *mouthparts*, surrounding the mouth. Following structures are present on head:

Antennae. The antennae are filiform. Each antenna consists of a small piece called the *scape*, an undifferentiated *pedicel* and a sufficently long *flagellum* composed of about twenty-five segments. Sensory bristles, probably olfactory in nature, are present on the surface of antennae.

Compound eyes. Two compound eyes are placed dorso-laterally on head. These are sessile as stalks are absent. Each compound eye is covered by transparent cuticle, the *cornea* which is divided into a large number of hexagonal facets. Each facet is the outer end of a unit known as an *ommatidium*. Such a structure gives mosaic vision.

Ocelli. Three simple eyes or ocelli are placed between the compound eyes. An ocellus consists of a group of visual cells the *retinulae* and a thick transparent *lens* which is the modification of cuticle.

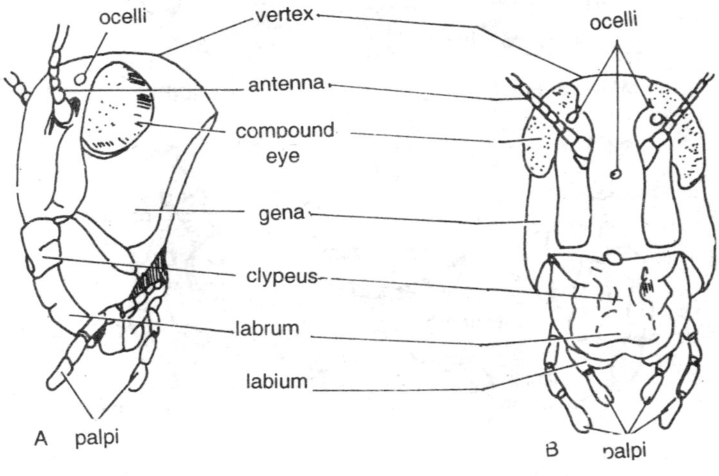

Fig. 7.2. *Grasshopper*. A—Head in lateral view, B—Head in frontal view.

Mouthparts. The mouthparts of grasshopper are *chewing type*. The mouthparts include the *labrum* or upper lip attached to the ventral edge of the clypeus. Beneath this is a membranous tongue-like organ, the *hypopharynx.* On either side is a single, hard jaw or *mandible* with a toothed surface fitting for grinding. Beneath the mandibles are a pair of *maxillae*. Each maxilla consists of a basal *cardo,* central *stipes,* a long curved *lacinia,* a long rounded *galea* and a *maxillary palp* which arises from the *palpifer*. The *labium* or lower lip comprises a basal *submentum,* a central *mentum,* two movable flaps, the *ligulae* and a *labial palp* on either side.

Thorax. Thorax is the middle division of the body of grasshopper.

It consists of three well-defined segments: an anterior *prothorax,* a middle *mesothorax* and a posterior *metathorax.* The prothorax is separated from the mesothorax by an extensive inter-segmental membrane and is independently movable. The mesothorax and metathorax are immovable and fused together to form a box called *pterothorax.*

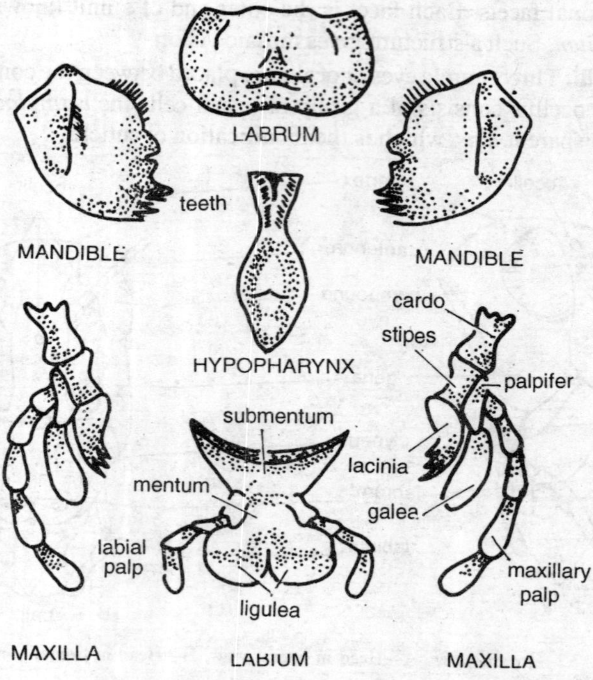

Fig. 7.3. *Grasshopper.* Mouthparts.

***(a)* Prothorax.** The tergum of the prothorax is very large and saddle-like. It is called the *protergum* or *pronotum.* It covers the neck in fornt, mesothorax behind and pleura on the sides. It is divided by three transverse furrows or sulci into four parts. The sulci mark the position of internal ridges.

The pleura of the prothorax are called *propleura.* These are almost fully covered by the lateral extensions of the pronotum. Each propleuron comprises a large anterior *episternum* and a small posterior *epimeron,* the two being separated by a pleural suture.

The sternum of the porthorax is known as the *prosternum*. It is very small and bears a small median spine-like projection, the *prosternal spine*.

(b) **Mesothorax.** The tergum of the mesothorax, called the *mesotergum*, is small and largely concealed under the pronotum. Its sternum, termed the *mesosternum*, is quite large. The *mesopleura*, i.e. pleura of mesothorax, are very prominent, each divided by a sub-vertical suture into an *episternum* and a posterior *epimeron*.

(c) **Metathorax.** The metathorax resembles the mesothorax in its sclerites except that its tergum is uncovered.

Legs. Each thoracic segment has a pair of jointed legs attached ventro-laeterally. Each leg consists in a linear series of 5 segments: *(i)* a short basal *coxa* which articulates with thorax, *(ii)* A small *trochanter* which is fused, *(iii)* a large and stout *femur*, (iv) an elongated, slender, spiny *tibia*, and *(v) tarsus* made of 3 short segments or tarsomeres bearing tiny pads (*plantulae*) on ventral side. Terminal tarsomere also bears a median fleshsy pad (*pulvilus* or *arolium*) between two lateral claws.

Wings. In grasshopper, each of the mesothorax and metathorax bears a pair of wings. The forewings are narrow and mostly parchment-like. These may be coloured uniformly or with deepening shades towards the bases or may be spotted. This pair is also named a *tegmina* because in the position of rest it covers the abdomen and the hind pair of wings. The hind-wings are broad and membranous and kept folded in fan-like manner having a number of longitudinal folds in alternating directions. Each wing develops as a sac-like projection of the body covering and flattens to a thin double membrane that encloses tracheae, nerves and blood sinuses. The cuticle thickens along the sinuses to form strengthening *nervures* or *veins*. Although these veins vary in their patterns amoung the different species, they are constant in individuals of certain species, where they serve for classification.

Abdomen. The abdomen is the posterior division of the body in grasshopper. It is cylindrical in form and is longer and more slender than the thorax. It consists of eleven segments, each partly telescoped into the posterior end of the anterior segement. They are connected to one another by soft inter-segemental membranes which allow movement. The first abdominal segment is immovable connected with the thorax.

(a) Tympanum. On each lateral side of 1st abdominal segment is an oval tympanic membrane that covers an auditory sac or organ of hearing.

(b) Spiracles. 8 pairs of smalll spiracles are present. Those of the first pair are larger than the rest and lie in first segment in front of tympanum, remaining 7 pair of spriacles are situated antero-laterally on segments 2nd to 8th. In segments 9 and10, terga are partially fused, while sterna are completely fused. Segment 11 consists of a tergum only which forms the *supra-anal plate* over anus. A small process behind 10th segment on either side forms a small process, the *anal cercus*.

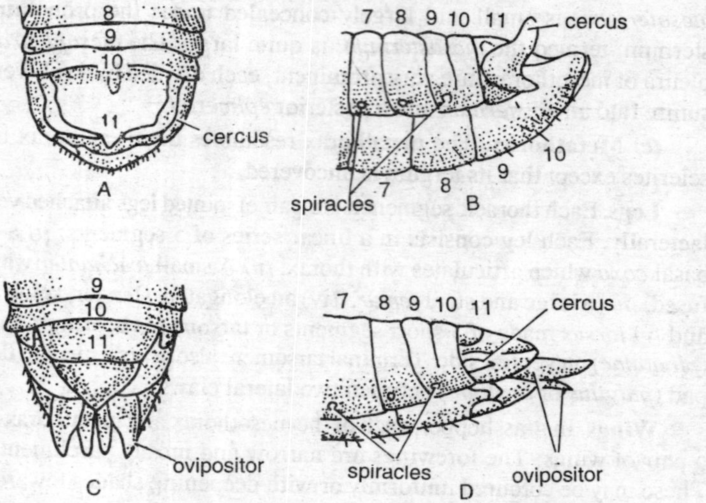

Fig. 7.4. *Grasshopper.* A—Male abdomen (Dorsal view); B—Male abdomen (Lateral view); C—Female abdomen (Dorsal view); D—Female abdomen (Lateral view).

(c) External genitalia. The terminal segments are modified in both the sexes for copulation and oviposition.

In male the 9th and 10th terga are as in the female. The 9th sternum is the last visible sternum. It is very large and boat-shaped. It is known as the *hypandrium*. It comprises two parts: the anterior 9th sternum proper and the posterior sternal lobe or *subgenital plate*. There is hardly any line of demarcation between these parts. The inter-segmental membrane at the posterior margin of 9th sternum is reflected forwards to form the base of a membranous pouch, the *ginital chamber*.

The 11the segment is similar in both sexes. It is represented by a shield-like plate, the *epiproct* or *subanal plate*, above two lateral lobes, the *paraprocts* or *podical plates.* A pair fo small, conical, unjointed processes arise from the inter-segmental membrane behind the 10th segment and extend backwards. These are termed as *cerci*.

In female the tergum of the 8th segment resembles the terga of the preceding segments. The 8th sternum is, however, the last visible sternum. It extends backwards as a large subgenital plate which is produces posteriorly into a small, median, triangular process, the *egg guide*. The inter- segmental membrane at the posterior margin of the 8th sternum is reflected forwards to form the base of a membranous pouch, the *genital chamber*. The 9th and 10th terga are much smaller and partly fused on the sides.

Integument

It is made up of cuticle, epidermis or hypodermis.

Cuticle. The cuticle is differentiated into an outer thin *epicuticle*. It is made up of lipids. It is impermeable to water. Middle *exocuticle* made up of hard, pigmented, thick protin. Inner *endocuticle* is thick and soft.

Epidermis or hypodermis. It consists of a signh layer of columner cells resting on a basement membrane. Specialized gland cells are scattered among other cells. The epidermis secretes the cuticle.

Body cavity

The body cavity of grasshopper is known as *haemocoel* filled with *haemolymph* or *blood*. All the systems of body lie in haemocoel. In adults, coelom is represented within the gonads.

MUSCULAR SYSTEM

The muscles are of striated type, very soft and delicate but strong. The number of muscles is very large. They are segmentally arranged in the abdomen but not in the head and thorax. The most conspicuous muscles are those that move the mandibles, the wings, the metathoracic legs and the ovipositor.

DIGESTIVE SYSTEM

The digestive system comprises an alimentary canal and digestive glands.

(I) Alimentary Canal

The alimentay canal consists of three principle parts: foregut, midgut and hindgut. Foregut and hindgut are lined with cuticle.

(a) Fore-gut. The fore-gut includes the preoral cavity, pharynx, oesophagus, crop and gizzard or proventriculus.

***(i)* Preoral cavity.** The preoral cavity is bounded anteriorly by labrum and clypeus, posteriorly by labium and laterally by mandibles and maxillae. It has in it the hypopharynx that hangs just in front of the labium and bears opening of salivary duct. The mouth lies at the top of the preoral cavity between the epipharynx and the hypopharynx.

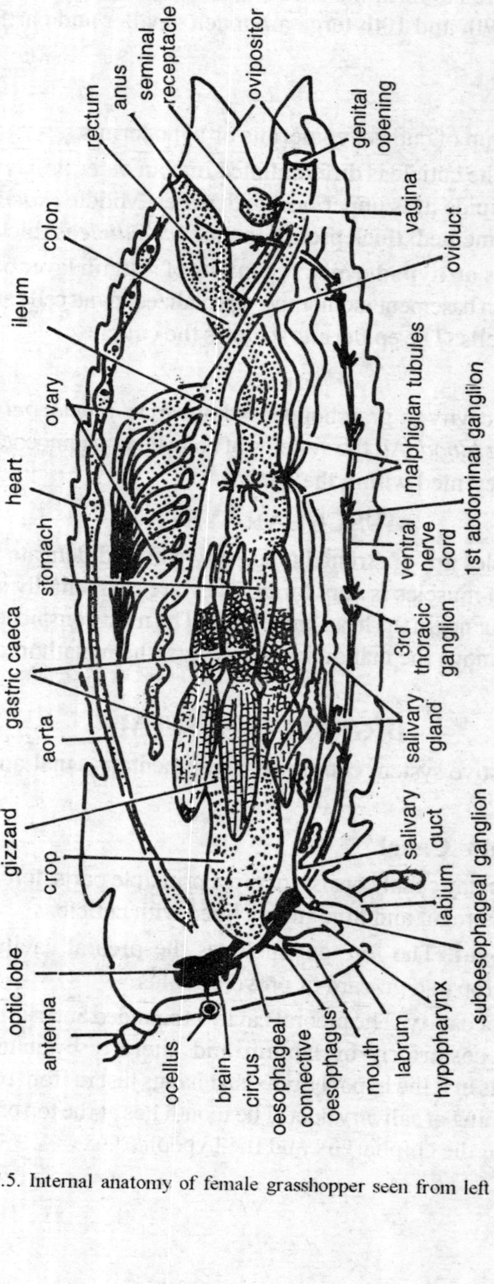

Fig. 7.5. Internal anatomy of female grasshopper seen from left side.

***(ii)* Pharynx.** The mouth opens into the pharynx. It is a very short tube that ascends vertically upwards and merges with the oesophagus without any external demarcation.

***(iii)* Oesophagus.** The oesophagus is a narrow tube that soon bends backwards. Passing through the neck, it reaches the thorax where it expands to form the crop.

***(iv)* Crop.** The crop is a large thin-walled sac. It occupies the anterior half of the thorax. It is mainly meant for the storage of food, though some digestion also occurs in it.

***(v)* Gizzard.** The gizzard follows the crop from which it is hardly distinguishable externally. It lies in the posterior half of the thorax and has thick muscular wall. The anterior part of gizzard is modified into a ring of teeth to masticate the solid food coming form the crop. Two rows of hairly cushions or *pulvilli,* which act as strainers, are present behind teeth. The posterior part of gizzard projects into the mid-gut.

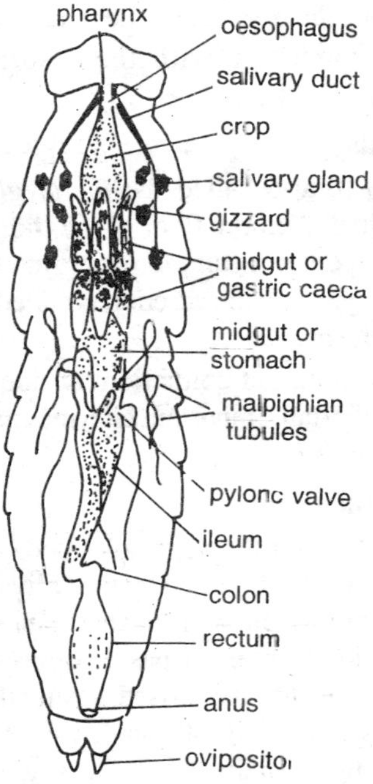

Fig. 7.6. *Grasshopper*. Alimentary canal in dorsal view.

***(b)* Midgut.** Next to gizzard is the midgut. It includes the ventriculus or stomach which reaches up to the middle of abdomen. It is thin- walled, devoid of a cuticular lining and sphinctered at both ends. Opening into the anterior end of stomach is a ring of six pairs of elongated, cone-shaped pouches called *hepatic* or *gastric caeca*. One pouch of each pair extends anteriorly over proventriculus and the other posteriorly over ventriculus. Gastric caeca produce digestive enzymes and facilitate absorption of digested foods.

***(c)* Hind-gut.** The hind-gut is differentiated into three regions: ileum, colon and rectum.

***(i)* Ileum.** The ileum is the anterior broad part of the hind-gut. A number of fine, pale-yellow, thread-like processes, the *Malpighian tubules,* open into the beginning of the ileum. These are excretory organs of grasshopper.

***(ii)* Colon.** The colon is the narrow middle part of the hind-gut. It first runs upwards and then bends backwards.

***(iii)* Rectum.** The rectum is the swollen terminal part of the hind-gut. It has thin wall and opens out by anus. The anus lies between the podical plates and under the base of the subanal plate.

The hind-gut is concerned with the absorption of water and elimination of faeces.

(II) Digestive Glands

The *salivary glands*, also called the *labial glands*, are a pair of branched structures situated on the floor of the thorax beneath the crop and the gizzard. They open by a common duct into the preoral cavity at the base of the hypopharynx. Their secretion is called saliva and contains a starch-splitting enzyme *amylase*.

Hepatic caeca and lining of epithelium of mid-gut, both secrate a number of enzymes in mid-gut. Important enzymes are *amylase*, *maltase*, *invertase*, *tryptase* and *lipase*.

Feeding and digestion

The grasshoppers feed on the vegetable food. Food is held by the forelegs, labrum and labium, lubricated by the salivary secretion (which contains some enzymes) and chewed by the mandibles and maxillae. Chewed food is stored in the crop. It passes onwards into the gizzard little by little, where it is further pulverised, strained and passed on into the stomach. The glands in the walls of stomach and hepatic caeca secrete a few enzymes which bring about digestion. The slightly alkaline or acidic secretions of midgut contain maltase, lipase, lactase, protease, tryptase and erepsin. The absorption of food material takes place in the mid-gut. By the time the food material reaches the rectum, the maximum nutritive material has been availed and the excess water is absorbed in the rectum. The undigested material or residue is transformed into slender faecal pellets to be ejected out through the anus.

CIRCULATORY SYSTEM

Circulatory or blood-vascular system is of *open* or *lacunar type*. It bears a close resemblance to that of cockroach. Blood is confined to vessels (heart and aorta). Only a portion of blood circulates through them and remaining takes its journey within body cavity or haemocoel.

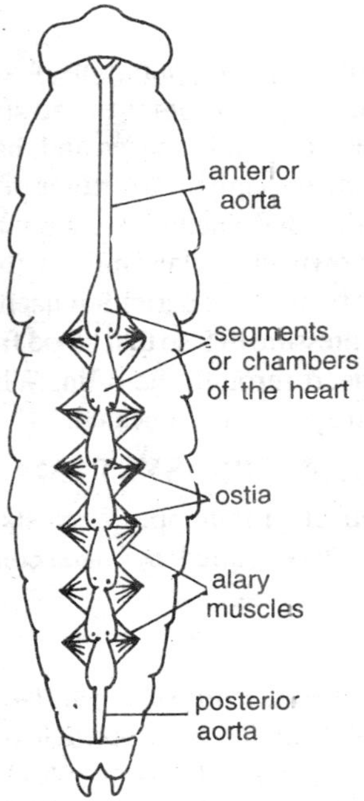

Fig. 7.7. *Grasshopper*. Circulatory system.

Heart and Aorta

There is a single, slender, tubular and pulsatile *heart* lying mid-dorsally in the abdomen. It is divided into a linear row of chambers, one in each segment. The heart has a number of side-openings named *ostia* which are provided with valves to permit the flow of blood in one direction only. A number of muscular strands, the *alary muscles* are divergently spread out fan-wise over the diaphragm to enlarge and reduce the cavity of pericardium by their contraction and relaxation. The heart is suspended in a shallow pericardial sinus which is separated from the main body cavity (*perivisceral sinus*) by a thin, delicate and perforated transverse partition known as *dorsal septum* or *diaphragm*. A similar *ventral diaphragm* separates a narrow *ventral sinus* or *perineural sinus* (surrounding nerve cord) from perivisceral sinus.

Haemolymph

The blood plasma contains colourless blood cells which act as phagocytes to remove foreign organisms. The blood serves mainly to transport food and waste materials.

Circulation of Blood

The heart, by contraction of its wall, pumps its blood into the anterior aorta that takes it to the head sinuses. From the head sinuses, blood flows backwards through the haemocoelomic spaces and, bathing the tissues, reaches the abdomen. Passsing through the apertures of ventral and dorsal diaphragms, it fills the pericardial haemocoel. Now the heart expands due to the elasticity of its own muscular wall, at the same time, the contraction of alary muscles reduces the pericrdial haemocoel. Expansion of the heart creates negative pressure in it so that blood from the pericardial haemocoel flows into the heart through the ostia. When full, the heart again contracts and the entire process is repeated.

RESPIRATORY SYSTEM

The respiratory system of grasshopper consists of a network of ectodermal white, shining air tubes called, *tracheae* communicating with exterior by *spiracles*.

1. Spiracles

There are 10 pairs of spiracles in grasshopper, two pairs of thoracic and eight pairs of abdominal spiracles. Each spiracle of the first pair lies in a sclerotized plate called *peritreme*. It remains covered by pronotum. It is grarded by two valves- anterior and posterior. The posterior valve is movable. The spiracle is closed and opened by a pair of muscles, stratching between the peritreme and the posterior lip on the inner side. Each spiracle of the second pair lies in a small membranous area between the meso-and the metapleura, a little above the base of the second leg. It is guraded by two movable lips which are controlled by a single internal muscle. This spiracle leads into a single tracheal trunk.

The first pair of abdominal spiracles are larger than others and lie close in front of the tympana. The remaining spiracles lie in the anteriro-ventro-lateral angles of the terga of segments 2-8. All abdominal spiracles are alike. Each leads into a very small chamber, the *atrium*, whose anterior wall is movable to close or open the spiracle.

The first four pairs of spiracles (two thoracic and two abdominal) open during inspiration and are called the *inspiratory spiracles*. The last six pairs of spiracles (all abdominal) open during expiration and are termed the *expiratory spiracles*

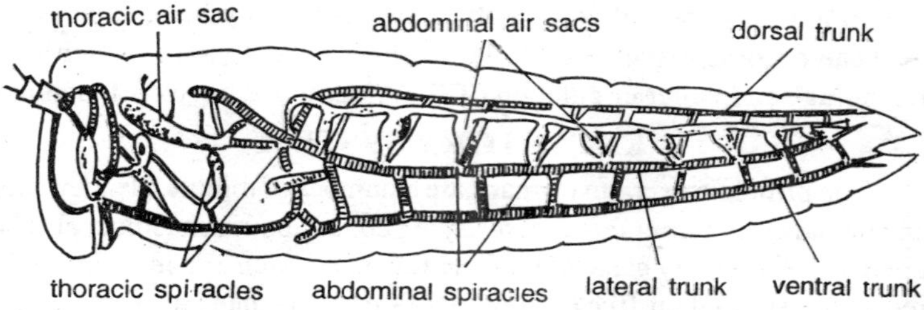

Fig. 7.8. *Grasshopper*. Respiratory system.

2. Tracheae

Each spiracle opens into a small chamber, the *atrium*, which opens into a network of fine, elastic, ectodermal air tubes, the *tracheae*, which branch and ramify to all parts of body. Main tracheal tubes form longitudinal tracheal trunks in the dorsal, ventral and lateral regions of body. A trachea consists of a single layer of squamous epithelial cells (ectotrachea) which secretes a chitinous cuticular lining on the inner side

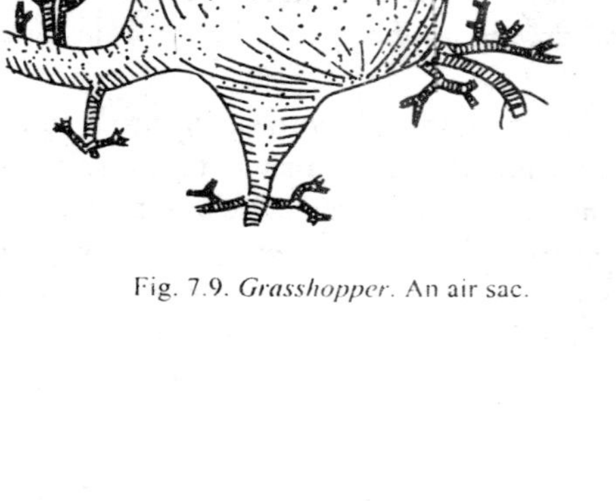

Fig. 7.9. *Grasshopper*. An air sac.

(endotrachea) and is shed at moulting. It forms a spiral thickening, known as *intima* or *taenidia*, to prevent collapsing of tracheal walls.

Finest branches of tracheae are called *air capillaries* or *tracheoles*. These are only 1 μ in diameter and are single celled. These have no spiral thickenings and are usually filled with tissue fluid which serves in internal respiration, like blood in other animals. Oxygen gets dissolved in this fluid and passes by diffusion to the tissue cells in exchange for the unwanted carbon dioxide.

In grasshopper and certain other insects (Honey bee), some of the tracheae become expanded into large thin-walled air sacs, which serve for storage and better circulation of air. Air sacs are devoid of taenidia.

EXCRETORY SYSTEM

The principle excretory organs are a number of minute, slender, blind and thread-like *Malpighian tubules*, lying coiled about in haemocoel. They open into alimentary canal at the junction of stomach and intestine, while remain closed at their free ends. Wall of tubules is made of a single layer of cells covered externally by a basement membrane and internally with numerous microvilli, forming a characteristic bursh border. Cells extract metabolic wastes such as urates from haemolymph, convert them into uric acid and pass on to hindgut for elimination with faeces. Water is reabsorbed by the proximal absorptive part of the Malpighian tubule. Removal of dry excretory wastes is characteristic of small land animals that have only a limited water supply.

NERVOUS SYSTEM

In grasshopper the nervous system consists of:

(i) Central nervous system
(ii) Peripheral nervous system
(iii) Autonomic nervous system.

(i) Central nervous system

It includes a *brain* or *cerebral ganglion*, a pair of *sub-oesophageal ganglion*, a pair of *circum-oesophageal connectives* and a *ventral nerve cord.*

(a) *Brain.* Brain lies dorsally above oesophagus. It consists of 3 pairs of fused ganglia. Brain is distinguished into: (i) a large anterior *protocerebrum* consisting of two large optic lobes, (ii) a *deutocerebrum* comprising of two small lobes situated ventrally on the posterior side of protocerebrum, and (iii) a small *tritocerebrum* of two lobes behind deutocerebrum. A loop-like *tritocerebral commissure* runs along the inner sides of circum-oesophageal connectives and connects the two sides (lobes) of tritocerebrum. Brain chiefly serves as a relay centre which receives stimuli from sense organs and thus directs the movements of body.

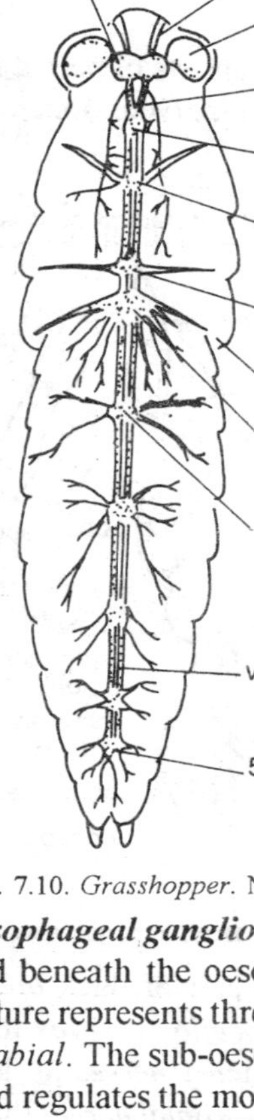

Fig. 7.10. *Grasshopper*. Nervous system in dorsal view.

(b) Sub-oesophageal ganglion. The sub-oesophageal ganglion also lies in the head beneath the oesophagus. Like the brain, it is also a composite structure represents three pairs of fused, ganglia-*mandibular*, *maxillary* and *labial*. The sub-oesophageal ganglion primarily acts as a motor centre and regulates the movements of the mouthparts.

(c) Circum-oesophageal connectives. Brain is connected by a pair of circum-oesophageal connectives around oesophagus to a ventral sub-oesophageal ganglion.

(d) Ventral nerve cord. A double ventral nerve cord arises form the hind end of sub- oesophageal ganglion and extends down the mid-ventral region of body. It bears paired or biloded ganglia, 3 thoracic and 5 abdominal. Third or metathoracic ganglion is the largest as the first three abdominal ganglia are also fused with it. Similarly, 5th abdominal ganglion is the largest representing last 4 pairs of abdominal ganglia fused together. All segmental ganglia are coordinated by nerve fibres running in the ventral cord. Besides, each ganglion independently controls the movements of its respective segment (or segments) including appendages.

(ii) Peripheral nervous system

The peripheral nervous system consists of paired nerves that arise from the various ganglia of the central nervous system to innervate the body.

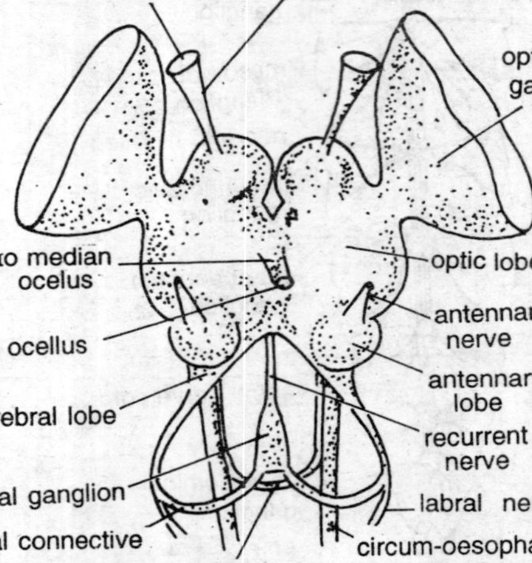

Fig. 7.11. *Grasshopper*. Brain in dorsal view.

The protocerebrum gives off a pair of thick *optic nerves* to the compound eyes and three slender *ocellary nerves* to the ocelli. The deutocerebrum sends a pair of *antennary nerves* to the antennae. The tritocerebrum gives off a pair of *labro-frontal nerves* that divide into two each: a *labral nerve* which innervates the labrum and a *frontal connective* that joins the frontal ganglion.

The sub-oesophageal ganglion gives off from its ventral surface several pairs of nerves. These include *hypopharyngeal nerves* to the hypopharynx, *mandibular nerves* to mandibles, *maxillary nerves* to maxillae, *labial nerves* to the labium and *cervical nerves* to the neck.

Several pairs of nerves leave each of the three thoracic ganglia and innervate mainly the parts of their own segment. The metathoracic ganglion, however, sends two paris of nerves to each of the first three abdominal segments also.

The first, second, third and fourth abdominal ganglia provide nerve to the fourth, fifth, sixth and seventh abdominal segments. The fifth or last abdominal ganglion sends nerves to the eighth and the remaining posterior abdominal segments.

(iii) Sympathetic nervous system

Sympathetic, autonomic, visceral or stomato-gastric nervous system includes certain ganglia and nerves connected to brain. Ganglia include: (i) a small *frontal ganglion* lying above pharynx in front of brain, (ii) an *occipital ganglion* above oesophagus behind brain and (iii) a pair of *ingluvial ganglia* in front of gastric caeca. Sympathetic nervous system controls the involuntary activities of alimentary canal, heart, aorta and repoductive organs.

SENSE ORGANS

The sense organs of grasshopper are adapted for receiving stimuli from the air and other environment in which it lives. The following sense organs are found in grasshopper.

1. Tactile organs. These are sensitive to touch. These are in the form of setae, hairs, spines etc. located in various parts of body such as antennae, mouthparts, cirri, legs, wings etc.

2. Olfactory organs. They resemble the tactile organs and are sensitive to chemicals. The antennae are supplied with the organs of smell.

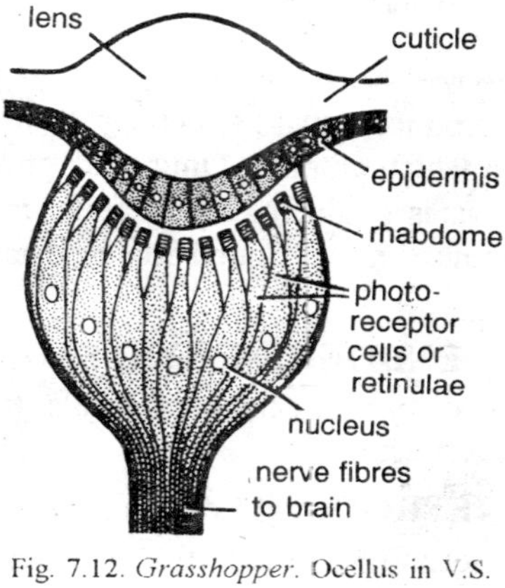

Fig. 7.12. *Grasshopper*. Ocellus in V.S.

3. Gustatoreceptors. Theses are the organs of taste. They are present on mouthparts, particularly palps.

4. Organs of sight. Three *ocelli* are sensitive to light but may form crude images at close range. An ocellus consists of a group of *photoreceptor cells* or *retinulae,* each ending in nerve fibre which leads to the brain. Outer end of each photoreceptor cell forms a *rhabdome.* Cuticle covering the group of photoreceptor cells forms a thick, biconvex, transparent lens. Real function of ocelli is not clearly known. *Compound eyes* are similar to those of prawn or cockroach in sturcture as well as function.

5. Auditory organs. There are a pair of auditory organs or *stridulatory organs*, one on either side of first abdominal tergum. It consists essentially of a *tympanic membrane* stretched within an almost circular chitinous ring. Attached to inner surface of tympanum is a whitish, sensory *scolopidia,* which are connected with metathoracic gaglion by an *auditory nerve.* Sound waves cause the tympanum to vibrate and the resultant sensory impulses are conveyed through scolopidia and auditory nerve to the metathoracic ganglion.

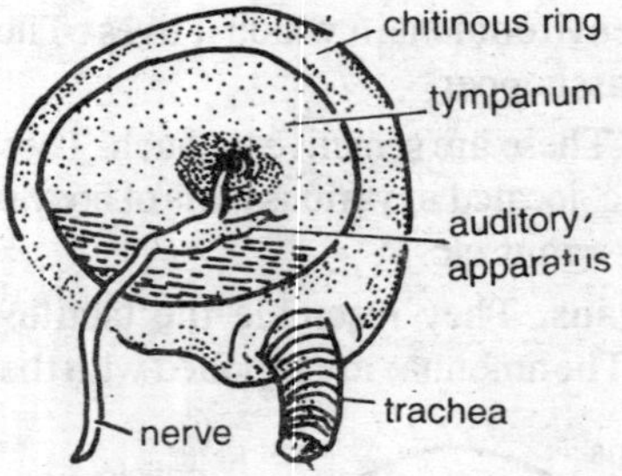

Fig. 7.13. *Grasshopper.* Interior view of tympanum and adjacent trachea.

Grasshoppers produce sound by rubbing tibial spines of hind legs against a wing vein which vibrates. Only male grasshoppers have the ability to stridulate, that is, producing sound. Some insects can hear sound beyound the range of human ears. Sound produced has some significance in mating.

ENDOCRINE SYSTEM

The endocrine system of grasshopper is similar to those of cockroach in structure as well as function.

REPRODUCTIVE SYSTEM

The sexual dimorphism exists between males and females. The

posterior end of abdomen is rounded in males where as in females it is pointed because of ovipositor.

Male reproductive organs

The male grasshopper has a pair of *testes* situated above intestine in segments 3rd, 4th and 5th of abdomen. Both the testes are oval stuctures and lie embeded in a mass of fat bodies. Each testis consists of a series of slender *tubules* or *follicles* in which spermatozoa develop. Each follicle opens by a *vas efferens,* and all the vasa effrentia open into *vas deferens.* Two vasa deferentia coming from testes meet below hindgut in 9th segment forming an *ejaculatory duct.* The ejaculatory duct opens out side through a large copulatory organ, the *penis* or *aedeagus.* Anteriorly, the ejaculatory duct receives two tufts of slender tubules, the *accessory glands.* In each tuft of accessory glands there is a clubs-shaped *seminal vesicle* which also opens into ejaculatory duct. The accessory glands secrete a fluid which help in transfer of sperms into female during copulation.

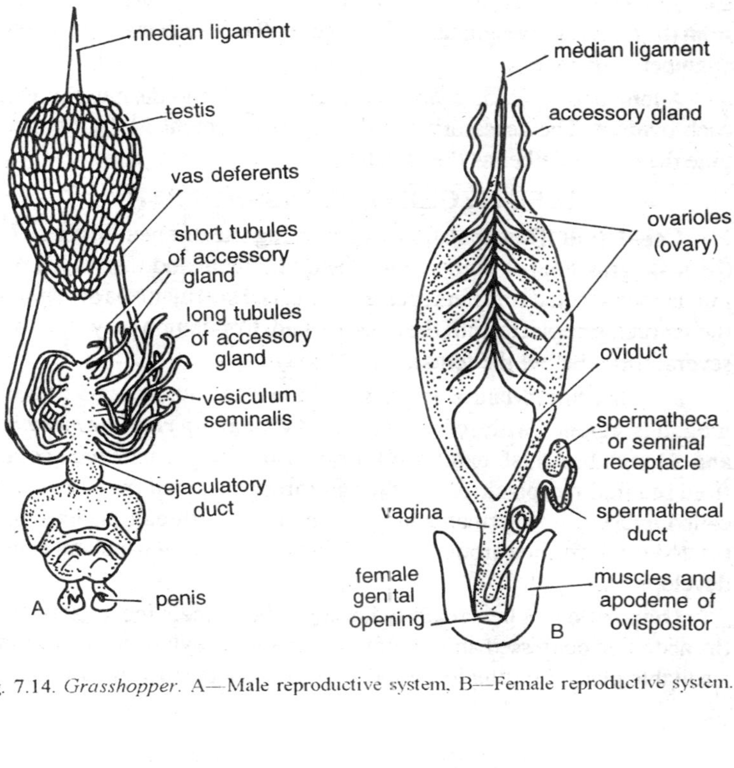

Fig. 7.14. *Grasshopper.* A—Male reproductive system. B—Female reproductive system.

Female reproductive organs

A pair of *ovaries* is present located over the intestine. These are connected to the dorsal body wall by a ligament of 6 to 8 tapering oblique ovarioles each containing a linear series of developing ova. As an oogonium decends down the ovariole, it enlarges and accumulates yolk and becomes a *primary oocyte*. It becomes surrounded by a *chorion* or *egg shell* secreted by epithelium of posterior end of ovariole. Shell has a minute pore, the *micropyle*, at its upper pole. After mating, sperm enters the ovum through micropyle to bring about fertilization.

Ovarioles of each ovary open into a wide *oviduct.* It runs along the outer margin of its ovary and then extends backwards and downwards to join its fellow of the opposite side in the 7th segment below the rectum. A *vagina* is formed by the union of oviducts. The vagina opens into the genital chamber at the base of egg guide. The genital chamber leads to outside at the base of long pair of valvulae.

An oval sac, the *spermatheca* or *seminal receptacle*, and its very long slender spermathecal duct lie coiled up in the 8th segment between the vagina and nerve cord. The spermatheca stores sprematozoa received from the male during copulation. The spermathecal duct opens into genital chamber at its base.

A long *accessory* or *collaterial gland* opens into the anterior end of each oviduct. The accessory glands secrete a cement-like substance to glue the eggs together as they are laid.

LIFE CYCLE OF GRASSHOPPER

Copulation. The copulation occurs during late summer. In copulation, the male grasshopper clings to the back of female and inserts his penis into her vagina and transfers spermatozoa. The spermatozoa are stored in the seminal receptacle until the eggs are laid. Copulation may takes place several times before the female starts to lay eggs.

Fertilization. Mature egg passing down the oviduct is 3 to 5 mm. long. It is supplied with yolk and covered with a thin *vitelline membrane* and a tough brownish outer shell or chorion. A spermatozoon coming from seminal receptacle enters the egg through a small pore in chorion, called *micropyle*. Spermatozoon nucleus fuses with egg nucleus and a blastoderm is formed around the periphery of egg from which an embryo develops.

Oviposition. When ready to lay eggs, the fremale moves about with the abdomen depressed and scratching the soil by ovipositor in search of a suitable place. Finding loose soil. she scoops a burrow in it by rotatory

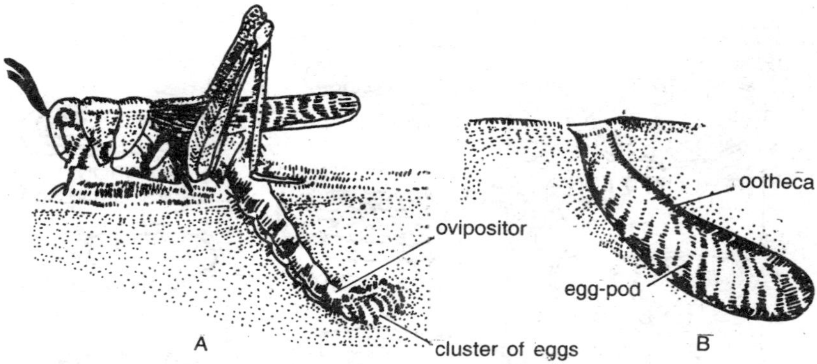

Fig.7.15. *Grasshopper*. Oviposition. A—Female grasshopper laying eggs in ground; B—Egg-pod of grasshopper.

movements of her abdomen. During this process, her abdomen bends verticlly downwards and stretches out greatly. The burrow may be upto 150 mm. deep. It is lined with a gelatinous secretion that hardens on drying and cements its wall. About 100 eggs are deposited in the burrow. They are surrounded by a sticky secretion that fastens them together to form an *egg-pod*. The burrow is closed with mud. Two to ten such egg-pods may be deposited by a single female during the breeding season that lasts from June to September.

Development. Embryonic development continues for about three weeks. It is then arrested and embryo enters into a rest period or *diapause* to tide over the adverse conditions of cold and food shortage in winter. With the onset of next spring, growth is resumed and the young grasshopper, hatching from egg, is called *nymph*. It is like a miniature adult but has relatively large head and lacks wings and reproductive organs. It feeds upon vegetation and grows rapidly. But the inflexible chitinous exoskeleton cannot sretch very much so that it is shed periodically. This complex process is termed *moulting* or *ecdysis*. Wings are developed gradually from external wing pads, and after 5 to 6 moults the young grasshopper becomes a full-fledged adult. This type of development is called *gradual metamorphosis*.

Economic Importance of grasshoppers

1. *As a pest.* Nymphs and adult grasshopprs eat several kinds of vegetation. Feeding is most active in the morning of sunny day. When food is scarce, they migrate into new feeding grounds and may damage farm and garden plantings.

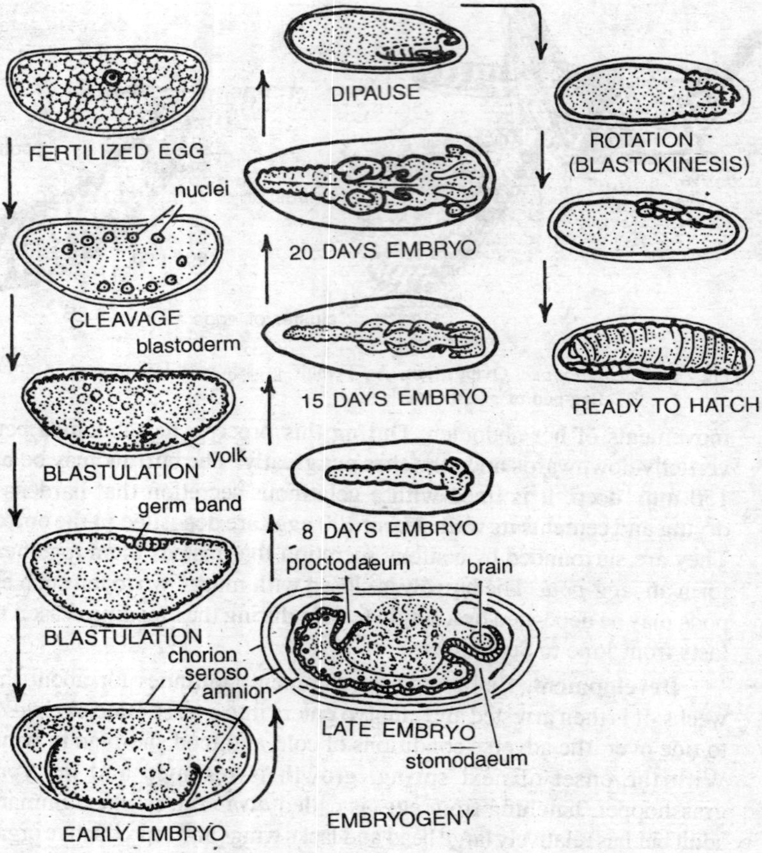

Fig. 7.16. *Grasshopper. Development*

Most African and Asian species are gregareous and migratory, the most common being *Schistocerca gregaria* and *S. migratoria* (desert locusts) extending from North Africa to North India. They have long wings and move in swarms from one region to another, feeding upon leafy vegetation, grasslands, trees and crops on the way. The swarms of this grasshopper were called *plague of grasshoppers* by Egyptians. The short-winged (or wingless) grasshoppers are solitary and confined to their respective ranges.

2. *As food.* Eggs, nymphs and adults of grasshoppers provide food for several predatory insects, frogs, reptiles, birds and mammals. They

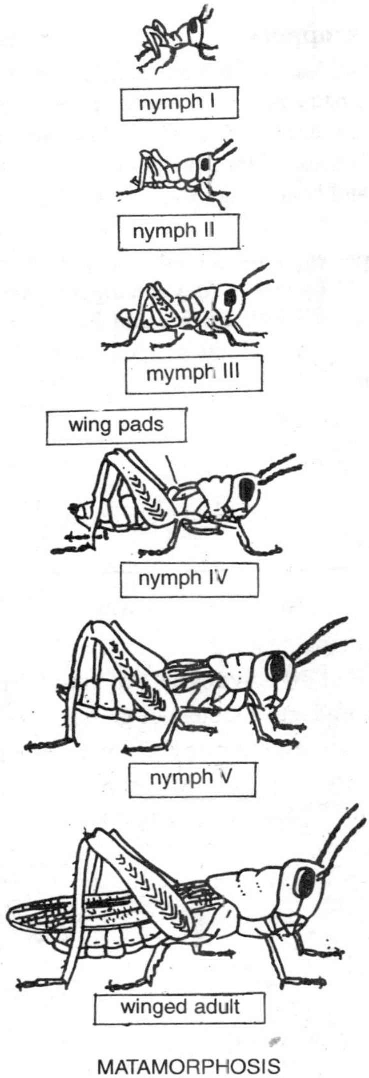

Fig. 7.17. *Grasshopper.* Life-cycle.

make good fish-bait. They are sometimes used even for human food. They continue to be used as food in Mexico, Japan and Philippines. They are commonly eaten by North American Indians and by primitive tribes in other parts of the world.

Control of Grasshoppers

Various insecticides are used either in the form of chemical sprays or dusts or poisoned baits that kill by contact or when eaten. Some recently employed insecticides are B.H.C., aldrin, chlordane, dieldrin, heptachlor and toxaphene. Methoxychlor leaves no residue harmful to man or domestic animals and hence is good for protecting fruits and vegetables and pasture lands.

The grasshopper eggs are eaten by some beetles, bee-flies, moles, and mice, the nymphs by robber flies and digger wasps and both nymphs and adults by large predatory insects and by frogs, reptiles, birds and mammals. Bacterial and fungal diseases also destory numbers of grasshoppers at time.

Revision Questions

1. Describe the external features of grasshopper.
2. Give an account of digestive system of grasshopper.
3. Describe the reproductive system of grasshopper.
4. Give an account of respiratory system in grasshopper.
5. Write short notes on : (a) Mouthparts of grasshopper, (b) Haemolymph, (c) Wing of grasshopper.

8

Musca

The houseflies of the genus *Musca* are very common in human dwellings. They are in plenty during the rainy season, though they appear early in spring and continue till late in winter. In winter, most of them die, but many survive in warm places. There are several species of housefly, *Musca domestica* is common in Europe, *M. nebulo* in tropical countries like India, Pakistan, Sri Lanka, Burma etc., *M. vicina* is common in all Oriental countries.

SYSTEMATIC POSITION

Phylum	-	Arthropoda
Class	-	Insecta
Subclass	-	Pterygota
Division	-	Endopterygota
Order	-	Diptera
Genus	-	*Musca*

Habits and Habitat

The housefly has a world wide distribution. They are abundantly around human habitation and filthy and dirty places.

The houseflies are very active and abundant during spring, summer and autumn. They like light, and rest at night on the ceilings, walls and fixture in the buildings. They become sluggish and decrease in number in November. The survivors pass winter in warm places. The houseflies can comfortably climb a vertical surface and walk upside down on a ceiling with their sticky feet. They are good fliers.

With the help of sponging type of mouthparts they take liquid food. They are unisexual. Their life-cycle show complete metamorphosis.

MORPHOLOGY

Shape, Size and Colour

It looks somewhat ovoid without wings and nearly triangular with wings at rest. It is about 1.92-2.2 mm. in width and 6-7 mm. in length. Its dark grey body, with yellowish tinge on the ventral side, has four longitudinal lines on the throax and one black streak on the abdomen.

Division of body

Body is bilaterally symmetrical and is divisible into three regions. (1) Head, (2) Thorax and (3) Abdomen

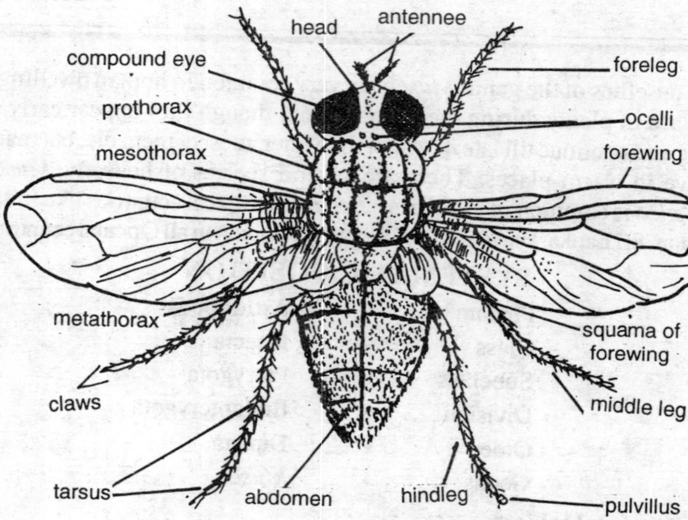

Fig. 8.1. *Musca nebulo.*

1. Head

The head is as wide as rest of the body. It is mobile and covered with hair and bristles. It bears following structures:

(i) *Compound eyes.* The compound eyes are large occupying almost the whole of hemispherical head, thus provides a wide field of vision. Each eye bears 4000 ommatidia.

(ii) *Ocelli*. The ocelli or simple eyes are 3 in number. They lie on the top of the head (vertex) medially between the compound eyes.

(iii) *Antennae*. The antennae are located in a depression in the middle of the front side of the head.

Each antenna is three-jointed; the proximal joint is called *scape*, the middle *pedicel* and the distal *flagellum* The flagellum is the largest and bears a three-jointed process called *arista.* The two basal segments of the arista are extremely small. The long distal segment bears two series of long hairs or *spinulae*. The antennae serve as sense organs.

(iv) *Mouth-parts*. The mouth-parts are sponging type adapted to suck a liquid food. It comprises the proboscis which is fleshy and retractile. The proboscis consists of three parts. The basal *rostrum,* middle *haustellum* and the distal *oral disc* or *sucker*. There is a movable joint between the rostrum and haustellum so that the haustellum can be folded back beneath the rostrum in resting condition. The rostrum bears a pair of dark brown or black, club-shaped *maxillary palps* on its front side. The haustellum has a groove on its front side which contains the *labrum-epipharynx* and the *hypophayrnx*. The edges of labrum-epipharynx are rolled to form a tube like *food channel*. The hypopharynx encloses the salivary duct. The oral disc consists of two oval *oral lobes* or *labella*. The oral lobes are traversed by a network of fine *grooves* or *channels,* called *pseudotracheae*. All pseudotracheae lead to the centrally located *mouth*, to which the lip of the food channel can be applied so that the liquid food collected by the pseudotracheae can be sucked up by the pumping action of the pharynx. The oral lobes are provided on the undersurface with minute, teeth-like structures for breaking up solid food.

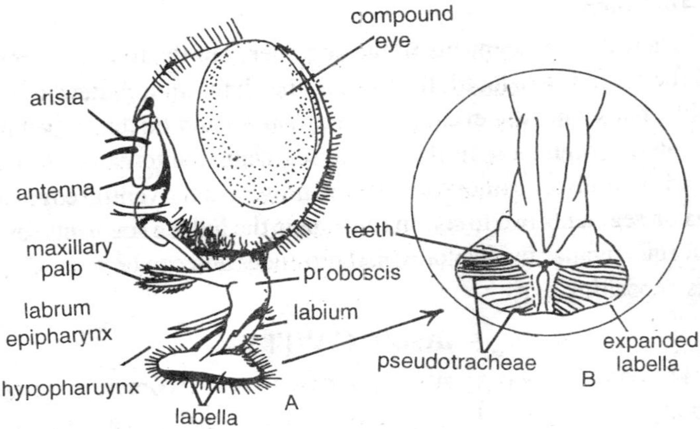

Fig. 8.2. *Musca*. A—Mouthparts, B—Proboscis enlarged.

For feeding, the proboscis is extended and its labella applied to the fluid food. The latter is drawn through the pores of the pseudotracheae into the food channel, whence it is sucked into the alimentary canal by the muscular pharynx.

2. Thorax

The thorax comprises indistinct three segments - *prothorax, mesothorax,* and *metathorax.* It bears three pairs of legs ventrally and two membranous wings dorso-laterally. The wings are present on mesothorax and a pair of knob-like structure called the *halteres* on the metathorax. The halteres bear sense organs and serve as balancing organs during flight. These also emit buzzing sound during flight.

The *legs* are thickly covered with hairs. Each leg consists of *coxa, trochanter, femur, tibia,* and five-jointed *tarsus,* terminating in a pair of curved *claws* with a pair of glandular pads, called *pulvilli* between them. The pulvillus bears a large number of tiny glandular hair which secrete tiny drops of a sticky liquid that sticks the fly to the object upon which it walks or sits These are called *tenant hair*. Between the pulvilli is a long tapering bristle the *empodium.*

The thorax has on its sides two pairs of *spiracles* in both the sexes. The first pair is between the prothorax and mesothorax and the other just below the halteres. They have thickened lips and may be opened or closed to a limited extent by the action of special muscles.

3. Abdomen

There are ten segments in the abdomen, but the first has atrophied and the second is reduced, the third to the sixth are well developed and visible, but segments seven to ten are reduced and lie telescoped inside the anterior segments. In the visible segments the terga are large and extend ventrally. A pair of spiracles are present in the ventral edges of the terga on segments two to six. In the female the hidden segments seven to ten form a tubular ovipositor which protrudes and can be seen when the fly is depositing eggs.

BODY CAVITY

The body cavity is greatly reduced by air-sacs, thoracic muscles and fat body. It is represented by narrow spaces, the *sinuses*, among the viscera. It is not lines by epithelium and contains blood. It is, therefore, called the *haemocoel*. It is incompletely divided by a horizontal partition, the *pericardium*, into two compartments; the small *dorsal sinus* that surrounds the heart, and the large *ventral sinus* that contains the remaining viscera. The two communicate through a passage bounded by pericardial cells.

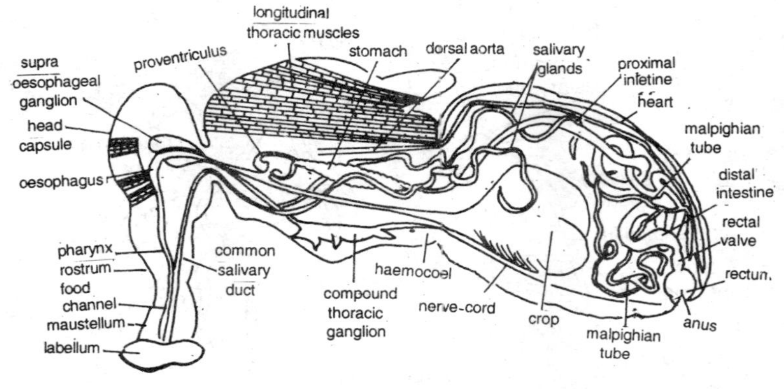

Fig. 8.3. *Musca*. General anatomy.

DIGESTIVE SYSTEM

The digestive system comprises the alimentary canal and digestive glands.

I. Alimentary Canal

The alimentary canal is complete and comprising three regions: foregut, midgut and hindgut. The foregut and hindgut are ectodermal in origin and internally lined with cuticle.

(i) ***Foregut.*** It includes :

Mouth. It is situated in the middle of oral lobes, leads by way of a grooved channel into a pharyngeal bulb.

Pharynx. It is a tubular organ provided with muscles for sucking food. Just within the head capsule, it merges with oesophagus.

Oesophagus. It is also a tubular structure passing through the cerebral ganglion. After entering into thorax the oesophagus leads into the *gizzard* or *proventriculus*. Before opening into gizzard, the oesophagus gives off a straight, slender duct which extends the length of thorax into the abdomen, where it expands into a large bilobed muscular *crop*.

Proventriculus. It is a small, circular chitinized sac lying in the prothorax, receiving oesophgus on its ventral side.

(ii) ***Midgut.*** It includes *ventriculus* and *proximal intestine*.

Ventriculus or *chyli stomach*. The proventriculus opens into the ventriculus. It is a straight, spindle-like sac, lined with columnar epithelium. The epithelium is folded to form a large number of *sacculi*. The ventriculus is followed by a long, coiled tube known as *proximal intestine*.

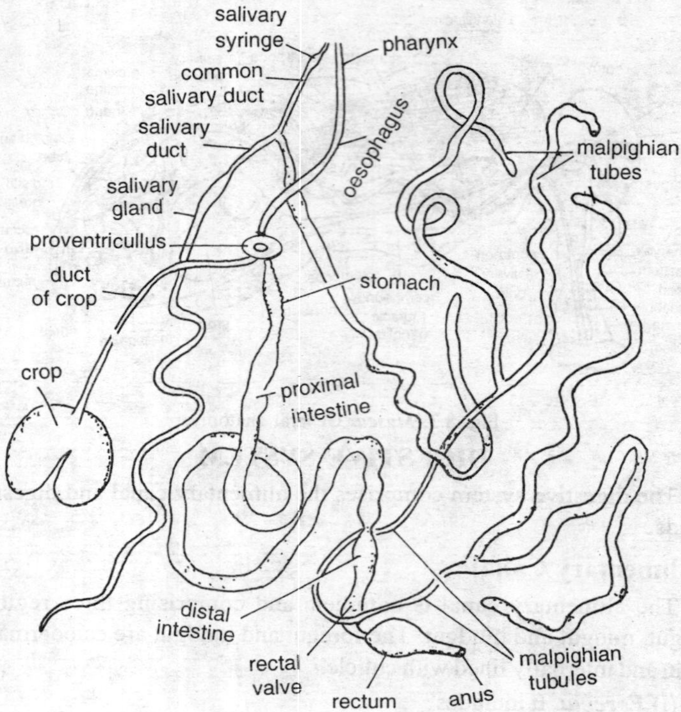

Fig. 8.4. *Musca.* Alimentary canal.

(iii) ***Hindgut.*** It comprises a *distal intestine* and *rectum.* The *distal intestine* is similar as proximal intestine but it is much shorter. It bears a rectal valve. The *rectum* is a dilated sac. It has tuberculated cuticle in anterior part, 4 rectal glands in middle part and thick muscular wall in terminal part.It opens out through the anus.

II. Digestive gland

There is a pair of salivary glands. These are wavy tube of uniform diameter. These are lined with cuboidal epithelium. Each gland gives off a salivary duct anteriorly. The two ducts unite to form a common salivary duct.The duct passes beneath the oesophagus. In the rostrum the common salivary duct is expanded to form the salivary syringe that pumps saliva onward to the hypopharynx.

Feeding and Digestion

During feeding the proboscis is extended and the two expanded oral lobes or the labellae are placed on the exposed liquid food. If the food material is solid the fly first pours a little saliva or regurgitated droplets of liquid from its crop to liquify it and then sucks the liquid. The liquid food fills the tubular pseudotracheae by capillary action. The pseudotracheae then converge into mouth, hence the food rises up the food channel due to the suction force set up in the muscular pharynx. The alimentary canal is long and coiled with the pharynx, oesophagus, crop-gizzard, stomach or midgut and rectum. The two small salivary glands pour saliva on food to hydrolyse starch. Digestion and absorption mainly takes place in the mid-gut.

RESPIRATORY SYSTEM

The respiratory system is highly developed. It consists of three components: the spiracles, air-sacs and tracheae.

Spiracles

There are 7 pairs of spiracles in a female and 9 pairs in the male. Of which first two pairs are thoracic and rest are abdominal. Each spiracle leads into a shallow vestibule which is partly separated by a valve-like mechanism from the atrium inside.

Air sacs and Tracheae

In the abdomen, the atria lead into the tracheae where as in the thorax the atria open into large thin walled *air sacs* from which small tracheae extends to muscles and other tissues. The thoracic air sacs send a number of cephalic air sacs in the head and a pair of very large *abdominal air sacs* in the abdomen. The cephalic and few thoracic air sacs remain expanded where as the others are collapsable. These are the reservoirs of air. The tracheae branch and rebranch to form the fine tubules, the trachioles.

The trachioles participate in the exchange of gases.

CIRCULATORY SYSTEM

The circulatory system of housefly is very simple and open type as found in other insects. There is a muscular tube or *heart* lying in a cavity i.e. the *pericardial sinus* along the mid-dorsal line.It extends the length of the abdomen and is divided into 4 chambers. Each chamber having a pair of opening through which the blood is sucked form the pericardial sinus. The anterior end is continued into a slender *dorsal aorta* or *anterior aorta*. The heart is supported by *alary muscles*. Besides heart there are 4 accessory pulsating organs called *auxillary hearts,* at the base of each wing and a few in the thorax which assist in circulation of blood.

The dorsal aorta continue along the stomach and appears to terminate in a mass of cells behind the proventriculus. It pours the blood into the anterior part of haemocoel. The haemocoel is incompletely divided into a large *ventral sinus* and a small *dorsal sinus* or *pericardial sinus*, by a horizontal partition, the *pericardium*.

The blood is colourless and is called the *haemolymph*. It contains a fluid plasma and numerous blood corpuscles, the *haemocytes*. It does not contain any respiratory pigment. It helps in physiological exchange of materials.

EXCRETORY SYSTEM

In housefly the excretory system consists of one pair of yellowish-green tube-like structures known as *Malpighian tubules*. These are present at the junction of midgut and hindgut. Each tubule is divided into two branches. The branches are very long, highly convoluted and cannot be separated from the fat-bodies.

NERVOUS SYSTEM

In housefly the *supra-oesophageal ganglion*, *sub-oesophageal ganglion* and *circum-oesophageal connectives* are fused to form a common mass which is perforated by an aperture for the passage of oesophagus. The ventral nerve cord bears a compound *thoracic ganglion* which is formed by the fusion of three thoracic ganglia.

From brain a median *ocellar nerve* originates and it is supplied to ocellus. A pair of *antennary*, paired *pharyngeal nerves*, paired *labial nerves* innervate their respective organs. Paired *cervical nerves* arise from the nerve cord before joining to the compound thoracic ganglion. It supplies to the muscles of neck. From compound thoracic ganglion a pair of each *prothoracic, mesothoracic* and *metathoracic dorsal nerves* arise which innervate the thoracic muscles. A pair of each of *accessory mesothoracic* and *metathoracic dorsal nerves* arise from the thoracic ganglion and suppliment the corresponding dorsal nerves in supply. To the legs, a pair of each *prothoracic, mesothoracic* and *metathoracic crural nerves* are supplied. Two pairs of *abdominal nerves* are given by the ventral nerve cord behind thoracic ganglion in the region of thorax. In the abdomen a series of abdominal nerves are given alternately on right and left to supply abdomen. The ventral nerve cord ends in the genital organs.

REPRODUCTIVE SYSTEM

The sexes are separate and are slightly different in size, the female being slightly larger than the male.

1. Male Reproductive Organs

The male reproductive organs consist of a pair of *testes*, *vasa deferentia*, *ejaculatory duct*, *ejacultory sac* and *penis*.

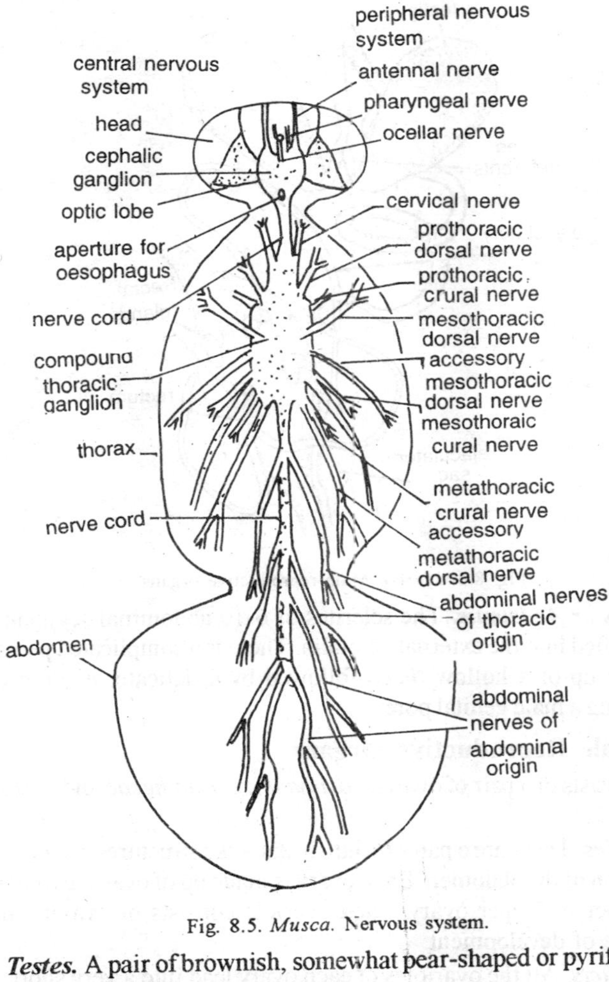

Fig. 8.5. *Musca*. Nervous system.

(a) ***Testes.*** A pair of brownish, somewhat pear-shaped or pyriform testes are present in 5th abdominal segment.

(b) ***Vasa deferentia.*** From each testis a duct arises known as *vas deferens* and extends forward. The two vasa deferentia unit to form a median tube, the *ejaculatory duct*.

(c) ***Ejaculatory duct.*** It is a much coiled structure, anteriorly it is wider.It enlarges posteriorly to form *ejaculatory sac*.

(d) ***Ejaculatory sac.*** It is a muscular sac-like structure supported by ejaculatory sclerites.It helps in ejaculation of semen.

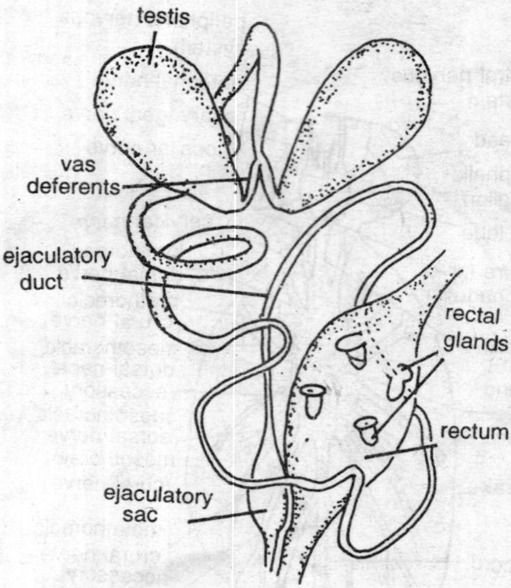

Fig. 8.6. *Musca*. Male reproductive organs.

(e) ***Penis or Aedeagus.*** The sclerites of 6-10 abdominal segments are modified into the external genitalia. There is a complicated penis.It is made up of a hollow *theca* followed by a delicate, *hyaline glans* bearing a male genital pore.

II. Female Reproductive Organs

It consists of a pair of *ovaries*, *oviducts*, *spermathecae* and *accessory glands*.

(a) ***Ovaries.*** These are a pair of whitish, disc-like structures occupy major portion of the abdomen. Each ovary is made up of ovarioles and their number is 70 per ovary. Each ovariole consists of ova at various stages of development.

(b) ***Oviducts.*** All the ovarioles of each ovary lead into a very short, thin walled tube, the *oviduct*. The two oviducts units to form a curved common oviduct. It dilates to form sacculus which is followed by a still larger vagina. The vagina opens out through female genital pore present in the 9th abdominal segment.

(c) ***Spermathecae.*** These are three (one right and two left) long tubular structures which open into the genital tract at the junction of sacculus and vagina.

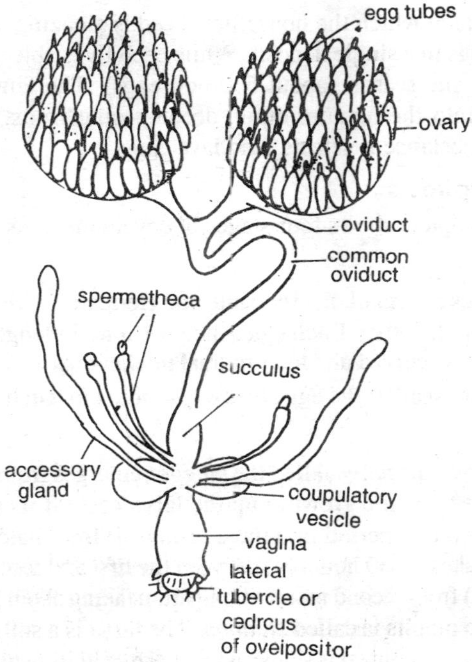

Fig. 8.7. *Musca.* Female reproductive organs.

(d) ***Accessory glands.*** A pair of long, tubular glands lying near spermathecae. They secrete a sticky fluid that cover the eggs to adhere them together and to the substratum.

(e) ***Copulatory vesicles.*** These are a pair of sac-like structures.

(f) ***Ovipositor.*** It consists of the abdominal segments 6-9 with long intersegmental membrane in between. It normally remains withdrawn into the 5th segment.

Life-history

Copulation

The flies breed from June to October when they become very active. The greatest breeding activity occurs in the warmest months of August and September. The pairing of the male and female takes place on the ground when at rest, and lasts for a few minutes. The male alights on the back of the female. The female inserts its ovipositor into the genital atrium of the male to receive the spermatozoa. The eggs are laid 6-8 days after copulation.

Oviposition

The rate at which the houseflies breed is amazing. A single fly lays 100-150 eggs in a single deposit. While each fly is able to lay such 5 or 6 deposits in one season (about 500-600 eggs). The female extends her ovipositor into the manure dung, decomposing fruits, vegetables etc., whatever substance it selects, and lays eggs.

Metamorphosis

It is complete shows four stages of development as follows :

Eggs

The eggs are small, nearly white and elongated with one end slightly broader than the other. Each egg is about 1 mm. in length and bears two dorsal, rib-like, curved and longitudinal thickenings.

In warm weather the egg takes 8 - 24 hours to hatch.

Larva

The larva, specially called the *maggot,* emerges from a slit along one side of the egg, while growing up the larva cast off its skin twice, thus dividing the larval period into three *instars* (i) from hatching to the first moult that takes 20-30 hours (ii) between the first and second moult taking 24 hours (iii) from second moult till pupation taking about 3 days. The time between two moults is called *stadium*. The larva is a soft, legless, worm-like creature. It is conical is shape, being narrow in front and broad behind. The newly hatched maggot is 2 mm. long. Its body consists of 12 segments and an indistinct head. The head can be withdrawn under the first segment

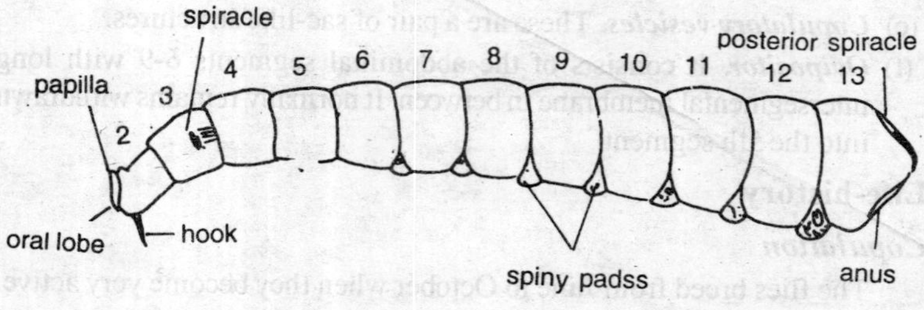

Fig. 8.8. *Musca*. Larva.

or *pseudocephalon*. It bears a pair of *oral lobe* bearing minute papillae-like *antennae*, a pair of *maxillae* with photo-receptive papillae, a pair of hook-like *mandibles* and a small *labium*. It is *metapneustic* having only one pair of D-shaped posterior abodominal spiracles with two slit-like apertures in each. The *anus* is located ventrally on the last segment.

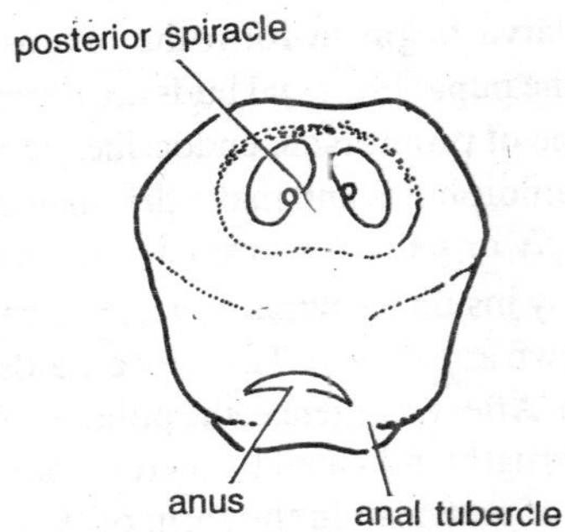

Fig. 8.9. *Musca*. Larval posterior end showing spiracles.

On the anterior border of each of the 6th-12th trunk segments, there is a semilunar pad beset with minute, stuff spines. These *spiniferous pads* are locomotory organs and help the larva in moving backwards and forwards. The body of the larva is covered externally by cuticle.

After first moult a second instar is formed. It is larger than first instar. Besides the posterior pair of spiracles which becomes larger, the 2nd instar acquires an anterior pair of spiracles also. Thus the larva is *amphipneustic*. After second moult, third instar is formed. The third instar is a full grown larva measuring 12 mm. in length.The third instar grows in size and ready to pupate within 3 days.

Pupa

When fully matured, the larva usually crawls away from its original moist place and travels to some dry and sheltered cervice. Here it rests for sometimes. The emergence of pupa is proceeded by a general contraction of body. Some of the anterior segments are even withdrawn within the body. Thus the larva changes into a pupa without moulting, the last larval skin hardens to form an outer covering of *puparium* which encloses the pupa. Such a pupa is called *coarctate*. It has no chitinous covering of its own but only a soft pupal skin, the outer puparium has been formed by the last larval skin. The puparium is barrel-shaped and it becomes dark brown, externally it is segmented and shows traces of larval spiracles and spiny pads which become non-functional. The pupa takes in air by means of a pair of spine-like pupal spiracles projecting between the fifth and sixth segments of the puparium. The pupa is absolutely immobile, and the pupal stage lasts from 4 to 5 days. During this time internal changes take place, the larval organs undergo degeneration and thus *histolysis* occurs. The *imaginal buds* of the larva begin to form the organs of the adult, the *histogenesis* occurs in the pupa. Imaginal buds are dormant cells, they are stimulated by a hormone of prothoracic endocrine glands which become active only during metamorphosis and make the imaginal buds grow. By these process the *adult fly* or *imago* is formed in the pupa.

When the fly is ready inside the puparium it has a sac-like structure in the front of the head known as *ptilinum*. This process aids in the emergence of imago from puparium. After emergence, the ptilinum is withdrawn in the head. Its position is externally indicated by a crescentic slit called *lunule* situated above the base of antennae in the form of 'U'.

Imago

It is somewhat smaller than the adult. The wings are less developed and the colour is light. In the contact of air the wings become hard and somewhat large. The body increases in size and within one or two weeks it becomes adult fly and ready for egg laying.

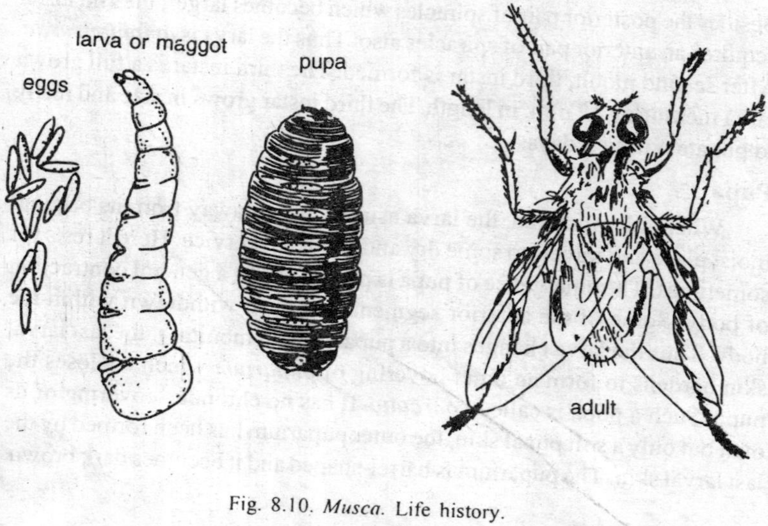

Fig. 8.10. *Musca*. Life history.

Economic Importance

The housefly is very harmful insect besides causing a lot of annoyance, it carries micro-organisms of several human diseases as it feeds on all eatable as well as manure, faecal matters and decomposing matters. A single fly carries about 33 million germs in its intestine and more than 100 million germs on its legs etc.

1. Dissemination of germs

It spreads various diseases by carrying them on the hairy surface of its body, legs etc. The germs are carried in droppings, vomit spots via saliva. The diseases are tuberculosis, diarrhoea, typhoid, cholera, leprosy, plague, anthrax, dysentery, gonorrhoea etc.

Larvae of housefly ingest eggs of a nematode, *Habronema*, and the infection is passed through pupae to adults which transmit it to the horses and to the eyes of children causing conjunctivites.

Sometimes the eggs are laid on open wounds, where they hatch out as larvae when infected fruits or vegetables are taken, larvae hatch out in the gut, perforate the mucous lining and may cause haemorrhage. This attack by larvae on human is called *myiasis*.

2. Contamination of food

The housefly contaminates our food material by transmitting the infective stages of *Taenia*, *Ascaris* etc.

CONTROL OF HOUSEFLIES

The houseflies are the dangerous enemies of human beings. Therefore, it is very essential that the growth and increase of houseflies and the infection by them must be stopped. It can be done in the following ways:

1. Preventing their breeding

Elimination of the breeding media is the first and most important step in fly control. This can be done by :

(a) *Physical method.* All the filth, where eggs can be laid, should be kept covered and periodically removed every week. Open toilets should discouraged.

(b) *Chemical method.* The manure or rubbish if accumulated for a longer time, it should be treated chemically with lime, copper-sulphate, crude oil etc. to prevent the breeding of flies.

2. Destruction of adult flies

(a) *Mechanical methods.* Flies can be collected by baits and then killed. Fly-papers, and fly-swatters can also be used.

(b) *Chemical methods.* Spraying of houses, barns etc. with DDT, benzene hexachloride etc. is useful. A few drops of 3% formaldehyde solution in a plate of milk attract and kills flies. Repellants like crezote oil and borax, keeps the flies away. A mixture of coal-tar and carbon disulphide can be applied to the wounds of animals. Fumigation by neem leaves is also useful.

(3) *Protection from infection by flies.* All types of filth should be kept away or duly covered.The houses must be properly screened and food, fruits and vegetables must not be exposed in houses or markets.

Revision Questions

1. Give an account of life-history of the housefly and suggest measures for checking of fly nuisance in the urban area.
2. Give an account of life-history of *Musca.* Mention the economic importance of this insect.
3. Give an account of morphology of common housefly.
4. Write short notes on : (a) Histolysis, (b) Maggots, (c) Halteres, (d) Ommatidium.

9

Mosquitoes

Everyone is familiar with mosquitoes. These are common in tropics and found in marshes and damp places. They are a nuisence in places where they occur in large number because of their biting habits. There are 29 gnera and 2000 species of mosquitoes in World. The important genera, as for as human health is concerned, are *Culex*, *Aedes*, and *Anopheles*.

SYSTEMATIC POSITION

Phylum	-	Arthropoda
Class	-	Insecta
Subclass	-	Pterygota
Division	-	Endopterygota
Order	-	Diptera

Habits & Habitat

Mosquitoes are found in damp and marshy places. They are abundant in tropics and sub-tropics. They are active in summer. Mosquitoes are nocturnal, as they come out in night for feeding. During day time they hide in hollows of trees, bushes, dark places of buildings etc. The female feed on the blood of man and other large animals where as male sucks the juice of flowers and fruits. They are oviparous and lay eggs in water. Most of the mosquitoes die in winter and extreme hot months of May and June. Some survive hiding in protected places. The mosquitoes can fly for 2-3 hours continuously and cover 20-30 km.

MORPHOLOGY

Shape, Size and Colouration

The mosquito has a slender, delicate and long legged body. It measures 3-6 mm. in length. The body colour is greyish black and the wings of *Anopheles* carry dark spots. *Aedes* has no spots on wings but the body has a black and white stripes.

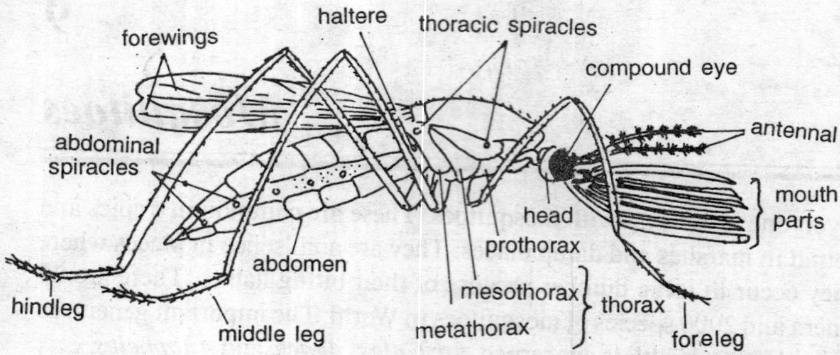

Fig. 9.1. *Culex.* Female.

Division of body

Like other insects, the body is divisible into three parts :

(1) Head

(2) Thorax

(3) Abdomen.

1. Head

The head is globular and highly mobile on a slender neck. It bears a pair of compound eyes, antennae and mouthparts. *The compound eyes* are large and kidney shaped. They occupy a considerable part of the head. The ocelli are absent. The top of the head has an epicranium below which is clypeus. *The antennae* are filiform, each with 15 joints. The basal segment is the *scape* which is concealed by a very large globular second segment, the *pedical* containing a *Johnson's organ* which is auditory in function, the other 13 joints form a *flagellum.* The flagellum has many bristles present in rings. The bristles are longer and much more numerous on the antennae of males giving them a bushy appearance. The antennae of females are with only a few, short bristles. *Mouthparts* are of *piercing* and *sucking type*, contained in a proboscis. These are :

(a) A pair of mandibles

(b) A pair of maxillae

(c) Labium

(d) Labrum and epiparynx

(e) Hypopharynx.

(a) *Mandibles.* The paired mandibles are long, thin and needle-like styles with fine blade-like distal edges. These act as piercing organs and puncture the host skin. The mandibles are, therefore, found only in females and are absent in males.

(b) *Maxillae.* These are also elongated needles, very much like the mandibles, but ending into saw-like blades. Each maxilla bears a pair of 5-segmented maxillary palp.

(c) *Labium.* It is formed by the fusion of 2nd pair of maxillae and is in the form of a median ventral, elongated and fleshy proboscis sheath. It is like an open gutter of half tube with a groove on its dorsal surface. The tube lodges the mouthparts and is roofed over by labrum and epipharynx. The free distal end of the proboscis ends in a pair of white pointed lobes known as *labellae*. The labellae are tactile organs clothed with numerous sensory hair.

(d) *Labrum and epipharynx.* The labrum and epipharynx are fused together forming a stiff structure with pointed tip. It lies dorsally above the groove of labium and roofing it. It bears ventrally a groove. The groove and hypopharynx forms a food channel.

(e) *Hypopharynx.* It is a long, pointed and flattened plate like double sword structure. A salivary duct opens at its tip.

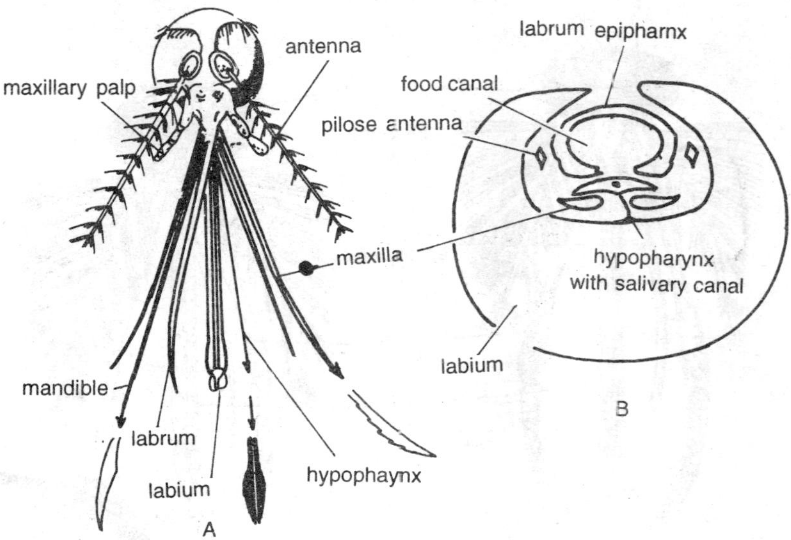

Fig. 9.2. Mouthparts of Mosquito. A-Frontal view. B-T.S. of same.

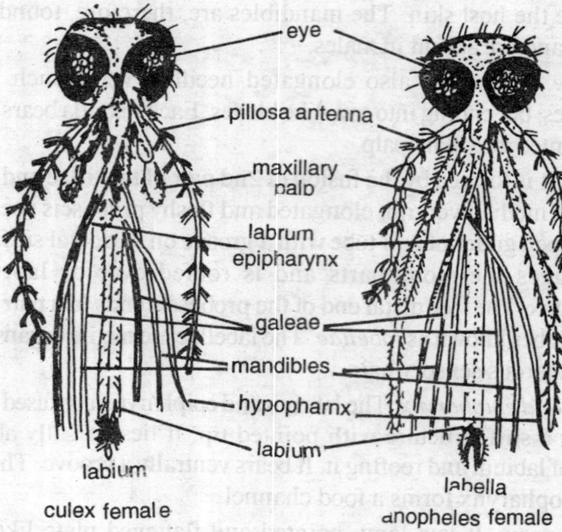

Fig. 9.3. Head and mouthparts of female *Culex* and *Anopheles*.

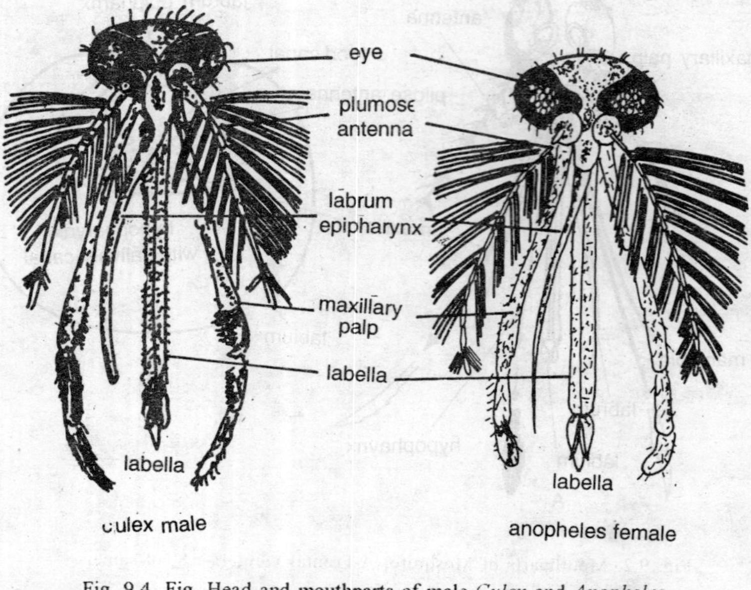

Fig. 9.4. Fig. Head and mouthparts of male *Culex* and *Anopheles*.

2. Thorax

The thorax is as usual made up of three segments. The prothorax and metathorax are small whereas the mesothorax is very large and its tergum consists of three sclerites : *scutum, scutellum* and *post-scutellum.* Each segment bears a pair of legs. The legs are long, slender and delicate. They have short coxae, long tarsi and simple claws.

One pair of wings are present on dorso-lateral sides of mesothorax. These are membranous. The metathorax bears a pair of *halteres* or *balancers.* These represent rudimentary hind-wings, useless for flight and probably sensory in function. These are supposed to produce characteristic sound.

3. Abdomen

The abdomen is slender and made up of 10 segments, the first one is fused with metathorax so that 9 segments are visible. In male last 3 segments are modified by genitalia. It bears an *anus* on the 8th segment and a *genital pore* on the 9th segment.

LIFE HISTORY

The life history in mosquitoes comprises four stages :

1. Egg
2. Larva
3. Pupa
4. Adult.

The female always takes a blood meal before egg-laying to have extra nourishment essential for this purpose.

1. Eggs

After mating the female lays eggs on still water, the eggs may be laid in ponds, or pools, or rain-filled containers. In *Culex,* the eggs are cigar-shaped and tapering at one end. The eggs laid at night and one *Culex* female may lay 300 eggs. The eggs are laid by side standing erect, and glued together by the legs to form boat-shaped rafts which float on water. Female *Anopheles* lays 40-100 eggs at one time. They float singly i.e., do not form a raft. The eggs are pointed at both the ends and have a pair of lateral air floats. In *Culex* the eggs have no air float. The eggs in the raft are vertical and, their heavier ends downwards. In *Aedes* the eggs are laid singly in moist soil. They hatch when flooded. In *Culex* the eggs lasts 2-3 days where as in *Anopheles* the eggs lasts for 24-48 hours.

2. Larva

In *Culex* the larva hatches out from the egg within 2-3 days. The larva

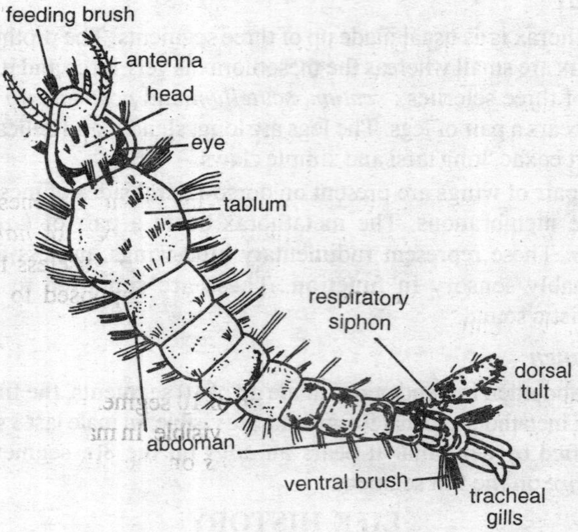

Fig. 9.5. Larva of *Culex*.

is small, transparent and wriggling creature. The body is distinct into three regions, the *head*, *thorax* and *abdomen*. The head is large and rounded structure bearing a pair of compound eyes, a pair of simple eyes or ocelli and complicated mouthparts.

The mouthparts include *labrum*, toothed *mandibles*, flattened *maxillae*, *labial plates* and two dense *feeding brushes*. The larva, thus, subsists on solid food. The thorax is unsegmented but bears three lateral pairs of unjointed tubercles, each bearing a tuft of bristles. The abdomen comprises nine segments, each bearing a tuft of bristles. The 8th segment of *Culex* larva bears a long, tubular, chitinous *respiratory siphon*. The siphon bears a *comb* of very small scales at the base, 2 large hair tufts and two rows of flattened spines called *pecten*. At the tip of the siphon there are two *spiracles* leading into the tracheae. Around the spiracles are five leaf-like lobes which can close over the spiracles to prevent water from entering. The respiratory system is *metapneustic* in which only the last pair of abdominal spiracles is open. At the end of the 9th segment is an *anus* surrounded by four leaf-like *tracheal gills*. The tracheal gills bear tracheae instead of blood vessels. The 9th segment bears a tuft of dorsal bristles at its tip and ventrally a bushy tuft of bristles called *ventral brush*.

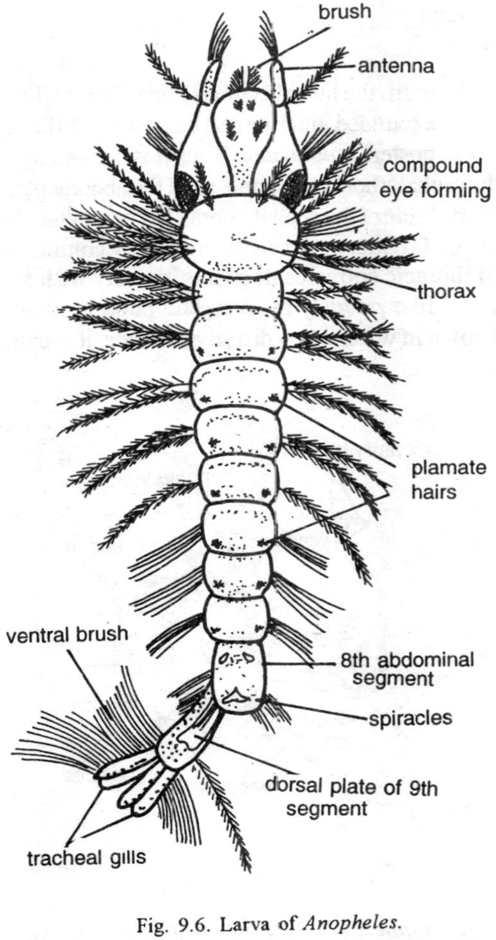

Fig. 9.6. Larva of *Anopheles*.

The larva a *Anopheles* is without a respiratory siphon instead the 8th segment bears a raised chitinous quadrilateral plate with two spiracles. The spiracles are surrounded by five small leaf like flaps. Near the spiracles, there are small bristles which form a pair of *pecten*. *Comb* is absent. The larva has *palmate bristles* or hairs forming a pair of tufts on each thoracic and most of the abdominal segments. A palmate bristle consists of a short stalk from which a number of slender and flat leaflets radiate. In *Anopheles* the larva can breathe and feed, lying parallel with the surface of water whereas in *Culex* it floats in water obliquely with head downwards.

During this period the larva moults for four times. A full grown larva is about 12 mm. long.

3. Pupa

After the last moult, the larva becomes pupa. This is the resting stage. The pupa has a large rounded anterior part consisting of the *cephalothorax* and a slender, bent posterior *abdomen*. Thus it is *comma-shaped*. It floats in water with the cephalothorax uppermost and the abdomen curved beneath it. On the mid-dorsal side of the cephalothorax is a pair of tubular *respiratory trumpets* which are broader at the distal end. They communicate with an anterior pair of thoracic spiracles. It swims actively with a pair of large, chitinous and leaf-like *paddles* or *fins*. The pupa is unable to feed. It flutters up and down in water when disturbed, hence, it is called a *tumbler*.

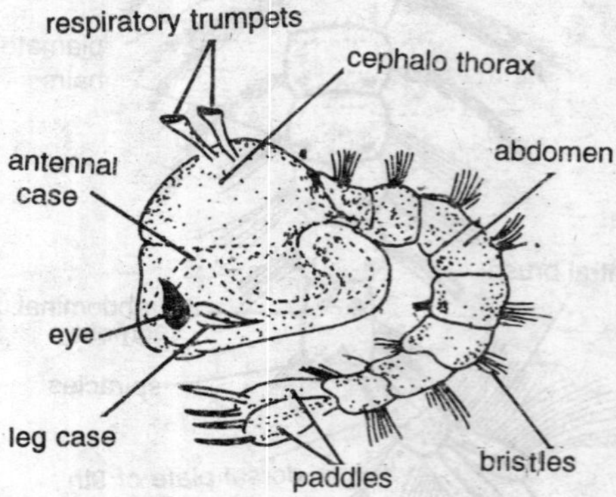

Fig. 9.7. Pupa of *Culex*.

In the pupa of *Anopheles* the respiratory trumpets are short and broad and the abdomen is more strongly curved than that of pupa of *Culex*.

The pupal period lasts for 1 to 3 days. During this period the pupa is transformed into a complete adult. Most of the larval organs in the pupa are broken down. The process of breaking down and disintegration of larval organs is called *histolysis.* The adult organs are formed. For the formation of adult organs groups of formative cells are set aside in the larva, they are called *histoblasts* or *imaginal buds.* This process is called

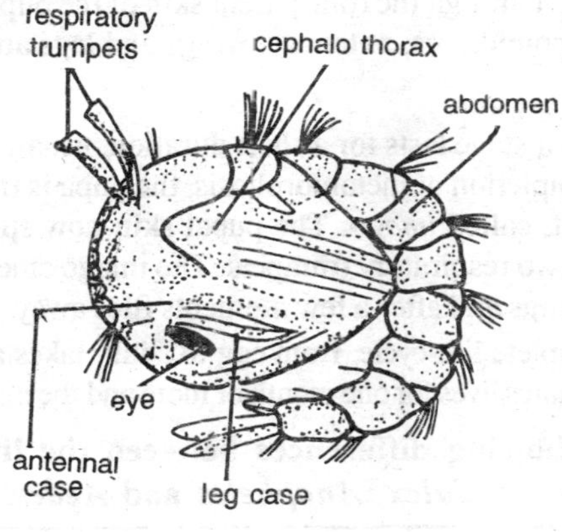

Fig. 9.8. Pupa of *Anopheles*.

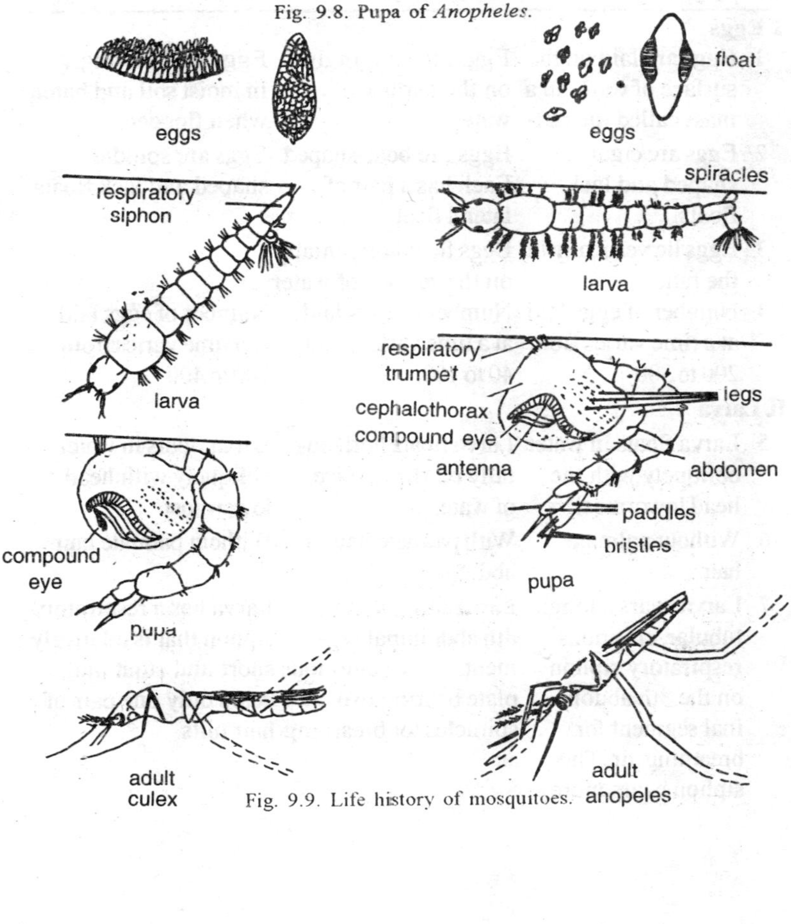

Fig. 9.9. Life history of mosquitoes.

histogenesis. Through the transparent skin of the pupa, the adult organs, such as compound eyes, antennae, wings and legs are visible.

4. Imago

The pupal stage lasts for a short duration, usually two or three days. After the completion of metamorphosis, the pupa is transformed into the perfect insect, called *imago.* The pupal skin now splits along the back between the two respiratory trumpets. The imago emerges out with well-developed wings and after a few moments flies away.

The complete life-cycle, from egg to adult, takes about ten to several days. The females lives for one mouth or more and the male for about a week.

Table showing differences between the life history of *Culex, Anopheles* and *Aedes.*

	Culex	*Anopheles*	*Aedes*
I. Eggs			
1.	Eggs are laid on the surface of water in a mass called the raft-	Eggs are laid singly on the surface of water	Eggs are laid singly in moist soil and hatch when flooded.
2.	Eggs are cigar-shaped and lack floats.	Eggs are boat-shaped. Each has a pair of lateral float	Eggs are spindle shaped and lack floats.
3.	Eggs lie vertically in the raft.	Eggs float horizontally on the surface of water	
4.	Number of eggs laid at a time varies from 200 to 300	Number of eggs laid at a time varies from 40 to 100.	Number of eggs laid at a time varies from 200 to 400.
II. Larva			
5.	Larva floats in water obliquely with the head lowermost.	Larva floats horizontally on the surface of water	Larva floats in water obliquely with head lowermost.
6.	Without palmate hair	With palmate hair on abdomen	Without palmate hair
7.	Larva bears a long, tubular, chitinous respiratory siphon on the 8th abdominal segment for breathing air. The siphon bears more	Larva has, on the 8th abdominal segment, a rised chitinous plate bearing two spiracles for breathing air.	Larva has a respiratory siphon that is relatively short and stout and bears only one pair of hair tufts.

	than one pair of hair tufts.		
III. Pupa			
8.	Pupa bears long respiratory siphons on the cephalothorax.	Pupa bears relatively short and broad respiratory siphons on the cephalothorax.	Pupa bears long respiratory siphons on the cephalothorax.
9.	It is colourless.	It is greenish	Pupa is colourless.
10.	Each paddle bears a single long bristle.	Each paddle bears one long and one short bristle.	As in *Culex*.
IV. Imago			
11.	Greyish body.	Greyish hairy body	Greyish black body.
12.	Wings lack dark spots.	Wings have dark spots.	Wings are without spots.
13.	Imago rests with the body parallel to the surface. Body hump backed when at rest.	Imago rests with the body making an actue angle with the surface. Body is not hump-backed when at rest.	Imago rests with the body parallel to the surface. Body is hump-backed when at rest
14.	Proboscis is bent downward, when imago is at rest.	Proboscis is in line with the body, when imago is at rest.	Proboscis is bent down when the imago is at rest.
15.	Abdomen bears scales on the sterna.	Abdominal sterna lack scales.	
16.	Maxillary palps are short in the female and longer than proboscis in the male but not clubbed.	Maxillary palps are as long as proboscis in both sexes and clubbed in the male.	Maxillary palps are short in the female and longer than the proboscis in the male but not clubbed.
17.	Scutellum is trilobed	Scutellum is evently rounded.	Scutellum is trilobed.
18.	Thorax is without markings.	Thorax lacks markings.	Thorax has silvery or white markings.
19.	Tip of a abdomen of female is blunt with cerci retracted.	Tip of abdomen of female is blunt with cerci retracted.	Tip of the abdomen of the female is pointed with cerci protruding.
20.	Spreads filariasis and encephalitis.	Spreads malaria.	Spreads yellow fever and dengue.

ECONOMIC IMPORTANCE

The mosquitoes are definit nuisence as they inflict painful and irritating bites besides sucking blood. Most of them act as transmitting agents or vectors for some diseases of man and other animals. The diseases caused by mosquitoes in man are malaria, yellow fever, filariasis, dengue, encephalities, leprosy etc.

(i) *Malaria*. Malaria a deadly human disease. It is caused by a protozoan, *Plasmodium*. It is transmitted by female *Anopheles*. In malaria there is high recurring fever accomplished by chills and shivering.

(ii) *Yellow fever*. Yellow fever is confined to Africa and south America. It is caused by a virus and the vector is *Aedes aegypti*. The symptoms include high fever, internal bleeding, severe headache, and pains in the bones, anaemia, severe jaundice and vomitting of blood and bile occur.

(iii) *Filariasis*. Filariasis is caused by a helminth parasite, *Wuchereria bancrofti*. Its vector is *Culex fatigans*. The disease begins with a slow inflammatory swelling of limbs, scrotum or mammae. Filariasis may eventually cause *Elephantiasis*, if left unattended.

(iv) *Dengue*. Also called *back bone fever* is caused by a virus and is carried by *Culex fatigans* and *Aedes aegypti*. The symptoms are sudden high fever with rash on the face and severe headache, pain in muscles and joints etc.

(v) *Encephalitis*. Encephalitis is caused by a virus and is spread by various species of *Culex* and *Aedes*. The symptoms are headache, high fever, drowsiness and inflammation of brain.

(vi) *Leprosy*. According to the scientists of JIPMER, leprosy can be spread by mosquitoes.

(vii) *Dermatobia*. Dermatobia is a disease of the skin of man and cattles of central and south America. *Psorophora*, a mosquito,carries the eggs of a bot-fly, *Dermatobia* (as it lays eggs on the body of mosquito). When the mosquito bites, the eggs of bot-fly quickly passed on the human skin and the larvae penetrate the skin causing swelling and cutaneous myasis.

Control of Mosquitoes

Following general methods may be used for the control of mosquitoes:

1. Elimination of breeding places

The larvae and pupae develop in the water, therefore, elimination of breeding places is important. Small ponds, ditches etc. can be filled up with mud. Stagnant water, swampy areas, marshes must be drained off. Water should not be allowed to be stored in open cans, barrels, buckets etc. Bushes, shrubs etc. should be cleared.

2. Destruction of larvae and pupae

It is easier to kill the aquatic larvae and pupae than the adults. It can be done by spreading over water surface with crude oil, kerosine oil, etc.. Oil emulsions of DDT, BHC etc. are effective larvicides. Biological methods can also be employed such as larvicidal fishes like *Gambusia*, *Guppy*, trouts, sticklebacks etc. can be introduced in the water reservoir. Aquatic insectivorous plants, such as bladderwort (*Utricularia*) can also devour the larvae and pupae of mosquitoes.

3. Destruction of adult mosquitoes

This can be done by hands, traps, fumigation, spraying etc. Fumigation with tarcamphor, cresol, sulphur or other derivatives of neptha burn to produce poisonous smoke which kill the adults. Spraying with DDT, gammexane, melathione etc. remain effective.

4. Sterlization

Sterlization of mosquitoes in Japan has been successfully done.

5. Personal protection

It can be done by; (i) The exposed parts of the body may be protected by the use of gloves, soxes, boots etc. (ii) Use mosquito repellents like odomos, mustard oil etc. (iii) Mosquito-nets should be used, especially at night (iv) The house should be built on a high grounds with proper drainage.

Revision Questions

1. Describe the mouthparts of a female *Anopheles* and compare them with those of a male.
2. Describe the life-history of mosquito that causes malaria in man.
3. What are the differences between the adults of *Culex* and *Anopheles*. Give an account of life-history of any one of them.
4. Describe the mouthparts of mosquitoes and show how they are adapted for the feeding habits of these insects.

10

Apis (Honey Bee)

Apis is a social insect living in colonies, build by themselves. A colony may have more than 50,000 individuals. They are semi-tropical insects that have adapted themselves in temperate climates, where they survive cold weather by forming themselves into winter balls within the hive.

There are five well recognized types of bees found in the World.

(i) *Apis dorsata* (Rock bee)
(ii) *Apis florea* (Little bee)
(iii) *Apis indica* (India bee)
(iv) *Apis adamsoni* (African bee)
(v) *Apis mellifica* (European bee)

Out of these, three are common in India. They are *A. dorsata, A. florea* and *A. indica. A. dorsata* is the largest variety with average size of about 20 mm. They build large combs on tree branches, high buildings etc. These are migratory species as during June and July they swarm to the hills, but in winter they come back. The workers build a fresh comb every time. A single comb may yield 60 pound of honey per year.

A. florea is a minature of the rock bee. They are plain species and build small combs on the tree branches, in bushes etc. They yield hardly a few ounces of honey per comb per year.

A. indica is of commonest occurrance on the plain and forests. They build several parallel combs in protected places like hollow of trees, caves, in rocks etc. The average output of honey is 6-7 pounds per colony per year.

SYSTEMATIC POSITION

Phylum	-	Arthropoda
Class	-	Insecta
Subclass	-	Pterygota
Division	-	Endopterygota
Order	-	Hymenoptera
Genus	-	*Apis*

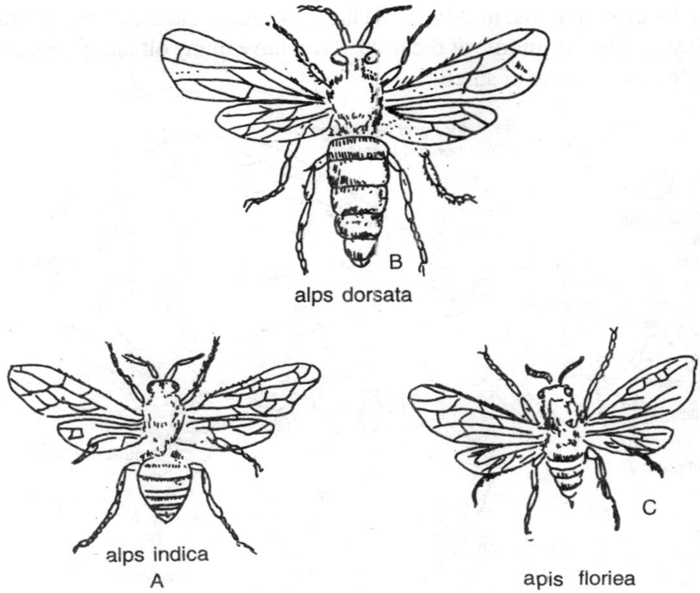

Fig. 10.1. Workers of three species of *Apis*.

Castes of honey bee

The nest of honey bee is known as *bee-hive*. A hive in summer consists of 32 - 50 thousand or more individuals. A colony may be *weak* or *strong* according to the number of worker bees. In a colony there are three types of individuals namely the *queen*, *workers* and *drones*. Due to the existence of several morphological forms, bees are said to be *polymorphic species*.

Structure of worker bee

The worker is the smallest member of the colony. Its body is densely covered with hairs. The colour is usually brown or black bearing several peculiar structures which help in the collection of pollen and nectar from the flowers.

The body is divisible into :

(1) Head, (2) Thorax, and (3) Abdomen.

1. Head

The head is flattened triangular widest cross-wise through the upper corners, which are capped by the large compound eyes. The compound eyes are present on dorso-lateral sides of the head. Each eye has several ommatidia (4900 in queen, 6300 in workers and 13000 in drones). Three ocelli are present in the middle of the face. The antennae are many-jointed and inserted in the middle of the head. They have many olfactory pits that provide a keen sense of smell.

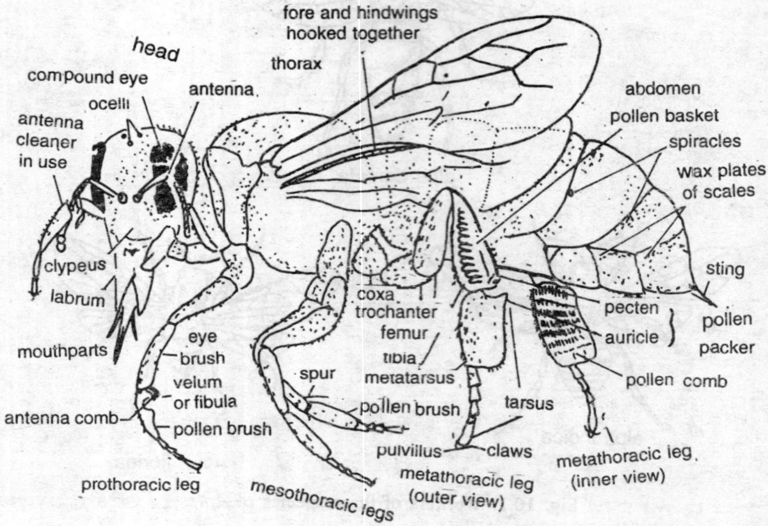

Fig. 10.2. *Apis.* Worker bee in lateral view.

Mouthparts. The mouthparts are attached to the lower part of the head and are of *chewing and lapping type*. These are modified for collecting the nectar and the pollens. Mouthparts consist of *labrum*, *epipharynx*, *mandibles*, *maxillae* and *labium*. Labrum lies below the clypeus. Below the labrum is fleshy epipharynx which is an organ of taste. At the sides of labrum are the two heavy mandibles which work sidewise. In the workers they are spoon-shaped, but they are sharp, pointed and toothed in the queens and drones. The function of mandibles are to mould wax and manipulating the pollens. In the maxilla the lacina is absent while *maxillary palps* are vestigial and the *galea* is elongated and blade-like. In the labium, the *paraglossae* are very much reduced and *glossae* are very much elongated. They are united, hairy and form a honey-spoon or *labellum* at the terminal part. The *labial palps* are well developed and they help to make the ligula up. The whole apparatus is protected by the galeae of maxillae. When the bee sucks up nectar, the terminal lobes of labium and maxillae are pressed close together so as to make a tube between them.

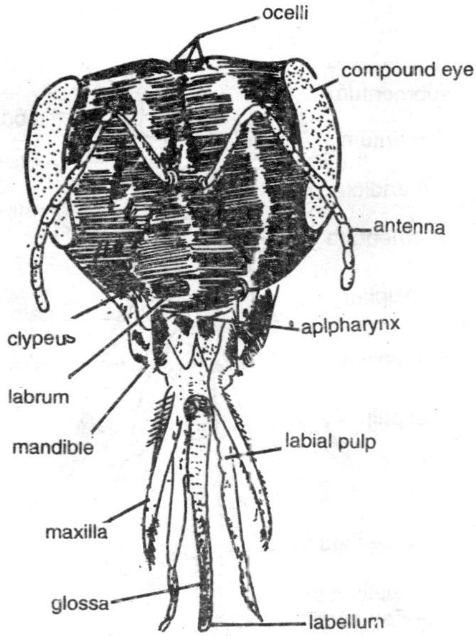

Fig. 10.3. *Apis*. Head.

The food is then taken in by a sucking action of the pharynx. At the tip of the tongue there occurs labellum and there is a groove running along the entire length on the ventral side, within the tongue this groove extends into a double barrel tube. A flexible chitinous rod lies along the back wall of this channel which is itself provided with a still finer groove along its ventral surface.

2. *Thorax*

The thorax is large and strong. It bears two pairs of *wings* and three pairs of *legs*.

The wings are small, narrow, membranous and transparent. They lie flat over the back at rest. They have a much modified and reduced venation. The fore- and hind-wings of each side are interlocked by hooks held in a groove, so as to work together during flight. Wings may vibrate over 400 times per second during flight. A worker is capable of long flight, even 12 km.

There are three pairs of legs, which are densely covered with hairs. The fore-legs are called prothoracic, mid-legs as mesothoracic and hind-legs are metathoracic. Each leg consists of five parts, *coxa*, *trochanter*,

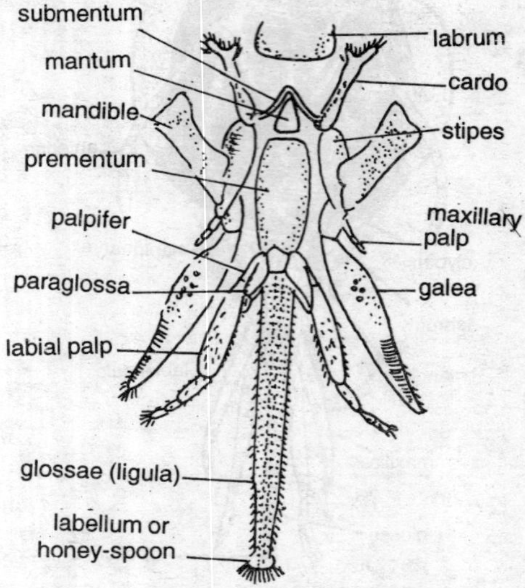

Fig. 10.4. *Apis*.Mouthparts.

femur, *tibia* and *tarsus*.The tarsus is five jointed and terminating into the *claws* and *pulvillus*.

(i) *Prothoracic legs*. Tibia of the fore-legs possesses distally at its anterior face a number of stiff bristles which forms pollen brush. From its posterior face there is also found a movable plate-like process or *velum*, which fits over a circular notch in the upper part of the first tarsal segment. The velum and *antenna-comb* together serve as antenna cleaners. On the anterior surface of the first tarsal segment there are some bristles which are termed as *eye-brushes* and serve to remove particles and pollen from the outer surface of the compound eyes.

(ii) *Mesothoracic legs*. The mid-leg has all the segments as seen in the fore-leg. The tibia bears a brush on its inner surface and a spine-like *pollen-spur* on its distal end. The spurs of both the mid-legs are used to remove pollen from the pollen baskets of hind-legs and to dislodge wax from wax pockets on the ventral surface of the abdomen.

(iii) *Metathoracic legs*. Each metathoracic leg has a large tibia with a cavity with bristles forming a *pollen basket* or *corbicula* used for storing pollens during collection, at the distal end the tibia has a row of stiff bristles called *pectines* below which is a flat plate, the *auricle*. The pectines and auricle form a *pollen packer* to convey packed pollens into the pollen-basket.

3. *Abdomen*

The abdomen is oval and less hairy than the rest of the body. It bears six distinct segments. It possesses wax glands and sting.

(i) *Wax glands*. The glandular area secreting wax lies on the ventral surface of the last four visible segments of the abdomen. The wax is secreted through minute pores in the form of flat scales. The wax is masticated by the mandibles before it is used for building the cells of the honeycomb.

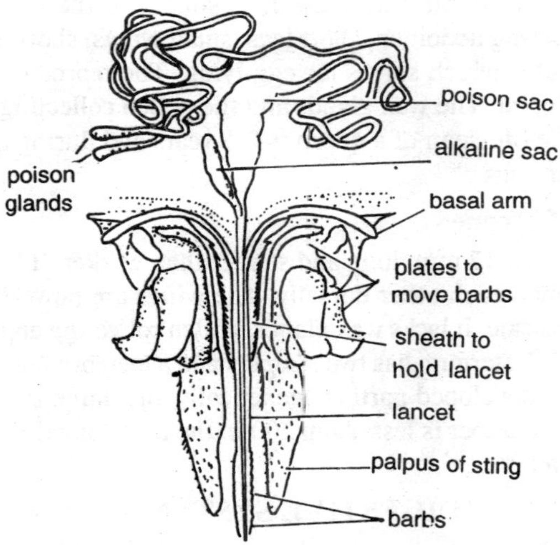

Fig. 10.5. *Apis*. A Sting.

(ii) *Sting or terebra*. In the worker bee the ovipositor changes so as to take the function of sting, i.e. for injecting poison. The actual sting or terebra consists of a central *poison-canal* enclosed by a *stylet sheath* and two *ventral stylets* or *lancets*. Stylets are grooved on dorsal side and form two ventral guide sails. The distal free termination of stylet and sheath are covered by spines or barbs. At the base, there is an expansion, which is called a *bulb*, while sting is continued into a pair of lateral basal arms. The sting is attached to a set of three plates :

(a) The inner most pair is the *oblong plates* posterior in position and representing the divided 9th sternum. To each is attached the basal arm of the stylet sheath of its side. (b) Two *triangular* or *fulcral plates* represent the reduced sterum of 8th segment and to each is attached the corresponding arm of the stylet. (c) The large *quadrate plates* lie dorsally to the triangular plates and at the posterior angle. These three pairs of plates act as the sternum of 9th segment and also function as levers. Two glands are also associated with the sting. A filiform poison gland pours its acidic secretion into a large poison-sac, which discharges its contents into the bulb. There is also an alkaline gland, which opens separately into the sting bulb.

Structure of Queen

The queen is 15-20 mm. long. It is similar to the worker except elongated, tapering abdomen, short legs, small wings, short mouthparts and an ovipositor which serves for egg-lying. The reproductive organs are well developed. The wax glands and the pollen collecting apparatus are absent. The life-span of a queen is 2-5 years and during this span it lays 1.5 million eggs.

Structure of Drone

A drone is 15-17 mm. long and stouter than worker. It has *holoptic eyes* which touch each other dorsally. The wings are powerful and the abdomen is truncate. It lacks wax glands, pollen collecting apparatus and the sting. The 9th sternum has two *claspers* and a membranous *aedeagus*. The drones are developed parthenogenetically, in a large special cell of the hive. Their number is less about 200 in the hive sometimes only one drone lives in a hive.

DIGESTIVE SYSTEM

The digestive system includes *alimentary canal* and *digestive glands*.

The alimentary canal comprises a *mouth* which opens into the *pharynx*. Pharynx is a muscular widened tube that opens into long slender *oesophagus*. The oesophagus running through the thorax and opens into an enlarged, thin walled sac, the *crop* or *honey sac*. The honey sac lies in the abdomen and it narrows posteriorly and opens into the *stomach*. The honey sac is of special interest in workers. The nectar collected from the flowers is held in it. From the sac the nectar is regurgitated into the cells of the comb or given up to another bee. The upper end of the stomach sticks up into the lower end of the honey sac as a small cone with an 'x' shaped valve in the summit. This opening is called *stomach-mouth* and is guarded by four lips. It opens whenever stomach requires food from the honey sac. Stomach leads into *small intestine* having a circlet of 100- 125 slender Malpighian tubules. Intestine opens into terminal part of alimentary canal, the *rectum* which opens out through *anus*.

The salivary glands are located on the back of the head in the front part of the thorax and open upon the upper part of the labium.

NERVOUS SYSTEM

It is made up of *brain*, *circum-oesophageal connectives* with double ganglionated *ventral nerve cord*. There are three thoracic ganglia. Last two are fused. 5-abdominal gangila are also fused in a mass.

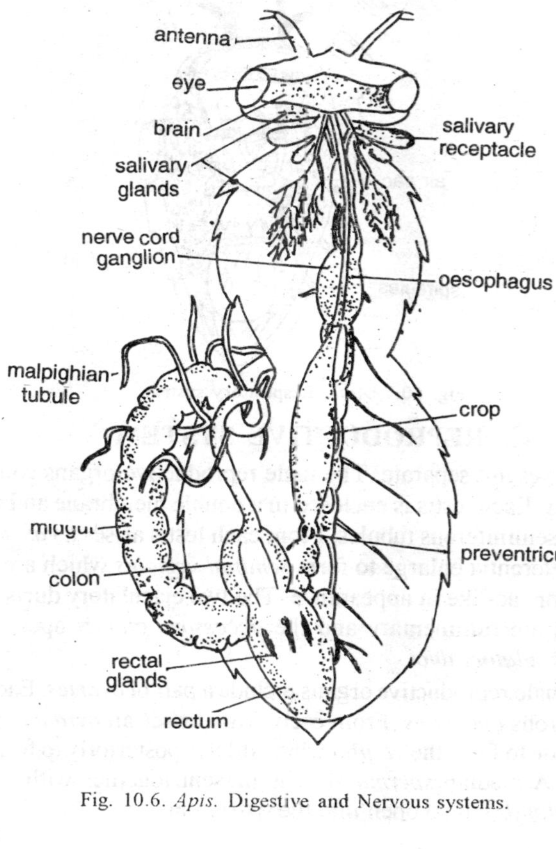

Fig. 10.6. *Apis.* Digestive and Nervous systems.

CIRCULATORY SYSTEM

It is similar to that present in cockroach. The heart consists of four chambers instead of 13 chambered heart of cockroach.

RESPIRATORY SYSTEM

The respiration is brought by the tracheal system as found in other insects. In all there are 10 pairs of spiracles present in thorax and abdomen. Two lateral tracheal trunks are enlarged in abdomen to form *air sacs*.

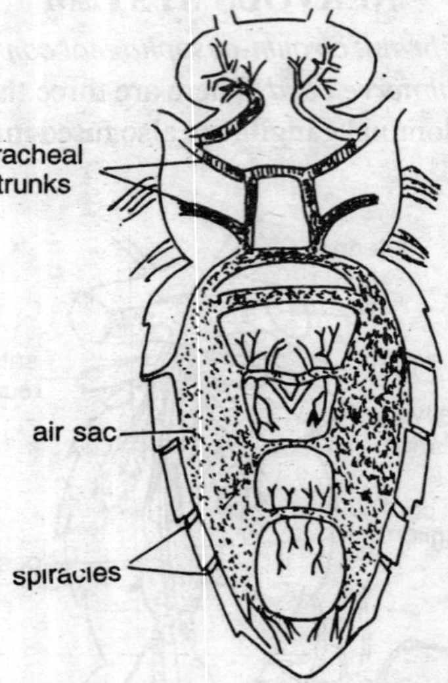

Fig. 10.7. *Apis*. Respiratory system.

REPRODUCTIVE SYSTEM

The sexes are separate. The male reproductive organs comprise a pair of *testes*. Each testis is enclosed in a double membrane and consists of 250-300 seminiferous tubules. From each testis arises a *vas deferens*. The vasa deferentia enlarge to form *seminal vesicles* which are usually cylindrical or sac-like in appearance. The two ejaculatory ducts leaving the vesicles are rudimentary and the *accessory glands* open into the common *ejaculatory duct*.

The female reproductive organs include a pair of *ovaries*. Each ovary bears numerous *ovarioles*. From each ovary arises an *oviduct*. The two oviducts unite to form the *vagina* which dilates posteriorly to form *bursa copulatrix*. A median *spermatheca* is present together with a pair of *collateral glands* which open into the spermatheca.

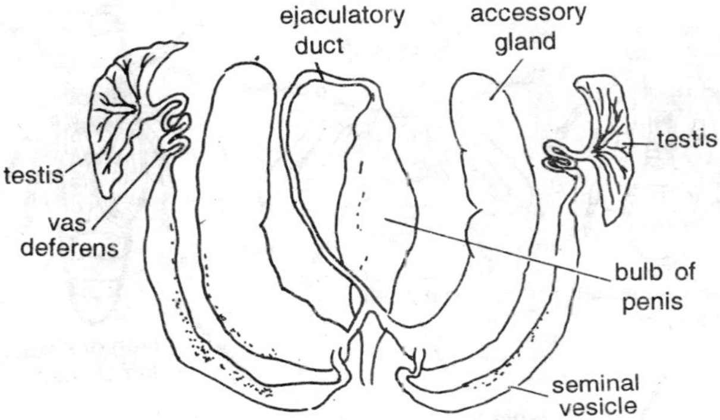

Fig. 10.8. *Apis*. Male reproductive organs.

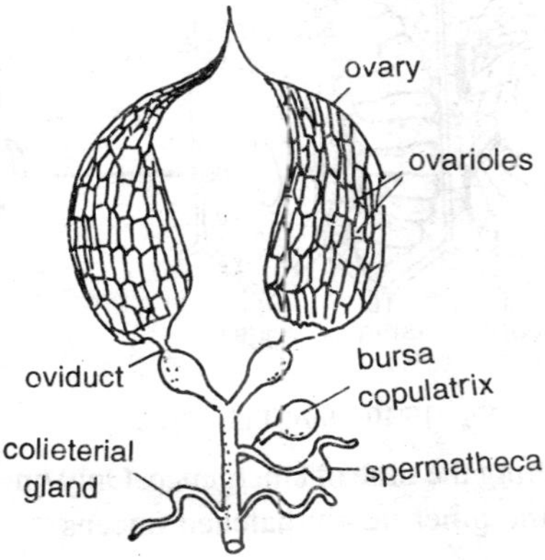

Fig. 10.9. *Apis*. Female reproductive organs.

LIFE HISTORY

The life-history of honey bee can be studied under following heads:

Swarming

Over crowding in the hive leads to swarming. A large number of bees come out of the hive during the spring or early summer.

On the fine forenoon, the old queen leaves the hive to establish a new one, accompanied by a large number of old workers and drones. Left behind in the hive are the young workers and several new queens still in

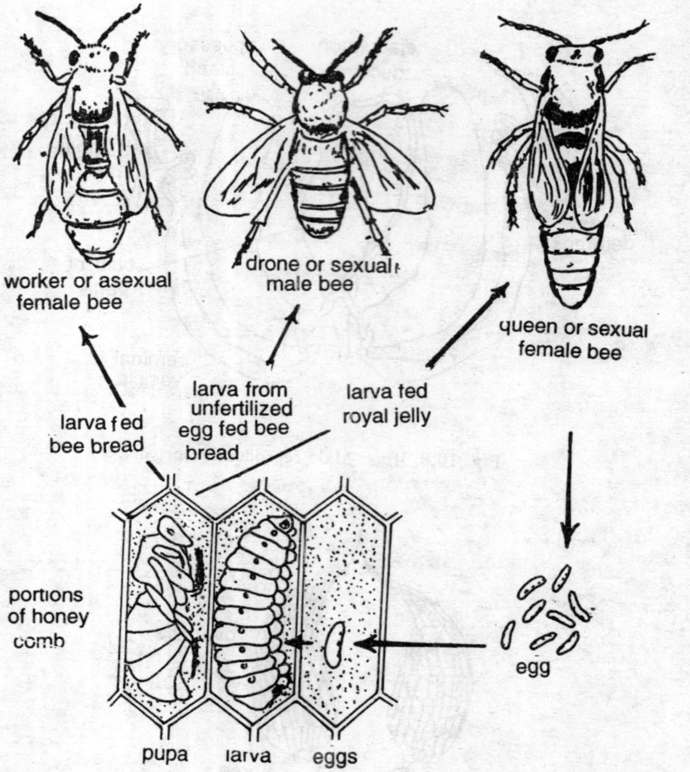

Fig. 10.10. *Apis.* Life Cycle.

their cells but approaching the time of emergence. Only one queen survives by stinging to death the other newly-hatched queens.

Supersedure

In supersedure the old queen with reduced egg laying capacity is replaced by a new, young and vigorous queen. The old and new queens may continue to live together or the old one may be dragged. Under unfavourable conditions or due to certain enemies the entire colony of bees migrates to a suitable place. The migration of the entire colony is known as *absconding*.

Nuptial Flight and Mating

About a week after emergence, the new queen takes her first aerial flight followed by a swarm of drones. The queen flies very high and the drones gradually drop out of the race.The last drone left in the race, mates with her. Mating occurs in mid-air, during which the queen receives spermatophores from the drone. The sperms are stored in spermatheca or sperm-reservoir of the queen to fertilize her eggs as long as she lives. The prime swarm is led by the old queen while the second swarm is accompanied by the newly-emerged virgin queen.

The queen after mating returns to the hive and does not leave it till become old. It was thought that a queen mated only once in its life. But it has now been known that out of experimented 110 queens, about 55 mated twice. This experiment was carried out very successfully in U.S.A.

After the swarming period the remaining drones are pushed out of the hive by the workers, if they survive.

Eggs and Egg-laying

Egg laying starts 3 days after copulation. The eggs are of two types-fertilized eggs developing into queens and workers and haploid eggs develop into drones, and are laid into separate chambers or cells. The fertilized eggs are laid into workers or queen cells, while the unfertilized into drone cells. In a season of abundant nectar, the queen lays upto 1000 eggs per day. Tle eggs are elongated and pinkish and are attached to the bottom of cells.

Larvae

After three days a tiny larva or *grub* is formed from each egg. There are no eyes and legs in the larva. For the first day all larvae are fed on *royal jelly* which is secreted by pharyngeal glands of young workers. Afterwards, larvae of the drones and workers are fed on honey and pollen, but the larvae of queen are continuously fed on royal jelly. In this way the food-supply causes them to develop differently. After several moults the larva grows, then its cell is sealed with a wax-cap.

Pupa

The larva changes into a pupa in the sealed chamber.

The wax-lid is porous and allows exchange of air for respiration. The pupa secretes a thin silken cocoon around itself and undergoes metamorphosis. It involves a change of legless, wingless and eyeless worm-like creature into a winged insect with legs and eyes. After the pupal period is over the lid, closing the cell, is cut off by the young bee with its jaws to become free.

The time of development for each caste is different which can be shown as:

	Caste	*Egg*	*Larve*	*Pupa*	*Total*
1.	Queen	3 days	$5^1/_3$ days	7 days	$15\,^1/_2$ days
2.	Worker	3 "	5 "	13 "	21 "
3.	Drone	3 "	6 "	15 "	24 "

The queen produces from her mandibular glands a special secretion called *queen substance*. It is picked up directly by a few workers that are in contact with the queen. These workers spread it to the other. The "queen substance" inhibits the workers from building brood cells for new queens.

The freshly emerged workers are first entrusted with the indoor duties for two to three weeks during which they act as nursing bees, dance attendance on the royalties, look after brood cells, build and repair the comb. Later on, they are put to outdoor duties and they are completely occupied in collection of nectar and pollen, guarding the hive, air conditioning, temperature regulation and ripening the honey, etc.

At the approach of winter, the drones are driven out of hive by the workers. They are either stung by workers or die of cold or hunger.

The queen and workers live there to pass over the unfavourable conditions, feeding over the stored honey and pollens. On the return of spring the activities are regained.

COMMUNICATION

The communication in insects is achieved by many ways such as visual stimuli, sound, chemical stimuli, body secretion etc. The honey bees have surprised by their peculiar movement, better known as *'bee dance'*. This peculiar sense of understanding and communication among bees has led *Carl Van Frisch* to be honoured with Nobel Prize in 1973.

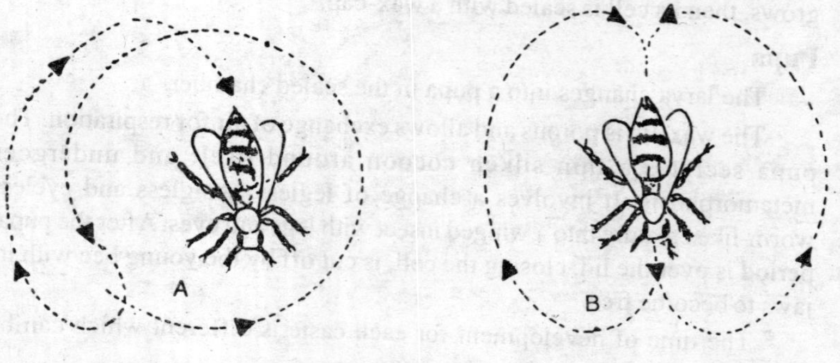

Fig. 10.11 *Apis*. Two different types of dances. A-Round dance, B- Tail wagging dance.

All the bees of a hive do not go out for the search of food. There are certain workers generally called *foragers* or *scout bees* who perform this duty and on their communication other bees move to the feeding place. These foragers collect the nectar and return back to the hive and communicate the source by performing the dance. The dance is performed in two manners, *round dance* and *tail wagging dance*. If the food source is less than 100 metres from the hive, it performs a circular dance. If the source is more than 100 metres, the dance performed is making a figure of 8. For example, if the source of food is 300 meters away, the worker makes 28 runs per minute, but when 3000 meters away, only 9 runs are made.

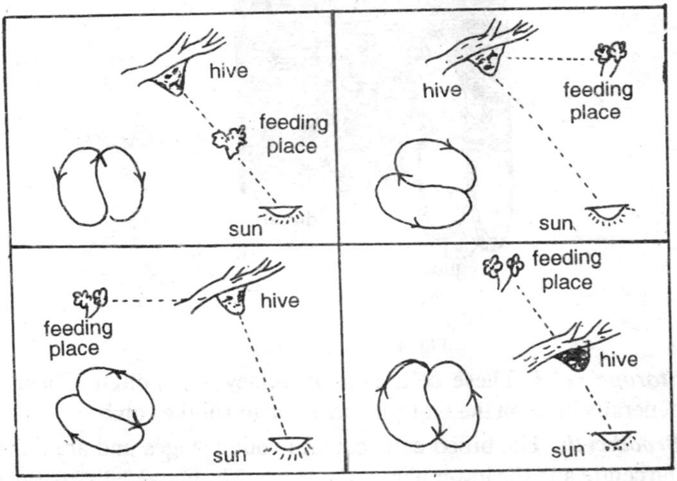

Fig. 10.12. *Apis*. Directional changes in bee dance to denote the feeding place with relation to the position of sun.

The round dance conveys the information about the source of food only where as the tail wagging dance indicates the source, distance, direction and the quantity of food. Between each semicircle, the bee moves along the vertical plane wagging its abdomen side to side. The frequency of wagging decreases with the increase of dance. The movement of sun has a definit relation with the direction of dance.

When forager performs its dance the other workers join the dance for a while and than leave the hive in the direction of food. *A. Wenner* (1964) claimed that foragers communicated to the other workers of their hive not only by their dance but also by producing peculiar sound which results due to the vibration of their wings. The intensity of sound is proportional to the distance of food from hive and also to the quantity of the food.

Structure of a Hive

The highest degree of nest construction among insects is found in bees. The hive is generally built hanging from a building, rock or branch of tree. It consists of a thick vertical sheet bearing two layers of hexagonal chambers known as *cells*. The cells are of following types:

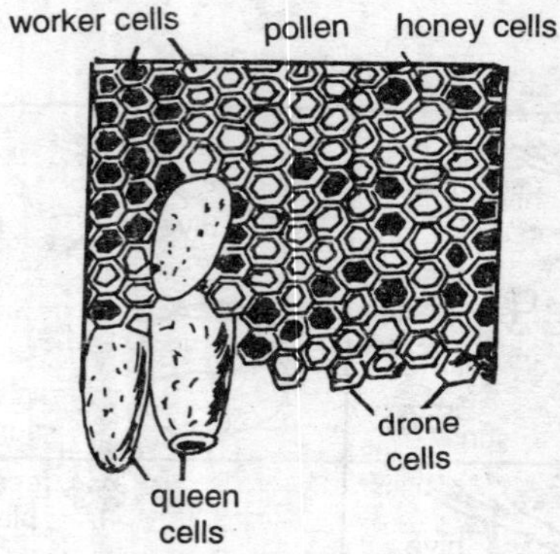

Fig. 10.13. Bee hive.

(i) ***Storage cells.*** These cells contain honey and pollen. These are generally built on the margin and at the top of the comb.

(ii) ***Brood cells.*** The brood cells contain young stages and are built in the centre and the lower part of the comb. The brood cells are further divided into three types:

(a) *Worker cells.* These are of the size of the storage cells and are used for rearing workers. They lie horizontally in the hive.

(b) *Drone cells.* These are about 6 mm. wide and are used for bringing up the drones. They are also horizontal in position.

(c) *Queen cells.* They are still larger and are used to rear the queens. They may be cylindrical or vase-like in form, and are vertical in position.

There are no special cells for adults, as they keep generally moving about on the surface cells of the comb or clustering cells which are situated on the upper part of the comb.

The wax used in budding of a comb is secreted from the *wax glands* present in the abdomen of workers. This wax has a melting point of 140°F.

Before use the wax is masticated and mixed with secretions of *cephalic glands* to convert into a plastic substance. The hive is polished with substance, *propolis*, prepared from pollens, is used for making the comb water proof.

Methods of bee keeping

The ultimate aim of bee keeping is to get more and more honey in pure form. The total method commonly used by old apiculaturists is very crude, cruel and of unplanned type. This old method is called as Indigenous method.

Indigenous method

1. ***Hives.*** Two types of hives are used in indigenous method of bee keeping e.g. wall or fixed hive and movable hive.

(a) *Wall or fixed hive*. It is purely natural type of comb because the bees themselves prepare the hive at any space on the wall or trees. There is an opening on one side through which bees come out the hive.

(b) *Movable hive*. It comprieses of hollow wood logs, empty boxes and earthen pots etc. placed in varandas of houses. There exist two holes, one is for entrance and the other for exit of the bees. The swarmed bees usually come to the box on their own accord. Some bee keepers use to take the clusters of the swarms from a tree and keep them in the hive.

2. ***Extraction of honey.*** For the extraction of honey, burning fire is brought near the bee hive at the night as a result of which bees are either killed or they escape off. Further the hive full of honey is being removed, cut into pieces and squeezed to get honey. Sometimes smoking is done so that the bees may escape from their hives.

3. ***Drawbacks of indigenous method.*** The indigenous method of bees keeping suffers from a number of drawbacks due to which it is not recommmended by present day panel. These drawbacks are:

(i) Honey becomes impure because at the time of squeezing, the brood cells, pollen cells, honey cells and larvae are also extracted.

(ii) The colony becomes weak due to killing of the eggs and the larvae at the time of squeezing.

(iii) Formation of new hive by the escaped bees requires extra energy which effects the yield.

(iv) The activities of the bees can be controlled.

(v) The hivation of bees on the same place is only matter of chance.

(vi) The honey robbers, like, rats, ants, wasps and monkeys may effect the hive easily.

(vii) The race improvement programme may not be applied, so there is no possibility for the selection of the best bees.

(viii) The hazards created by climatic factors can not be controlled.

Modern method of apiculture

To overcome the drawbacks of indigenous method an advanced method based on scientific facts has been developed. It has opened a new era for the cottage industry in India and has also given an opportunityfor lakhs of unemployed persons to keep them busy in this business. First of all care was taken to improve the texture of the hives and during this race hive patterns were introduced in India. The Newton model with 7 to 10 frames (21 x 14.5 cm) in the brood chamber with a shallow super (21 x 6.5 cm sized frames) has been most popular in South, East and Central India. Longstroth hive containing 10 frames (44.8 x 23 cm) has been used as a standard in Himachal Pradesh, Jammu and Kashmir, and Punjuab. In Uttar Pradesh another type of hive has been in use which was evolved at Jeolikote apiary and contained 8 frames (30 x 18cm). After gaining experience from the above metioned hives, Indian Standard Institute has standardized the hive of small and big sizes accommodating frames 21x 14.5 and 31x 20.4 cm respectively.

Now-a-days a typical type of movable hive is constructed which is capable of expansion or contraction according to the requirement of the place, season and climatic conditions.

Appliances for modern method

1. Typical movable hive

An artificial movable hive is constructed by wooden box base on bees space theory. The size and number of frames are variable from hive to hive according to the need. A small space is enough to permit the entrance and exit of workers and drones but queen once placed in hive never comes outside the hive. The perforation size on zinc sheet is only of 0.37 cm. but the thorax of the queen is 0.43 cm. to 0.45 cm. so the queen can never pass through this pore. This typical hive consists of 6 parts as given below:

(a) *Stand.* It is the basal part of the hive on which the whole hive is constructed. The stand is adjusted to make slope for the hive. Due to this slope rain comes down quickly.

(b) *Bottom board.* It is situated above the stand and forms the proper base for the hive having two gates in the front position. One gate functions as an enterance while the other as exit.

(c) *Brood chamber.* The bottom board carries the brood chamber which is the most important part of the bee hive. It is large in size provided with 5 to 10 frames. In each frame a wax sheet bearing hexagonal frames is held up by a couple of wires in a vertical position. Along with the margin of every hexagonal mark, the bees start making wall and ultimately the cells. Here every sheet of the wax is known as *comb foundation* which attracts the bees and provides the base for the comb preparation on both the sides.The frames are kept vertically in brood chamber which is covered over by other frames having a wire meshing through which the workers can easily pass.

(d) *Super.* It is also without cover and the base. Super is provided with many frames containing comb foundation to provide additional space for expansion of the hive.

(e) *Inner cover.* It is a wooden piece used for the covering of the super. It has many holes for proper ventilation.

(f) *Top cover.* It is ment for protecting the colony from rains. It is fitted with zinc sheet which is plain and sloping.

2. *Queen excluder*

It consists of a wire-gauze, extrans guards and drone traps with individual wires placed 0.37 cm apart. It readily permits the worker to pass through it but keeps back the queen in the brood chamber.

3. *Honey extractor*

It is used for the extraction of the honey from the comb and functions on principle of centrifugal force. When combs are centrifuged by this device the pure honey is thrown out without any damage to the comb.

4. *Uncaping knife*

When all of the combs are filled with honey they are sealed by capping with the wax. So, before such capped combs are placed in the honey extractor, the wax sealing has to be removed with the help of uncapping knife heated by steam before use.

5. *Other equipments*

Most of the useful equipments for the successful management of the bees are locally manufactured which are very cheap. As they are made locally, they may not be exactly similar to those made at other places. Thus, Indian Standard Institute has standardized some very common equipments for the production of unifrom interchangeable articles. Some materials like protective garments, gum cages, gloves, net veil, bee net, brush etc. are required for easy and well planned handling of the bees.

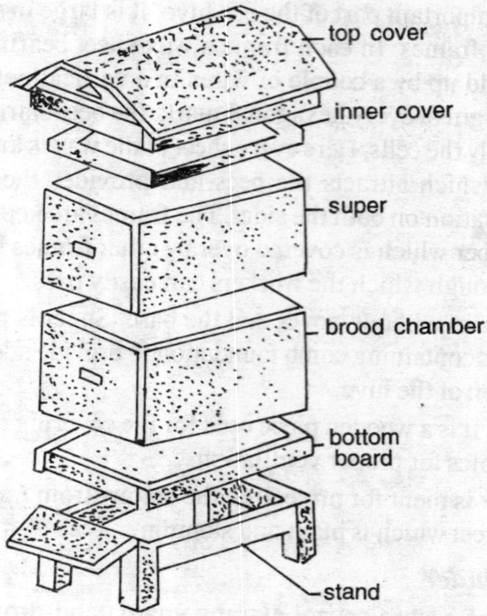

Fig. 10.14. Typical movable hive

Economic Importance

Honey bees are most valuable domesticated animal of man and have been associated with human race from time of Vedas and Ramayana. They provide us honey and wax and moreover they help in pollination.

Honey

The honey is used by man as the source of natural sweets for preparing cakes, breads etc. It has great food value as one pound of honey is equal to 4.5 pounds of grapes or 5 pounds of apples or 3.5 pounds of potatoes. It is a very powerful tonic containing vitamin B complex (Biotin, Choline, Thiamine, Riboflavin, Pyridoxin and Pantothenic acid) and vitamin C (ascorbic acid). It also contains calcium, iron, phosphate and manganese. Beside honey is an antiseptic and contains formic acid as preservative.

The chemical composition of one pound of honey is: 6 1/2 oz. Levulose (fruit-sugar), 5 1/2 oz. Dextrose (Glucose), 9 gm. Sucrose, 7 gm. Detrines and gums, 1 gm. of Fe^{++}, Ca^{++}, Na^{++} etc. and about 4% of undetermined sustances. The colour of the honey and chemical composition and flavour usually depend on the flowers visited by worker-bees.

Bee Wax

The wax is removed from the bee hives and is used in making polishes, candles, carbon papers, electrical and other products.

Pollination

Perhaps the greatest service of honey bee is the pollination of flowers of fruit plants and seed crops.

Bee Venom

It is secreted by poison-glands of stings. It is transparent and, contains formic acid, histamine, tryptophane, sulphur, many proteins, volatile oils, enzymes etc. Clinically it is used in the treatment of rheumatism, keratoconjunctivitis, iritis, iridocylitis and in the control of blood pressure.

Revision Questions

1. Give an account of external features of worker honey bee.
2. Honey bee is a social insect. Discuss.
3. Describe the life-history and economic importance of honey bee.

11

Cimex (Bedbug)

The bedbugs are domesticated insects and are intimately associated with man throughout the ages. They are sanguivorous as they suck the blood of man. Bedbug includes 30-40 species of wingless brownish blood sucking insects, mostly parasitic on birds and bats. Three species attack man these are *Cimex lectularius*, *C. hemipterus* and *C. rotundatus*. Out of these first and second are more common. The third one is found in Africa, Southern Asia and West Indies.

SYSTEMATIC POSITION

Phylum	-	Arthropoda
Class	-	Insecta
Subclass	-	Pterygota
Division	-	Exopterygota
Order	-	Hemiptera
Genus	-	*Cimex*

Habits and Habitat

They are cosmopolitan in distribution. They inhabit dark, damp human dwellings like houses, hotels, rest-houses, barracks, etc. They live in cracks in the walls and floor, in crevices in the beds and furniture, under mattresses, carpets and wall papers, and in similar places.

The bedbug is a temporary ectoparasite of man feeding on the blood. They are nocturnal, but may come out during the day. A well fed bug can do without food for several months or even a year. The starving bug may resorts to *cannibalism*. They are oviparous and undergo gradual metamorphosis. This bad-smelling pest lives for 5-6 months.

MORPHOLOGY

Shape, Size and Colour

The body is oval, dorseventrally flattened, measuring about 5 mm. in length and 3 mm. in width. The colour of the body is brown becoming deep purple or red after it has fed.

Division of body

Like other insects, the body of bedbugs is divisible into three parts :

(1) Head

(2) Thorax

(3) Abdomen.

1. Head

The head is short, broad and fits into a notch of the prothorax. It bears a pair of compound eyes, a pair of 4 jointed antennae, and mouthparts. The simple eyes or ocelli are absent.

The mouthparts are of the piercing and sucking type and form a hollow beak or proboscis in front of the head. The beak is formed by the *labium*, which bears a mid-dorsal groove for enclosing the four stylets of two *mandibles* and two *maxillae*. The saw-like mandibular stylets are used for making puncture in the skin whereas the maxillae for the inward flow of blood and downward flow of saliva. This is facilitated by two channels of the maxillae. The maxillary and labial palps are absent.

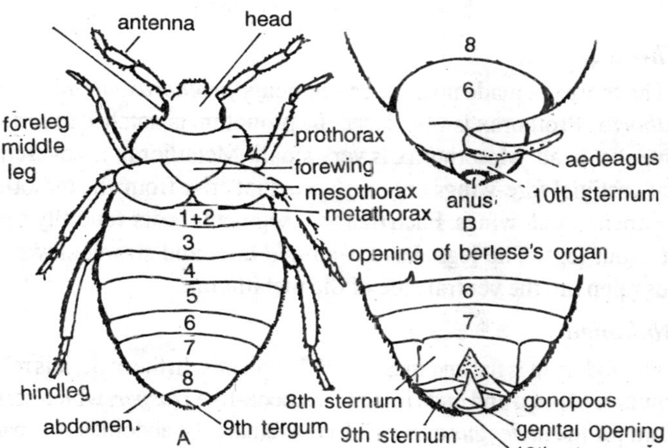

Fig. 11.1. *Cimex*. A—Adult male. B—Genitalia of male. C—Posterior end of female.

At rest, the rostrum bends beneath the thorax but while feeding, it is pointed downwards or forwards. The mandibular stylets make a puncture in the skin. The jointed rostrum bends and the maxillary stylets enter the wound. The saliva runs down the puncture and prevents the clotting of blood which is subsequently sucked up.

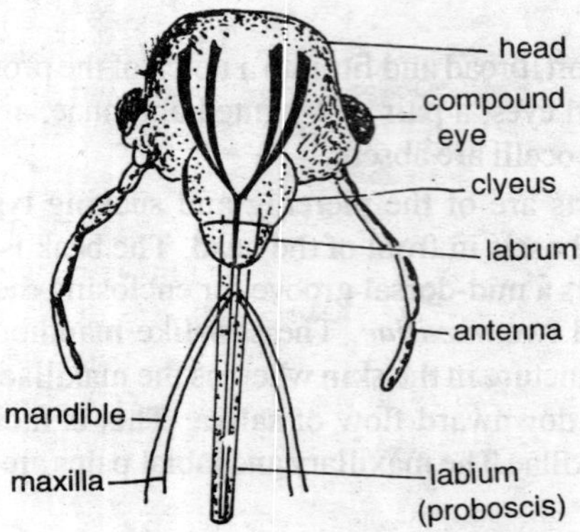

Fig. 11.2. *Cimex.* Mouthparts.

2. Thorax

The thorax is made up of three segments *prothorax*, *mesothorax* and *metathorax*. Prothorax is very large. Its pronotum is notched anteriorly to receive the head. Mesothorax is very small. Metathorax is covered by a pair of vestigial fore-wings or *hemelytrae* that arise from the mesothorax. There are no hind- wings. Each thoracic segments bears ventrally a pair of short, stout legs. The legs have 3-jointed tarsi and strong claws. Stink glands open on the ventral side of the metathorax.

3. Abdomen

The abdomen is flat and made up of 7 segments. In the male, it is relatevly narrow and pointed and ends in a curved hook-like *clasper*, which serves as a sheath for the *aedeagus* or *penis*. In the female the abdomen is broad and rounded and has a prominent notch in the 4th sternum, a little to the right of the middle line. The notch is the opening of a peculiar copulatory pouch, the *organ of Berlese* or *spermatophagous organ*, the cells of which are thought to ingest and destroy the excess sperms.

LIFE HISTORY

Oviposition

After copulation the female lays eggs in holes, cracks, cervices etc. A female deposits 200 to 500 eggs in a batch; the egg is one mm. long, yellowish-white in colour. It is barrel-shaped with a raised lid at one end containing a micropyle; eggs are laid throughout the year in warm countries.

Nymph

In about 6 to 10 days duration the young bug or nymph hatches out of the egg by pushing off the lid. The nymph is about the size of a pin head and is semitransparent. In its general appearance, the nymph resembles the adult except that it is smaller and paler.

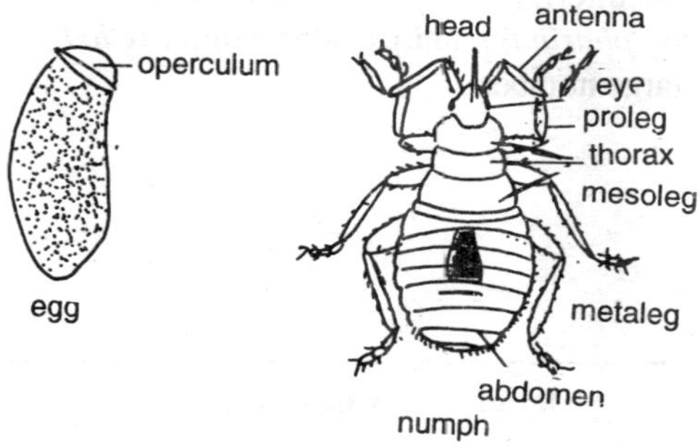

Fig. 11.3. *Cimex*. Life-history.

Few hours after hatching, the nymph becomes capable of piercing the host skin and sucking blood. It serches for man and feeds to its capacity in 3-4 minutes. It consumes blood three to four times of its own weight. If man is not available, it feeds on the blood of older bugs. After feeding it becomes rounded and bright red in colour. Now it settles down and in one to four weeks it moults into the 2nd stage nymph. The duration of Ist stage nymph depends upon temperature and other factors. The nymph undergoes five moults before changing into the adult. On the whole they take about 29 days to complete the life history. Therefore, the development of bedbugs is direct, i.e. without metamorphosis.

Economic Importance

The bedbug is a serious household pest. They feed on mammalian blood causing inflammation and irritation. This is due to the introduction of saliva into the human blood during feeding on it.

It is believed that the bedbug acts as a transmitting agent of kala azar (caused by *Leishmania donovani*), relapsing fever (spirochactosis), typhoid etc. They, however, do not cause the outbreak of these diseases in an epidemic form.

Control

The best method, to control them, is the spray of DDT (5% in kerosene). This can be sprayed on all the walls and furniture where the bugs rest. Fumigation with sulphur, sprinkling boiling water, spraying of emulsion of kerosene, benzene and petrol are useful in reducing infestation bedbugs.

Enemies of Bedbugs

Bedbugs have many natural enemies. They are eaten by large bugs like *Reduvius*, a Mediterranean spider called *Thanatos*, small red house-ant (*Monomorium pharanis*) and pseudoscorpion (*Chelifer*). Man also destroys them a large number.

Revision Questions

1. Describe the structure of *Cimex*.
2. Describe the structure and life-cycle of bed-bug.

12

Termites

Termites, commonly called *white-ants,* are social and polymorphic insects living in colonies. They have a well set caste system.

SYSTEMATIC POSITION

Phylum	-	Arthropoda
Class	-	Insecta
Subclass	-	Pterygota
Division	-	Exopterygota
Order	-	Isoptera
Type	-	Termites

Habits and Habitat

The termites occur throughout the tropical and temperate countries. About 1600 species are known. The most common are *Microtermes* and *Odontotermes*. These are colonial insects. They either burrow in wood or make subterranean nests or *termitaria.* They feed mainly on wood and the cellulose is digested with the help of symbiont flagellate, *Trichonympha campanula* living in the intestine of termites.

EXTERNAL FEATURES

These are soft-bodied small medium sized insects. The colour of the body is white or pale yellow. The body is elongated and flattened with smooth or hairy surface. The body is divisible into three parts :

(1) Head

(2) Thorax

(3) Abdomen.

1. Head

The head is free and well sclerotized. It bears a pair of short or long beaded antennae. The compound eyes may be well developed, vestigial or absent similarly ocelli may or may not be present. The mouthparts are chewing type.

2. Thorax

It is divisible into three segments: prothorax, mesothroax and metathorax. Each segment bears a pair of legs on ventral side. Each legs has 4 or 5 jointed tarsus and a pair of claws. The fertile form bears two pairs of wings on the dorso-lateral sides of mesothorax and metathoax. The wings are membranous and extend behind the body. They can be shed at basal sutures. The wings are absent in sterile forms.

3. Abdomen

The abdomen bears 10 segments, ending in a pair of short 2-6 jointed cerci. In male the 9th segment bears a pair of anal styles.

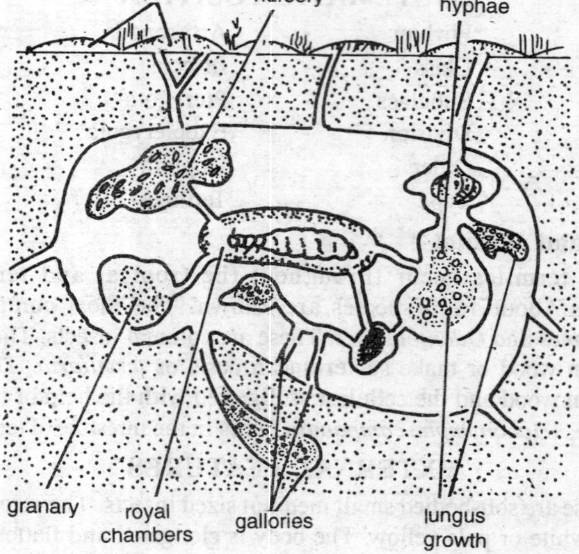

Fig.12.1. Termite nests.

POLYMORPHISM

The termites are polymorphic forms. A termite society has several castes, which are of two marine types :

1. Fertile caste or reproductive castes
2. Sterile castes

1. Fertile Castes

These are also called *reproductive* or *sexual castes.* The fertile individuals comprise the following forms:

(i) *Large winged or Macropterous forms.* These are true kings and queens. Their colour is yellow, brown or black. The wings extend for beyond the hind-end of the body. The compound eyes are widely separated and are large. The ocelli are usually present. They live in royal chambers. The sole duly of king is to fertilize queen and queen is to lay eggs.

(ii) *Short winged or Brachypterous forms.* These are supplimentary reproductive individuals. They have a less chitinized body of pale-yellow colour. The two pairs of wings are vestigial, short and stump-like. The compound eyes are small and the ocelli may or may not be present. These forms replace the true king and queen in case the king and queen die.

(iii) *Wingless or Apterous forms.* These are called *ergatoid kings* and *queens*. Their structure is similar to that of a worker. Their body is colourless and, wings are absent. The compound eyes are vestigial and the ocelli are absent. The sex organs are poorly developed.

2. Sterile Castes

These are wingless forms with rudimentary reproductive organs. These are blind or eyeless and colourless. The sterile castes are of three types :

(i) *Workers.* These are the most numerous of all castes. They are generally dimorphic, i.e., occur in two forms : small and large. In some cases they are trimorphic, being small, intermediate and large. The workers construct and repair the nest; collect and distribute the food; look after the eggs; and feed the nymphs, soldiers and the royal couples (true kings and queens). In some species, the workers cultivate a certain fungus on a finely chewed vegetable matter for feeding the royal couples and nymphs.

(ii) *Soldiers.* The soldiers are the most highly specialized. They defend the colony against enemies. These are pale coloured, pigmented forms with large chitinized big projecting mandibles. The brain is small and compound eyes are vestigial. In certain species there may be three grades of soldiers: large, medium and small sized.

(iii) *Nasutes.* In some higher genera the mandibulate soldiers are replaced by nasutes or *proboscideans* These forms are short with vestigial mandibles. Their head is prolonged into a rostrum, which bears at its tip the opening of *frontal gland.* They throw a jet of sticky secretion of the frontal gland on the invaders through their rostrum. This secretion immobalizes or kills the enemy. The secretion is also used to dissolve hard substances like stones, concrete etc. which are in the

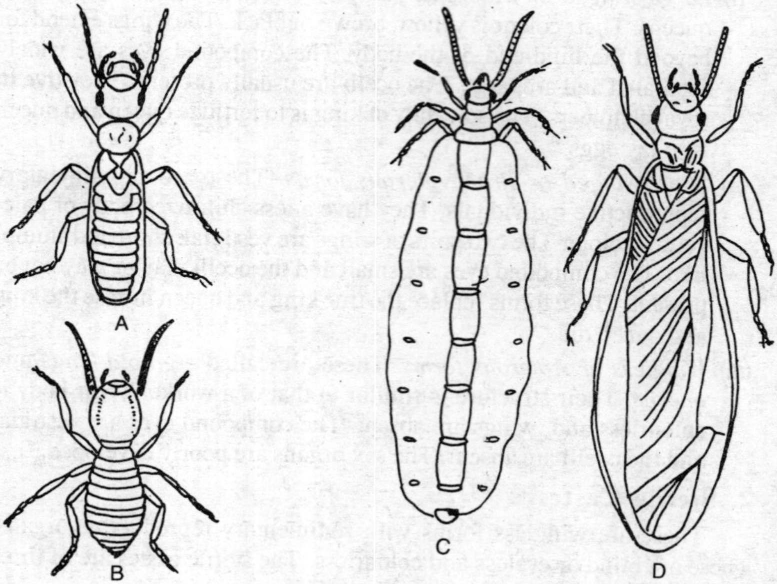

Fig. 12.2. A—Queen of termites without wings after nuptial; B—Wingless sterile worker; C—Mature queen; D—The winged male before nuptial.

way of the workers when they build the colony. They may be dimorphic or trimorphic i.e. small, medium and large sized.

Nests

Primitive termites live in galleries bored in wooden logs, fences, poles, furnitures etc. Some termites make tunnels in the ground, they destroy roots of grasses, vegetation and crops. Other construct a nest or termitarium. The nests consist of a labyrinth of irregular tunnels and chambers and may be entirely sub-terranean or with a mound above the surface. These mounds may be from few metres to nine metres. The natives of Belgium and Congo clear them of the termites and use them as their dwelling huts. There are other types of the termites which form termitaria of masticated wood. These are of the size of a football and are placed upon the trees. These nests are called carion nests.

LIFE HISTORY

Swarming and mating

In the rainy season, with the first shower of rain, a large number of

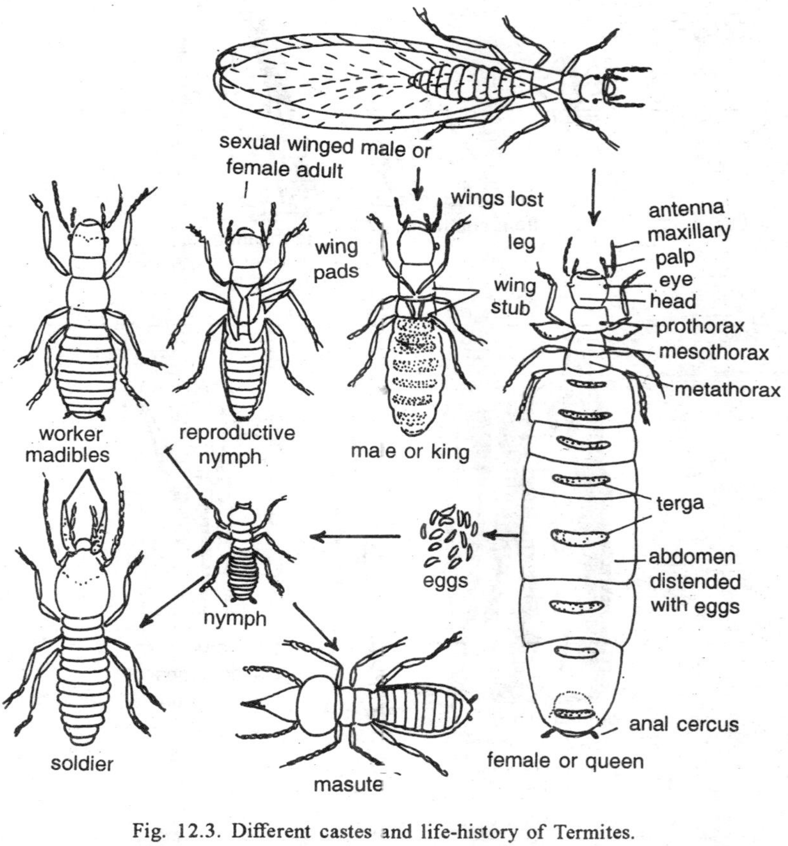

Fig. 12.3. Different castes and life-history of Termites.

macropterous forms (kings and queens) leave the termitaria through the holes made by soldiers and fly away to new sites. It occurs during night or day depending upon the species. Mortality is very high during and after swarming. A large number of these forms are eaten up by lizards, birds, toads, ants etc. Only a few lucky ones survive and found a new colony. At the new site the king and queen mate. Thus the flight is not a true nuptial flight but rather a dispersal flight.The king and queen together excavate a small burrow in the ground called the *nuptial chamber*. The queen lays eggs. These eggs are tended by the royal couple and hatch into workers, which soon take up their duties. The queen now undergoes drastic modifications. Her ovaries and fat bodies increase greatly while the jaw muscles degenerate.

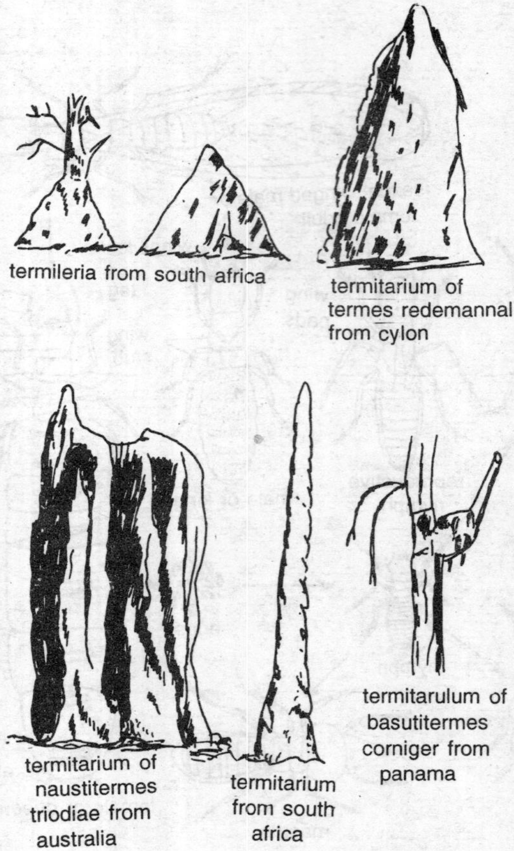

Fig.12.4. Termite nests.

The king lives with the queen for life, and they mate at regular interval. The queen grows her abdomen to an enormous size, 50-75 mm. Her legs being absurdly small, she is unable of the slightest movements and she lies passively in the royal chamber. She is attended by the workers and defended by the soldiers. On an average she lays 4000 eggs per 24 hours.The queen may live for 6-15 years and lays about a million eggs during her life.

Development

The eggs are pale, smooth, oval or elongated. They hatch into wingless nymphs, and resemble adults except in size, wings are absent and sexually immature. After several moults the nymph reaches the adult stage. Formation of different castes from similar eggs in perhaps determined by variable food supply, hormones, pheromones, social behaviours etc.

ECONOMIC IMPORTANCE

The termites are highly destructive insects. They attack and damage all sorts of plant materials, like wood used in buildings and bridges, furnitures, railway-sleepers, fences, crops, books and papers.

However, they are useful also in certain respects. They increase the fertility of the soil by making it porous with their burrows like the earthworms. They dispose off dead wood in the forests to make room for new vegetation. They are eaten by natives in some tropical countries, like South East Africa.

Control

They can be destroyed in the field by pouring kerosine and can be fumigated by Aresnic, Sulphar and Kerosin oil. Trees must be banded by coal-tar or 5% DDT emulsion in oil should be painted on the trunk to 0.45 metre from ground level. This treatment should be done after mansoons. A colony of termites can be eradicated only after findings out and killing the queen.

Revision Questions

1. Describe the external characters of a Termite.
2. Give an account of habitat, habits and economic importance of Termites.

13

Bombyx mori

Bombyx mori commonly called as *Chinese* or *mulberry silkworm moth* is a native of China. It is well known for the pure silk. The moths are reared for silk.

SYSTEMATIC POSITION

Phylum	-	Arthropoda
Class	-	Insecta
Subclass	-	Pterygota
Division	-	Endopterygota
Order	-	Lepidoptera
Genus	-	*Bombyx*
Species	-	*mori*

Types of Silkworms

1. *Mulberry silkworm.* The mulberry worm (*Bombx mori*) belongs to the family Bombysidae. It produces most of the raw silk we use. This silk is supposed to be superior in quality to the other types due to its shining and creamy white colour. The caterpillar feeds on mulberry leaves.
2. *Tasar silkworm.* A number of species of genus Antheraea produces tasar silk. These are *A. mylitta*, *A. paphia*, *A. royelet*, *A.peryni* etc. Their caterpillars feed on leaves of Arjan, Asan, Sal, Oak and various other secondary food plants.
3. *Eri silkworm.* The eri silk is produced by *Artacus ricinii.* Their caterpillars feed an castor leaves. The silk is creamy white but is less shining.
4. *Munga silkworm.* The munga silk is obtained from the caterpillars of *Antheraea assama* which feed on *Machilus* plants.
5. *Oak silkworm. Antheraea pernyi* of China and Japan, *A. roylet* of

Himalayas and *A. yamamai* of Japan have been reared for centuries. They produce good quality of silk.

6. *Giant silkworm. Attacus altas* found in India and Malaysia. It is largest of all living insects, reaching eleven inches in wing span.

The mulberry silkmoth never occures in the wild state, it is completely domesticated. *Bombyx* is extensively cultivated all over the world and usually produces a single brood in a year i.e. *univoltine*. Some strains, however, produce 2-7 broods a year (*polyvoltine*) and are cultivated in warm climate.

EXTERNAL CHARACTERS

The adult moths are robust and cream-white in colour with several lines across the fore-wings. The body is very heavy and hairy. It is about 25 mm. long and has a wing expanse of 40-45 mm. Though both the pairs of wings are apparently well developed, the moths fly very rarely. Females are longer than the males. The proboscis is absent, antennae are bipectinate or feathery in both the sexes. The life-span of adult is short usually ranging from 2-3 days. it does not take food during this period.

LIFE HISTORY

Male and female moths copulate in tail to tail position. Life-history comprises four stages.

1. Eggs
2. Larva
3. Pupa
4. Adult.

The metamorphosis is complete.

1. Eggs

Soon after fertilization, each female lays about 300-500 eggs in clusters upon the leaves of mulberry tree. They are fixed to mulberry leaves with a gelatinous secretion. The eggs are small, smooth and spherical. When laid they are yellowish white but become darker later on. The female dies within 3 or 4 days after egg-laying. In the univoltine race overwritering takes place in the egg stage, as such the eggs may hatch in months.In the polyvoltine race the eggs hatch in a few days.

2. Larva

The larva is called *caterpillar* or *silkworm*. The caterpillar larva, when hatches from the egg, measures 5-7 mm. in length. These caterpillars move on the leaves in a characteristic looping manner. Their body is rough, wrinkled and greyish in colour. They are made up of 12 segments which are distinct into three parts: head, thorax and abdomen. The *head* bears mandibulate mouthparts with which they feed upon the leaves.The thorax is 3 segmented and all the segments bear a pair of true 5 jointed legs ventrally. The *abdomen* is 10 segmented and bears 5 pairs of unjointed, stumpy *prolegs* or *pseudolegs*. These prolegs are present in each segment 3rd, 4th, 5th, 6th and 10th. A short dorsal *anal horn* is present on 8th segment. The prolegs of the first segment turn backwards to form the *claspers*. There is a row of small respiratory pores, the *spiracles*, on either side of the abdomen.

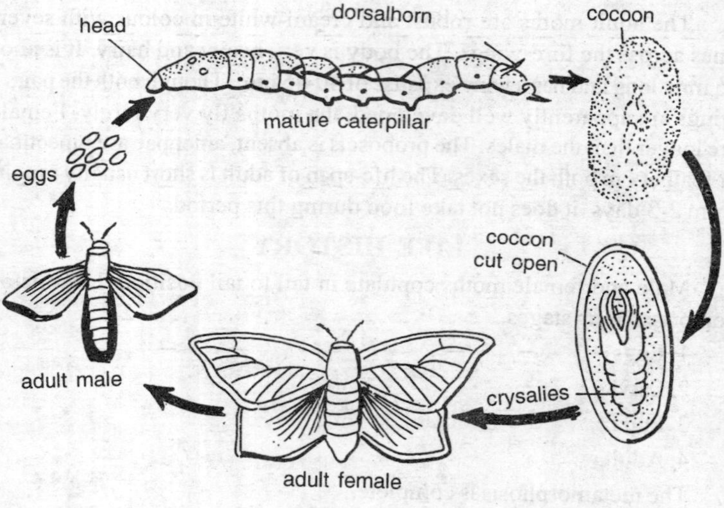

Fig. 13.1. *Bombyx mori*. Life history.

The larva has only three pairs of simple *eyes* or *ocelli*. The antennae, maxillae and labial palps are rudimentary.

The larvae feed voraciously upon the mulberry leaves and grow very quicky. They stop feeding, become inactive after 4-5 days, and than 1st moulting takes place. The 2nd stage larvae resemble the Ist stage larvae except that they are slightly bigger in size. They also eat voraciously for next seven days than 2nd moulting takes place and third stage larvae are formed. The larvae repeat this process for four times. The maturity is achieved in about 45 days since the time of hatching. The matured caterpillar now measures 7-10 cm. in length. By this time the formation of a pair of *salivary glands* is completed. Since these glands secrete silk they are called *silk glands*.

3. Pupa

The mature caterpillar stops feeding and excretes its last excreta and begin to secrete the sticky-secretion from the silk glands through a very fine pore called *spinneret* situated on the hypopharynx. The secretion is continuous and after coming in contact with the air the sticky secretion is converted into a fine, long and solid thread of silk. The thread is made up of fine filaments stuck together by *sericin* secreted by two other glands. The thread becomes wrapped around the body of larva forming a *cocoon*. This process continues for 3-4 days. The cocoon serves as comfortable house for the protection of the caterpillar for further development.

The cocoon is a thick, while or yellow and oval capsule. The outer or initial filaments of the cocoon are irregular, but the inner one that forms the actual bed of the pupa is one long continuous thread about 300-450 metres long wound in ring. The silk thread is secreted at the rate of 150 mm. per minute. Within 15 days the caterpillar is transformed into a brownish *pupa* or *chrysalis*. The pupa lies dormant but undergoes metamorphosis. The larval organs such as prolegs, claspers, anal horn and mouthparts are lost. The adult organs develop. Within 12-15 days the adult is formed.

4. Adult (Imago)

The adult secrets an alkaline fluid that moistens and softens one end of cocoon. The cocoon breaks at this end and the adult comes out. Soon after the emergence, the silk moths mate, lay eggs and die.

ECONOMIC IMPORTANCE

The silkmoth is a very valuable insect as it gives us silk.

Sericulture

Sericulture is the production of silk by rearing silkworms. The silk industry originated in China about 3000 B.C. A Chinese Princesses *Siling Shi* was the first to discover the art of reeling an unbroken filament from a cocoon. This art was kept a close secret for nearly 3000 years. This art later on spread to the rest of the world.

Scriculture is a regular industry. India has silk-producing centres in Assam, Bengal, Madras, Mysore, Punjab and Kashmir. Healthy eggs of high-yielding strains are procured from scriculture research stations. The eggs are placed in paper-lined trays made of split bamboo. The trays are kept out of the reach of ants. The eggs are periodically stirred by a feather. larvae are given chopped leaves of a good quality mulberry trees 5 to 9 times a day. The pupae are not allowed to become adults, as the latter rupture the cocoon to escape, and thereby cut the silk thread of cocoon into pieces.

Spinning of cocoons

This is the period when the caterpillar stops feeding and starts to secrete a pasty substance from the silk gland. In this condition worms should be picked up and transferred to the spinning trays and kept in a position of slope (slanting) to the sun for a short period. Within three days spinning is over and the cocoon is formed and this is the last phase of the rearing of silk worm.

Quality of cocoons

The quality of cocoons is dependent on the raw silk yield, filament length, reliability and splitting.

Marketing of cocoons

The price of cocoon is fixed during every season of the rearing. This price is, however, watched by the Government and cocoons are purchased by the rearers.

Post-cocoon processing

The method of obtaining silk thread from cocoons is known as post-cocoon processing. This includes *stifling* and *reeling*.

1. Stifling

The process of killing the cocoons is termed as stifling. Sericulturists should be very much careful that before the emergence of silk worm (the cocoons which do not have cut holes) good sized cocoons of 8 to 10 days old are selected for further processing. The cocoons are dipped in 0.5% Na_2CO_3 solution for 18 hours than they are dropped into hot water or subjected to steam or dry heat, sun exposure for 3 days or fumigation. In this way pupae or cocoons are killed. The killing of the cocoon in boiling water helps in softening the adhesion of the silk threads among themselves and loosening of the outer threads to separate freely, facilitating the unbinding of silk threads. The cocoons are further treated with 0.5 Na_2CO_3 solution for 15 minutes.

2. Reeling and spinning

The process of removing the threads from the killed cocoon is called as reeling. Four or five free ends of the threads of these cocoons are passed through eyelets and guides to twist into one thread and wound round a large wheel from which it is transferred to spools. Thus the silk obtained on the spool is known as *Raw silk* or *Reeled silk*. The waste outer layer or damaged cocoons and threads are separated, teased and then the filaments are spun. This spunned silk is called as *spun silk*.

The raw silk is further boiled, stretched and purified by acid or by fermentation and then carefully washed over again and again to bring about the well-known lustre on the thread.

The modernization of the reeling and spinning process by autolization and various labour saving processes has opened a new way to this cottage industry in the world. One autolizer can yield 3.715 kg silk per basin in 8 hours in Japan.

Silk

Silk is a pasty secretion of the silk worm produced by the silk glands. The silk glands are actually modified salivary glands which are long and sac-like. As this pasty secretion comes in contact with air, it becomes hard and forms strong and pliable silk strands. This secretion forms two cores of fibroin : (i) a tough elastic insoluble protein consisting of 75% of the fiber's weight and cemented together with sericin from the middle region of the silk glands at the time of secretion, and (ii) a gelatinous protein which is easily soluble in warm water. Some quantity of wax and carotenoid pigments are also detected. The diameter of the silk fibers is 0.0045 to 0.0082 cm. Its elasticity is found to be 20%.

Economic Importance of Silk

The raw silk is used in the manufacture of woven materials and the knitted fabrics for the preparation of garments, parachutes, parachute cords, fishing lines, sieve for flour mills, insulation coil for telephones and wireless receivers, and types of racing cars. Fabrics for garments in various weaves, plain, twill, stain, crepe, georgette and velvet, knitted goods such as vests, gloves, socks, stocking, dyed and printed ornamented fabrics for saries, jacketes, shawls and wrappers are made out of this material.

Revision Questions

1. What is metamorphosis ? Describe it with reference to the life-history of silkmoth.
2. Give an account of structure and life-history of *Bombyx mori*.

14

Mouthparts in Insects

Every organism should possess, in the view of feeding, organs which are either priinaries or secondaries modified structures which enable them to feed and live as long as they are alive are known as mouthparts. The insects shows various specializations in their mouthparts. No doubt the insects are most successful forms among invertebrates and are adapted to the nature or environment so intimately as that their number is increasing constantly. The success of this group in all respective reveals vivid variations in mouthparts that are also being modified according with their modes of lives.

According to the food and manner of feeding the mouthparts are of following types:

1. Biting and chewing type
2. Chewing and lapping type
3. Rasping and sucking type
4. Piercing and sucking type
5. Sponging type
6. Siphoning type

Out of these the rasping and sucking type is considered to be less important because it is radiating to the piercing and sucking type.

1. Biting and chewing type

The biting and chewing type of mouthparts are considered to be most primitive and they are said to be *mandibulate* type of mouthparts. These mouthparts are found in groups such as Orthoptera, Isoptera, some Coleoptera etc. The mechanism of feeding is simple because that are meant for pinch off, chewing up and swallowing the pieces of plant or animal tissues.

The mouthparts consist of :

(i) Paired mandibles

(ii) Paired maxillae
(iii) Labium or lower lip.
(iv) Labrum or upper lip
(v) Hypopharynx
(vi) Epipharynx

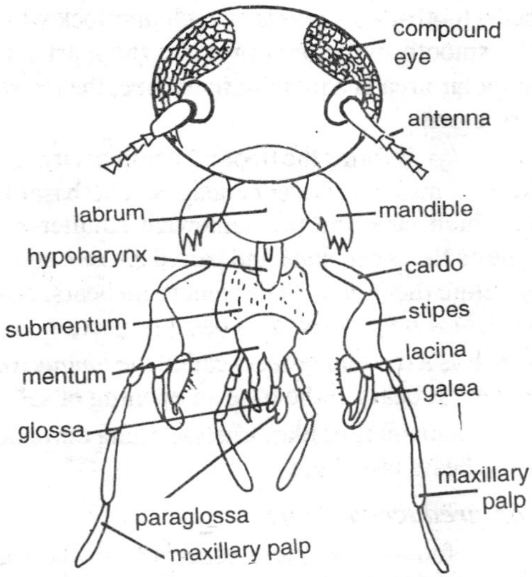

Fig. 14.1. Mouthparts of Cockroach (Biting and chewing type).

These mouthparts are further divided into three sub-divisions :

(a) Grinding type
(b) Grasping or Predaceous type
(c) Mandible-suctorial type.

(a) Grinding type

Especially this type is found among cockroaches and grasshoppers. Because of being considered them as generalized one so we have to describe the various parts of mouthparts first:

(i) *Labrum or upper lip*. It forms the a roof of the mouth cavity, hanging from clypeus and variable in shape and size.

(ii) *Maxillae*. These are paired and attached laterally. Each maxilla consists of a basal part, the *protopodite* which is made up of two segments, *cardo* and *stipes*. The stipe bears two branches distally. the inner *endopodite* and an outer *exopodite*. The endopodite consists of two lobes, inner *lacinia* and outer *galea,* which bears the taste buds. The exopodite is a branch of stipes possessing 5-jointed *maxillary palp* which acts as tactile organ.

(iii) *Mandibles.* A pair of mandibles lie in the mouth cavity and they work transversely to masticate or grind the food. Each one is an unjointed and triangular piece usually, which is very stout. The inner edge of mandible bears tooth-like *denticles* which interlock while capturing the prey. The smooth molar area provides the masticatory surface. Proximal to molar area is a membranous lobe, the *prostheca* which bears sensory hairs.

(iv) *Labium* or *lower lip*. It forms the floor of mouth cavity. It is formed by the fusion of two maxilla-like appendages. The basal large part is *submentum,* which supports the somewhat smaller *mentum*. The mentum supports the *prementum* and paired *glossae* and *paraglossae* together constitute the *ligula*. The prementum bears, on each side, a three-jointed *labial palp* borne on a basal projection, the *palpiger.*

(v) *Hypopharynx*. It is a median process called the *tongue*, occupying in the buccal cavity. It bears at the base an opening of salivary duct.

(vi) *Epipharynx*. It is a membrane like structure lining the under surface of the labrum and bears taste buds.

(b) Grasping or predaceous type

This category is found in Isoptera especially in the soldier castes which are provided with strong mandibles.

(c) Mandible-suctorial type

It is common in ants. Their mandibles are sickle-shaped possessing a groove ventrally and are provided with sucking tube in order to suck up the juice of the prey. The maxillae are slender and fit into the groove of mandible of their side.

2. Chewing and lapping type

This type of mouthparts are seen in the group Hymenoptera such as bees and wasps. Here the mouthparts are modified for collecting the nectar and pollens of the flowers and moulding the wax and serve for both chewing and licking of the food. The mouthparts are the same in all respect as that found in generalized type except in a few modifications. The *epipharynx* projects below the labrum and the *mandibles* are soft or smooth and spatulate. In workers these are utilized in building the honey combs. *Galea* is elongated and blade-like but the *maxillary palps* are vestigial. The *labium* is articulated with basal triangular *postmentum* or *submentum* and *mantum* which is having muscles. *Paraglossae* are greatly reduced where

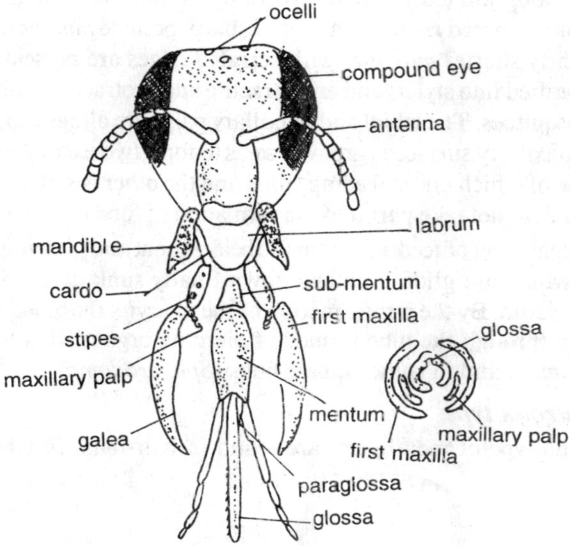

Fig. 14.2. Mouthparts of *Apis* (Chewing and lapping type).

as the glossae are much elongated constituting the retractile ligular tongue which terminated distally as a honeyspoon or *labellum*. The labial palps are well developed when the animal wishes to take food, the terminal lobes or maxillary galeae and labial palps are brought close together to form a groove or hollow tube which can be inserted into the flower where the labellum moves forwards and backwards to collect pollens and to suck up the nectar of flower. There are two views either the liquid food can be drawn by automatic capillary action or it is the action of pharyngeal muscles.

3. Rasping and sucking type

Such type of mouthparts are seen in Thrips. These are adapted for lacerating and sucking. One of the mandible is reduced markedly but there is nothing like piercing. The stylet is used for escaping the surface of the prey.

4. Piercing and sucking type

This type of mouthparts are common in bugs, mosquitoes, aphids, scale insects, flea and some other forms in which the mouthparts are modified with regards to diet. These are modified in four subtypes.

(a) Hemipterous type

Here the four needle-like stylets, which are grooved length-wise and lying in the long jointed lower-lip rostrum, are present. The labium is modified into 3 jointed *rostrum*. A pair of sharp, pointed *mandibles* and a pair of slightly shorter *maxillae* with serrated edges are present. These four are modified into stylets and are retractile and protractile, unlike that those of mosquitoes. The labial and maxillary palps are altogether absent. The inner maxillary surface is grooved so as to form two exceedingly fine channels out of which one is feeding canal and the other is salivary canal. The labrum does not take part into the formation of food channel.

The mechanism of feeding is that the beak is bent and piercing organs are stylets which are gliding up and down finally sunk deeply into the flesh of the victim. By the strong motion of the pharynx the blood or juice is sucked up through the tube formed of piercing organs. It is found in bedbugs (*Cimex*) and in some aquatic bugs (*Sphaerodoma*).

(b) Anoplurous type

Such sub-type of mouthparts are found in hair-louse in which the

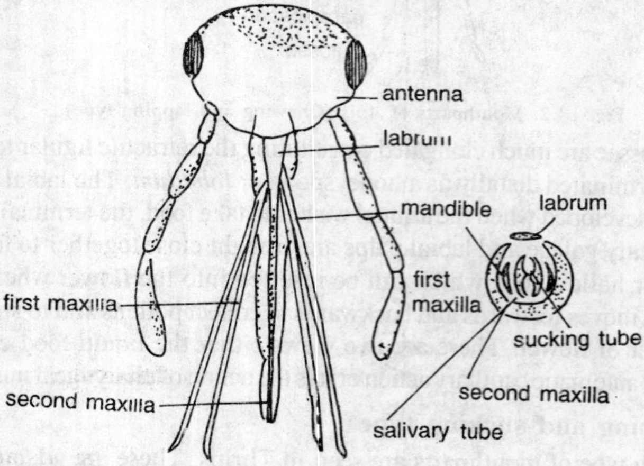

Fig. 14.3. Mouthparts of *Cimex*.

mouthparts are withdrawn into a pouch under the pharynx, if not in use. The pouch is said to be stylet sac, the rest all the structures are same as we found in bugs. Besides a minute tooth-like structure on the head called *buccal denticle* is present. The latter enables the louse to anchor its head to the skin of the host thereby it starts feeding.

(c) Flea type

This sub-type of mouthparts are well marked in fleas. The fleas are parasites on the domesticated animals like dogs, rats etc. The mouthparts are characterized by the presence of a pair of long blade-like *mandibles* with serrated edges. *Labium* consists of two halves fused at their bases so as to form a hollow *proboscis* enclosing the *mandibles*, the *labrum* and the *epipharynx*, which are fused to cover up the mid-dorsal opening of the labium. These mouthparts are adapted to pierce and suck the blood of the host.

(d) Dipterous type

This sub-type is found in mosquitoes. The mouthparts are adopted for piercing and sucking. The *mandibles* and *maxillae* are modified into fine rod-like stylets which are used to pierce the skin of host. The labrum is long needle-like and fused with the epipharynx and covers the mid-dorsal opening of the labium. The *hypopharynx* is double edged, served with a mid-dorsal groove, which forms the salivary passage for the secretion of salivary glands. The labial palps are absent, but maxillary palps are present.

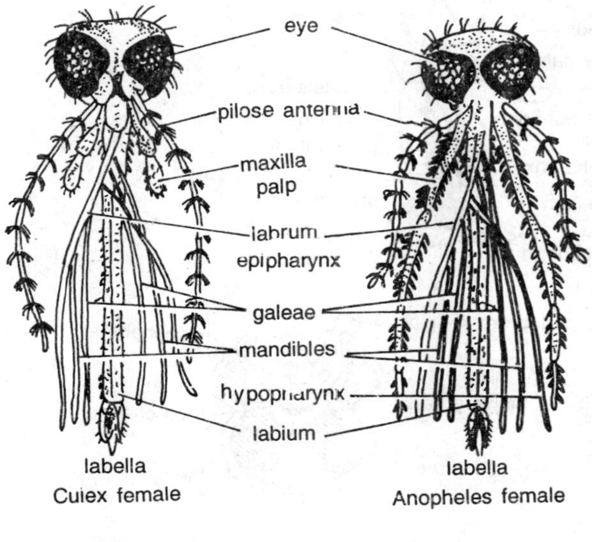

Fig. 14.4. Mouthparts of female *Culex* and *Anopheles*.

While feeding, the hypopharynx and labrum-epipharynx become closely applied to enclose a tubular passage for the ascend of the blood through this tube. The salivary secretion is poured through the groove of hypopharynx in order to check the clotting of blood.

In male mosquitoes, the mandibles are totally absent so that they suck plant juices, whereas the female sucks the blood of human beings and other cattles because of piercing the skin by powerful mandibles.

5. Sponging type

This type of mouthparts are found in Diptera such as house-flies, *Calliphora* and others.The adaptability of the mouthparts suited alone to suck as the liquid and not piercing the tissues of enemy. These forms are capable of sucking even solid food which has to be converted into liquid by their saliva before sucking.

The *labium* is modified into so called *proboscis* which has got a groove with membranous sides and the groove is covered by *labrum*,

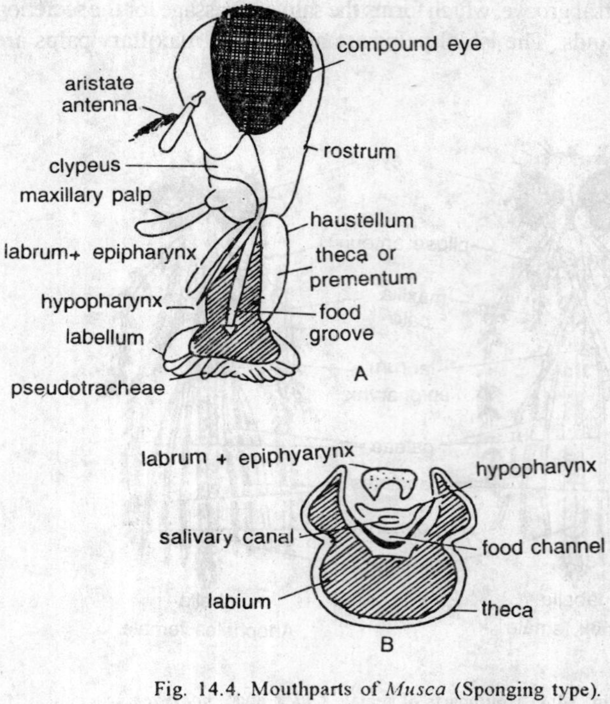

Fig. 14.4. Mouthparts of *Musca* (Sponging type).

epipharynx and *hypopharynx* dorsally.The proboscis expands dorsally into a oral-disc or *labellum* which possesses a number of pseudotracheae. Its proximal part is cone-like known as *rostrum* bearing *maxillary palp,* and the middle is *haustellum* with a mid-dorsal groove serving as a food passage. All the pseudotracheae are converging into the central aperture.

the *mouth*, which leads into the food passage. The mandibles are absent while the maxillae are represented by two maxillary palps, each made up of a single piece.

Generally the two labellae are kept retracted upwards and backwards during rest, but these labellae are protracted and expanded and placed over the liquid food. The fly can consumes solid food particles of an easily soluble substance like sugar. The fly, extrude a drop of saliva to make the solid in dissolved form. The liquid food is drawn by the pseudotracheae and then passed on to the food-canal which is formed by labrum-epitharynx and hypopharynx lying in the mid-dorsal groove of the haustellum.

6. Siphoning type

This type of mouthparts are seen in lepidopterans (butterflies and moths). These are highly specialized for sucking up the juice of flowers and fruits.

The mandibles are usually absent but if present they are very much reduced.The labium is too much reduced. The maxillary palps are considerably vestigial. The labium is represented by a triangular plate bearing the labial palps. The galeae of maxillae are modified into a proboscis through which liquid is drawn up. The proboscis is kept coiled when it is in rest.

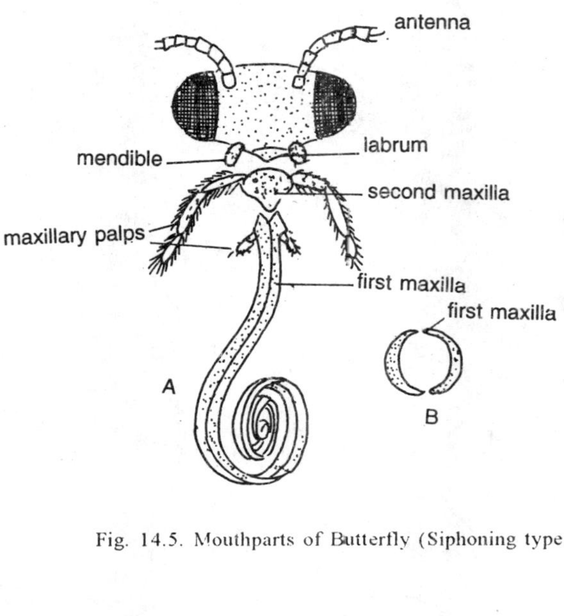

Fig. 14.5. Mouthparts of Butterfly (Siphoning type).

When the insect wants to feed than only the watch spring-like proboscis becomes uncoiled and extended to reach the spot where its diet is available.

In Hexapoda the function of the mouth elements is seen in two quite different lines upon which the insects are grouped as the mandibulate possessing the biting and chewing type and other in Hasutellata consisting of sucking. From the evolutionary point of view it is decided that the biting and chewing parts are generalized primititve kind from which other types have been originated.

15

Respiration in Insects

The respiration includes both physical and chemical processes. The physical envolves the transportation of oxygen to the tissues and removal of carbon dioxide from tissues where as chemical process envolves the oxidation of carbohydrates with the formation of energy, carbon dioxide and water. In the case of vertebrates the oxygen is transported to the tissues by blood with the help of a respiratory pigment, the *haemoglobin*, but in the case of insects the blood plays very little role in transportation of oxygen to the tissues except the larvae of *Chironomus* which have haemoglobin in blood plasma.

Various types of respiratory organs are present among the insects according to their mode of lives.

I. Tracheal respiration in terrestrial insects

In the insects, particularly in terrestrial mode of life, the respiratory organs are *tracheal tubes*. The tracheal tubes are invaginations of the ectoderm and are internally lined by cuticular intima which is thickened by spiral filaments called *taenidia*. It not only gives the support to tracheal branches but also prevents the collapse of wall. These tracheal trunks divide and redivide and ultimately ends into a cluster of tubes called *tracheoles*. The tracheoles are devoid of taenidia. The tracheoles are lines by a protein called *trachein*. The ultimate trachioles enter the individual cells of thc body, and are partly filled with a fluid called *tracheole fluid* at their distal end. By the reduction in the amount of fluid, more of the surface walls of the tubules are exposed to oxygen and, more oxygen is immediately supplied to tissues.

In certain insects like cicadas, honey bees, butterflies etc. the tracheae dilate into *air sacs* of different sizes. The air sacs lack the spiral thickening, allow an increased supply of oxygen and afford a greater breathing capacity of the insect.

The tracheal branches open to outside by many apertures called *spiracles*. The spiracles are elongated or circular apertures and situated on the lateral sides of the body. Usually these are 10 pairs in number, 2 pairs are thoracic and 8 pairs abdominal. Each spiracle is supported by a small annular sclerite called *peritreme.* The spiracles can be closed or opened by spiracular muscles. In some cases the spiracles are absent so that on the basis of presence or absence of spiracels the system is divided into *open type* or *closed type*.

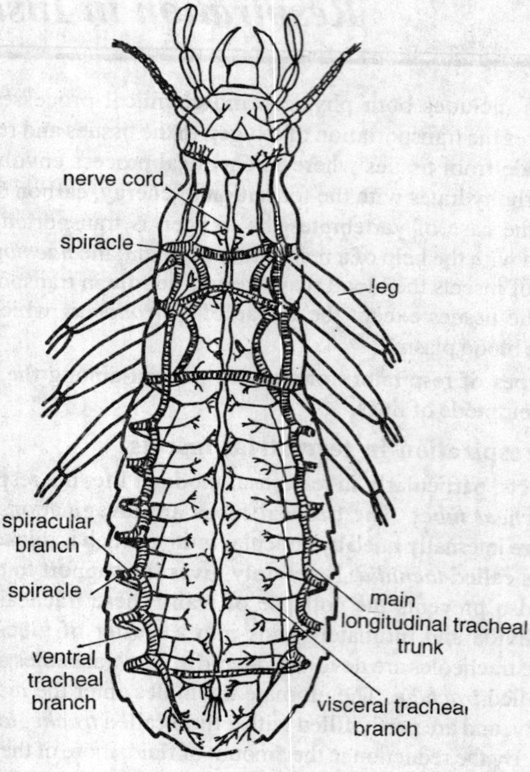

Fig. 15.1. Tracheal Systems. Generalized insect, showing the arrangement of the tracheae.

Keilin (1944) and *Hinton* (1947) have proposed the clasification of the tracheal system on the basis of functional spiracles as follows :

1. *Polypneustic tracheal system.* It represents the open tracheal system and can be divided into following sub-types.

(a) *Holopneustic tracheal system.* It consists of all 10 pairs of functional spiracles found in nymphs and adults of large number of

(b) *Hemipneustic tracheal system.* Here only 8 pairs of spiracles (the mesothoracic and first 7 abdominal) are functional whereas other pairs are non-functional as mostly found in larvae.

(c) *Peripneustic tracheal system.* Here only nine pairs of spiracles are functional. Only a single pair located on the meetathorax is non-functional. Such a system is seen in terrestrial larvae of Neuroptera, Mecoptera, Lepidoptera etc.

2. *Oligopneustic tracheal system.* It represents partial open tracheal system. It is divided into following sub-types :

(a) *Amphipneustic tracheal system.* Here only two pairs of spiracles i.e. first thoracic and last abdominal are functional as found in the larvae of Diptera.

(b) *Metapneustic tracheal system.* Here only one pair of spiracles i.e. last abdominal is functional. It is seen in larvae of Cutcidae, Tipulidae.

(c) *Propneustic tracheal ststem.* Here only first pair of mesothoracic spiracles are open e.g. pupae of Diptera.

3. *Apneustic tracheal system.* It represents the closed respiratory system. All pairs of spiracles are closed and non-functional. It is present in various aquatic and endoparasitic larvae.

4. *Hypopneustic tracheal system.* It represents partially degenerated tracheal system in which one or two pairs of spiracles disappear completely so that the insect bears less than 10 pairs of spiracles are found as in Mallophaga, and Siphunculata (only one pair of thoracic and 6 pairs of abdominal spiracles are present)

5. *Hyperpeneustic tracheal system.* In such a system more than 10 pairs of spiracles. Diplura like *Japyx* bears 4 pairs on the thorax and 7 pairs on abdomen, in all there are 11 pairs of spiracles are present.

II. Closed type of respiration in aquatic insects

The closed type is present in some aquatic insects and endoparasitic larvae. Insects living in aquatic medium have different adaptations. Insects of order Odonata, Plecoptera and Ephemeroptera are primarily aquatic where as insects of order Hemiptera, Coleoptera, Lepidoptera, Diptera and Neuroptera are secondarily aquatic. Most of the forms are aquatic only during larval stages.

The immature stages of Odonata, Plecoptera and Ephemeroptera are aquatic and draw oxygen through general body surface or specialized structures, the *gills*, so that they have two types of respiration :

1. Skin or cutaneous respiration.
2. Gill respiration.

1. Cutaneous respiration

In those, insects where cutaneous respiration occurs, the tracheal system is fairly developed without spiracles. The gaseous exchanges takes place through general body surface by diffusion. The larvae of aquatic lepidopterans show this type of respiration.

2. Gill respiration

It provides an enlarged respiratory surface formed by evagination of integument. The gills are usually thin walled. In aquatic forms three types of gills are present :

(i) Tracheal gills

(ii) Spiracular gills

(iii) Blood or anal gills.

(i) Tracheal gills

It is formed by hallow evagination of integument or intestinal wall. These are supplied by fine tracheal branches and so that known as tracheal gills. The tracheal gills may be filamentous (as present in the case of

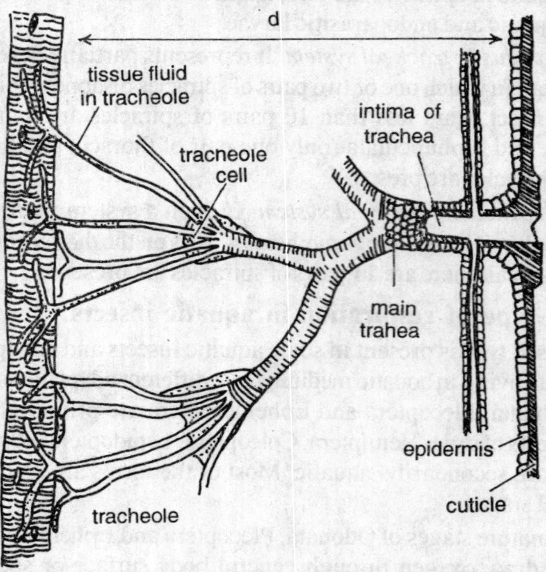

Fig. 15.2. Tracheal Systems. Scheme of tracheal systems operating by simple diffusion. The internal branches are, of course, greatly simplified. The tracheae are continuations of the body wall, and the body wall cuticle is continues with the tracheal lining. In the tracheols cells and are without cuticular supports.

Plecoptera, some Ephemeroptera and Trichoptera) or plate-like (as in larval Odonata and some Ephemeroptera). In some odonates peculiar *rectal gills* are present. These are six folds formed by the rectal wall. They are well supplied with tracheal branches and hence sometimes they are referred as *branchial basket*. The insect draws water in the rectum through the anus and exchange of gases takes place here.

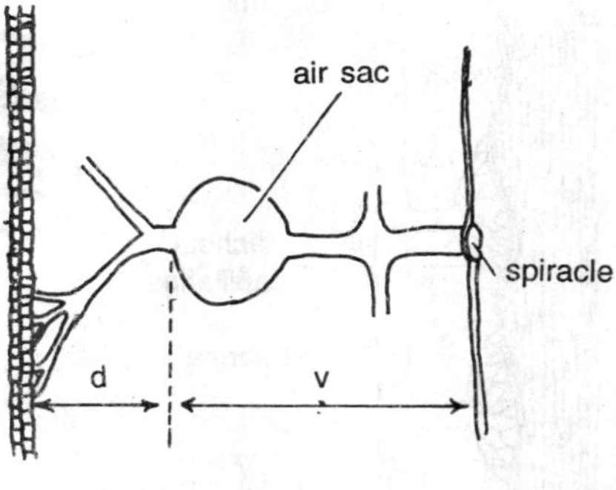

Fig. 15.3. Scheme of tracheal systems in which active ventilation augments the movements of gases in the tracheae. The intervation of air sacs or other very spacious tracheal trunks, which can be collapsed or expanded as a result of body activity or special musculature, makes the outer part of the tracheal system a ventilating region.

(ii) Spiracular gills

These gills are found in some Coleopteran and Dipteran aquatic pupae. Here 1 or 2 pairs of spiracles are drawn out to long processes which have named as spiracular gills by *Hinton* (1953). Some of the pupae can live in damp climate. So that the gills are considered to adopted both for aerial as well as aquatic respiration through gill surface.

(iii) Blood or anal gills

They are thin diverticulae of integument or of the proctodaeum. They are present in aquatic larvae of beetle (*Hydrobia*). These gills are supplied with blood hence they are known as blood gills but in the case of larvae of *Tipulid* the tracheal branches are also present so that they function as the tracheal gills are well as blood gills. In the case of blood gills the respiratory function has not been demonstrated. *Wigglesworth* (1953) has shown that in the mosquitoes larvae the blood gills are concerned with the regulation of movement of ions absorbed from the water.

III. Open type of respiration in aquatic insects

This type of system is found in secondary aquatic forms belonging to order Hemiptera and Coleoptera. They have spiracles through which tracheal branches drawn in atmospheric air. Since the insects live in water they have developed certain adaptations to ensure the supply of air while they are in submerged stage. The various modifications are :

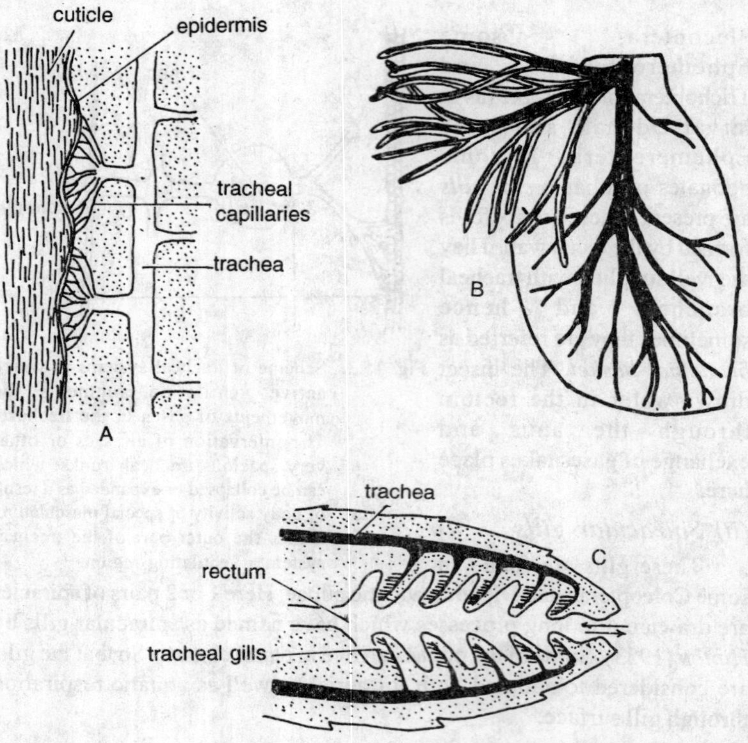

Fig. 15.4. Respiratory Adaptations. A. Scheme of surface trachea in aquatic insects with closed tracheal systems. Tracheal capillaries immediately below the epidermis are the site of respiratory exchange. Oxygen absorbed here moves into the main tracheal trunks which branch through the body tissues. B. A tracheal gill from a mayfly nymph. C. Rectal gills of a dragon fly nymph.

1. Air stores

Some of the nymphs and adults of water-boatman and the back-swimmers and also the nymphs of *Spherodema,* certain amount of air is carried by the insect attached to some part of body when the insect dips in water. In the case of water-boatman and back-swimmer air is carried on ventral surface of body. The air is kept by hairs called *hydrofuge hairs,* which resist penetration of the air filled by water. In adults of some beetles the air is carried by wing cover or elytra. The spiracles remain in contact with air and draws it but the air is not available for a long time so that the insect have to come again and again to take fresh air bubbles to the surface. Besides in respiration the air bubbles also has hydrostatic function.

2. Gaseous plastron

This plastron is a permanent bubble held by various means around the body by certain insects. The plastron communicates with spiracles so that the spiracles are able to draw air from water. Thus it has permanent nature. The insect need not to come on surface again and again for fresh air. *Thorpe* (1950) described such type of respiration in certain aquatic insects.

3. Air tubes

Many insects which remains submerged in water acquire and obtain air through one or two tubes which can break the water surface. In such insects the spiracles are situated at the tip of the tube and the rest of spiracles are non-functional and hence either may closed or under-developed (larvae of mosquitoes). In the *Culex* larvae a tube projects above the surface of water to draw air in. The rat-tiled (dipteran larvae) have the tube of about 3-4 inches long. When the larva remains submerged in water the long tube projects outside to draw atmospheric air. While this type of tube present in different forms in some aquatic Hemiptera such as *Nepa, Belostoma* and *Spherodema,* there is a long respiratory spihon through which air is drawn.

4. Plant perforating spiracles

In some insects the spiracles are situated on sharp-pointed processes which are penetrated by insect into the air cavities of hydrophytes for the purpose of getting air. This is found in the larvae and pupae of certain colepterans and dipterans.

Thus from the above account, it is clear that aquatic insects have assumed several modifications in the respiratory system to enable them to live in water. Without such modifications it would not have possible to them to lead an aquatic life.

16

Metamorphosis in Insects

Insects develop from the primordial cells in the ovaries of females. The eggs are laid and hatched out side the body after a variable periods of time. The development within the egg is called *embryonic development* and period after hatching of egg is called *post-embryonic development.* One of the most characteristic features of insects is the fact that they almost always hatch in a condition morphologically different from that of the imago. In order to reach the latter instar they consequently have to pass through changes of forms which are collectively termed *metamorphosis,* these are usually most pronounced towards the end of post-embryonic development and they are accompnied by physiological and biochemical changes. During metamorphosis the exoskeleton or outer skin is cast-off periodically. This process is known as *moulting* or *ecdysis.* The larval stage between two moultings is called *stadium.* Thus the insects growes by a series of changes and these growing stages are called *instars.* The number of moults vary but is constant for a given species. On the basis of these changes the metamorphosis is distinguished into following types:

1. Ametabolous development or no metamorphosis
2. Hemimetabolous development or incomplete metamorphosis
3. Paurometabolous development or gradual metamorphosis
4. Holometabolous development or complete metamorphosis
5. Hypermetabolous development

1. Ametabolous development or development with no metamorphosis

In some primitive forms of insects the egg hatches in a form that resembles the adult in all morphological details except in size and colour and in the absence of wings, reproductive organs and external genitalia (styles etc.). The stages of development are : *Egg-Young-Adult.* Ametabolous development is seen in *Lepisma*, spring tails (*Thysanura, Collembola* and *Protura* i.e. in apterous insects).

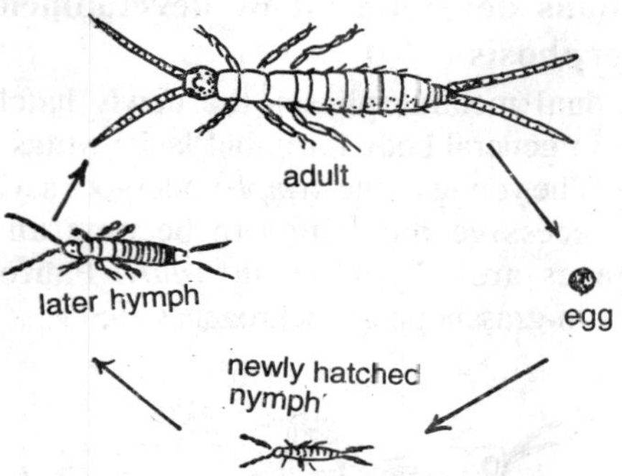

Fig. 16.1. *Lepisma*. Ametabolous development.

2. Hemimetabolous development or development with incomplete metamorphosis

This type of life-cycle includes gradual metamorphosis as in *paurometabolous* type except that the eggs are laid in water and develop in aquatic nymphs or *naiads*. These naiads posses tracheal gills but adults are formed from naiads by repeated moults and maturity. Final moult results in the winged adult. The developmental-stages are : *Egg-Naiad-Adult*. Hemimetabolous development is seen in May-flies, Dragon-flies, Stone-flies etc.

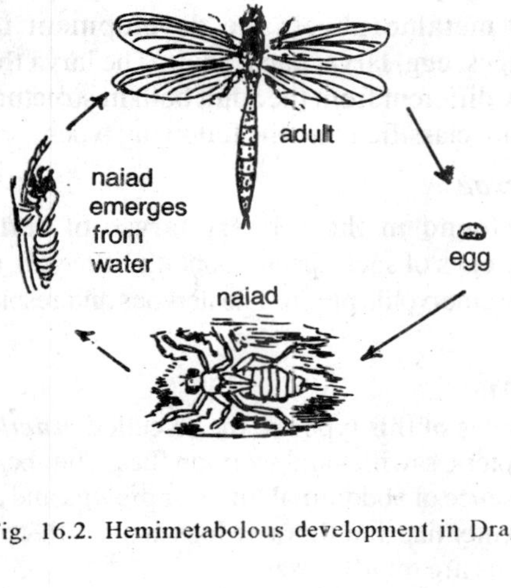

Fig. 16.2. Hemimetabolous development in Dragon-fly.

3. Paurometabolous development or development with gradual metamorphosis

In case of gradual metamorphosis, the newly hatched creature resembles an adult in general body form, but lacks wings and external genital appendages. The young or the *nymph* undergoes several nymphal stages through successive moultings to become an adult. The developmental-stages are : *Egg-Nymph-Adult*. Paurometabolous development occurs in grasshopper, cockroaches etc.

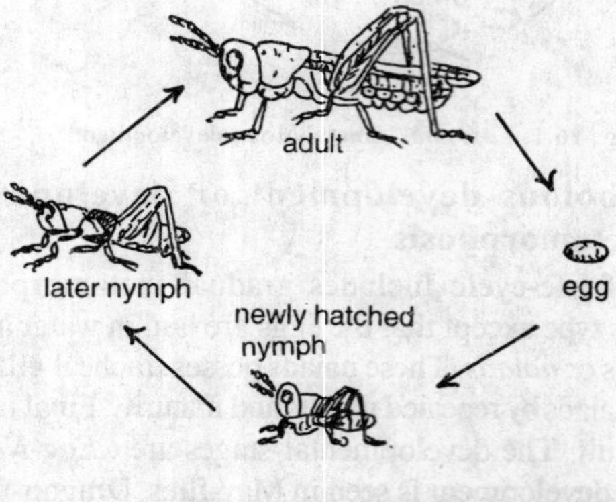

Fig. 16.3. Paurometabolous development in grasshopper.

4. Holometabolous develpment or development with complete metamorphosis

In complete metamorphosis the development includes four developmental stages, egg, larva, pupa, adult. The larva that hatches out of the egg is totally different from the adult both in structure as well as in habit. The larvae are classified into the following types :

(i) Protopod larvae

This type is found in the primary larvae of certain parasitic Hymenoptera. The eggs of such species contain little yolk and the larvae emerge are an early embryonic phase. The nervous and respiratory systems are undeveloped.

(ii) Polypod larvae

Typical examples of this type are the so called *eruciform larvae* of most of the Lepidoptera, sawflies and scorpion flies. They bear well defined segmentation, presence of abdominal limbs or prolegs and a *peripneustic* tracheal system. Antennae and thoracic legs are small. compound eyes. ocelli and wing pads are totally absent.

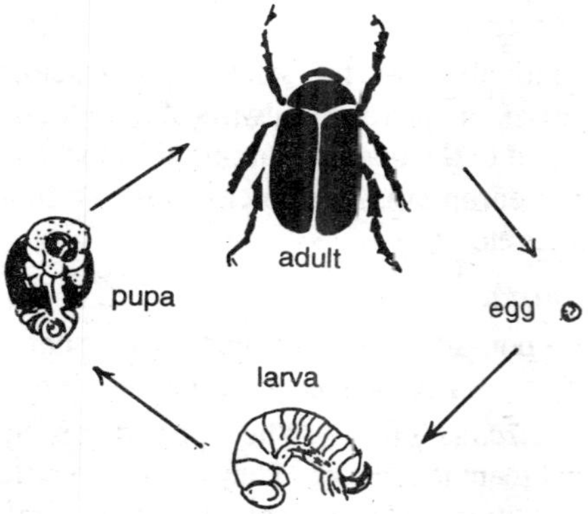

Fig. 16.4. Holometabolous development.

(iii) Oligopod Larvae

These are characterized by the presence of more or less well-developed thoracic legs and absence of abdominal appendages except sometimes for a pair of similar caudal process. Two common types can be distinguished :

(a) Campodeiform larva

They are so called from their resemblance to *Campodea* (Thysanuran insect). The larva bears an elongated more or less fusiform somewhat depressed body which is often well sclerotized and markedly *prognathous* head, long thoracic legs and usually a pair of terminal abdominal processes. Compound eyes, dorsal ocelli and wing pads are absent. Campodeiform larvae occur in the Neuroptera, some Coleoptera, Trichoptera and Strepsiptera.

(b) Scarabeaeiform larva

These are subcylindrical, C-shaped larvae with shorter thoracic legs, a soft fleshy body and no caudal process. The larvae are seen in Scarabaeoidea, in Coleopteran families etc.

(iv) Apodous larvae

In this type the trunk appendages are completely absent and most cases are probably derived from oligopod type. Such a larvae are seen is several families of Coleoptera.

After several changes the larva turns into pupa. The pupa appears to be inactive externally but internally it is very active and forms the imago. The following types of pupae are distinguished by *Hinton* (1946) :

(a) Decticous pupae

These have relatively powerful, sclerotized, articulated mandibles. This is the primitive type of pupa and is always of exarate i.e. the appendage, are not closely applied to the body but are used in locomotion. This type of pupa occurs in Neuroptera, Macoptera most Trichoptera, in some Lapidopteran families etc.

(b) Adecticous pupae

This type have a non-articulate mandibles and are not used in escaping from the cocoon. Three forms of adecticous pupae are recognizable.

(i) *Exarate* or *decticous pupae*. The appendages are free of any secondary attachment to the body. They are found in the Siphonoptera and Strepsiptera, most Coleoptera and Hymenoptera.

(ii) *Obtect adecticous pupae*. Here the appendages are firmly pressed against its body and are soldered down to it by a secretion produced at the last larval moult. These pupae occur in all higher Lepidoptera, some Coleoptera etc.

(iii) *Coarctate pupae*. This is an adecticous as well as exarate type of pupa which is enclosed in puparium. The puparium is formed from preceeding larval cuticle. It does not show any changes in external appearance. It is seen in Cyclorrhaphan Diptera.

5. Hypermetabolous development or development with hypermetamorphosis.

When a developing insect passes through two or more markedly different larval instars it is said to undergo *hypermetamorphosis*. The last or final larval instar changes into non feeding, inactive stage called *prepupa*. It is enclosed in the cuticle of preceeding larval stage. Such a metamorphosis occurs in some Neuroptera, Coccidals and many endoparasitic Hymenoptera.

Internal Metamorphosis

In those cases where the metamorphosis is gradual or incomplete, the internal organs do not undergo any major modifications but some very minor changes are observed. In the case of holometabolous and hypermetabolous devlopment, major changes take place in internal organs because of different mode of life. All these modifications of internal organs take place within the pupa. There may be other structural or physiological internal changes, so that the pupa is inactive externally but it is active internally. During early pupal stages internal organs break down or dissolved (probably as a result of lysosomal breakdown) or by *phagocytes*. This process is known as *histolysis*. This phenomenon differs in different insects. In some cases it is minor or may be major. After histolysis new tissues are formed from pad-like structures known as *immaginal pads* or *histoblasts*. This process is known as *histogenesis*. It is ultimately followed by a process of *organogenesis*. The internal changes within the pupa differ in different groups of insects. In Coleoptera, Lepidoptera, Hymenoptera and some Diptera intestine and other larval parts are greatly modified without general histolysis. Similarly in certain dipterans the general histolysis is noted so that all the organs are dissolved and reformed except the nervous system, heart, reproductive organs and thoracic appendages. The study of internal metamorphosis is thoroughly studied in all groups of insects but still it is unknown in some of the groups of insects.

Hormones and their role in Metamorphosis

Insects have a rigid exoskeleton, because the exoskeleton do not expand to allow for enlargement, insect growth is associated with the

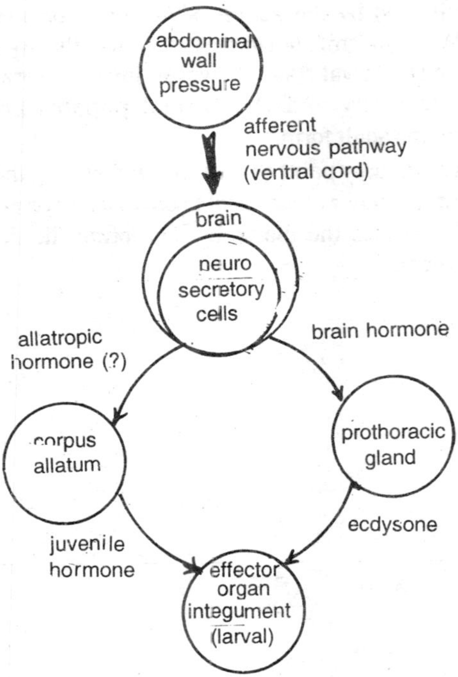

Fig. 16.5. Control of moult and metamorphosis in insects.

periodic shedding of the exoskeleton in a process known as *ecdysis*. Ecdysis is followed by the expansion and hardening of a new, soft, larger exoskeleton that has devloped underneath the old exoskeleton. Thus insect development includes a series of ecdysis or moults, and the stages between the moults, are known as *instars*.

In insect development, *metamorphosis* is the change in the body form that converts an immature individual into an adult. Metamorphosis is either incomplete or complete comprising eggs-nymphs and adult in former case and eggs-larvae-pupae-adults in latter case. Several hormones regulate insect growth, moulting and metamorphosis. The neurosecretory cells of the brain produce a *brain hormone* that travels along axons to a structure called *corpus cadiacum*, which is attached to the brain. From the corpus cardiacum, the brain hormone is released and carried to the *prothoracic glands*, which is stimulated to release *ecdysone*. Ecdysone is also called the *moulting hormone* or *moult and maturation hormone* because it promotes moulting and tend to spur the developmental process along toward pupation and adulthood. During larval stage the effect of ecdysone is countered by the action of *juvenile hormone* secreted by *corpra allata*. When juvenile hormone is secreted, the organism remains immature. In the last larval stage, however, juvenile hormone secretion decreases dramatically and the animal pupates and begins its transformation to the adult form.

From an evolutionary point of view, it is interesting that some plants produce compounds similar or identical to ecdysone or juvenile hormones. This apparently protects the plants by disrupting the development of insects that eat them.

17

Trilobites

Trilobites are an extinct group of marine arthropods that were once abundant and widely distributed in Palaeozic Seas. They reached their hight of distribution and abundance during the Cambrian and Ordovician periods and disappeared at the end of Palaeozoic Era. Over 3900 species have been described from the fossils obtained so far.

Habits and Habitat

The trilobites were exclusively marine since they were found only in association with the remains of marine fossils. They were mostly bottom dwellers, crawling over sand and mud with the help of legs. They probably swallowed mud and silt and used the organic particles as food, like many annelids.

EXTERNAL FEATURES

Size

Most trilobite species ranged from 3- 10 cm. in length, although some planktonic forms were only 0.5 mm. in length. Some species attained a length of 67.5 cm. (*Terataspis*).

Division of Body

The body was oval dorso-ventrally flattened and divided into three more or less equal sections:

1. Cephalon or Head
2. Thorax or Trunk
3. Pygidium.

Each of the body division was in turn divided into three regions by a pair of furrows running from anterior to posterior region and forming a median axial lobe, flanked on either side by a *lateral* or *pleural lobes*. In some cases the thorax and abdomen formed the *opisthosoma*.

1. Head or cephalon

It was distinct, solid, unsegmented, semicircular and often prolonged backwards at the angles. The head was covered by a carapace or cephalic shield with the posterior lateral margins projecting backward. The head was composed of atlest five fused segments. The 5th occipital segment was most distinct. The original head segmentation was frequently indicated by transverse grooves on the carapace. The pleural skeleton was reduced to thin membranes to which the appendages were attached. The lateral portions of the carapace were known as *cheeks* or *genae*. The cheek of either side was traversed by an oblique frontal, facial or occular suture. On each side of the middle of the carapace there was a pair of eyes that varies considerably in size, depending upon the species. X-ray sudies indicate a structure similar to the compound eyes of living arthropods. These eyes are called *maculae* by *Lindstrom*. In some species of *Triarthrus* and *Amphyx* there was a single simple eye and its structure is unknown. In some trilobites the eyes were absent. Three types of compound eyes were recognized in trilobites:

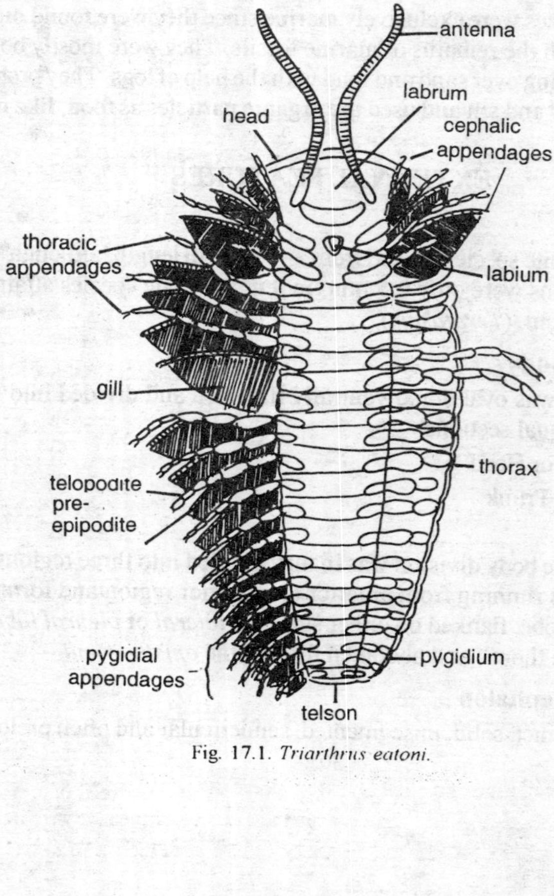

Fig. 17.1. *Triarthrus eatoni.*

(i) ***Holochroal.*** It was the first type of eyes and had globular or biconvex lens. The two eyes were very close to each other.

(ii) ***Presmatic.*** It was characterized by the presence of plano-convex lens.

(iii) ***Schizochroal.*** The lens were biconvex and the two eyes were widely separated.

The *mouth* was located in the middle of the underside of the head just behind and beneath a lip-like flap called the *labrum*. On each side of the labrum was a long sensory *antenna*. Behind antennae were located four pairs of similar appendages. They were similar to the appendages of

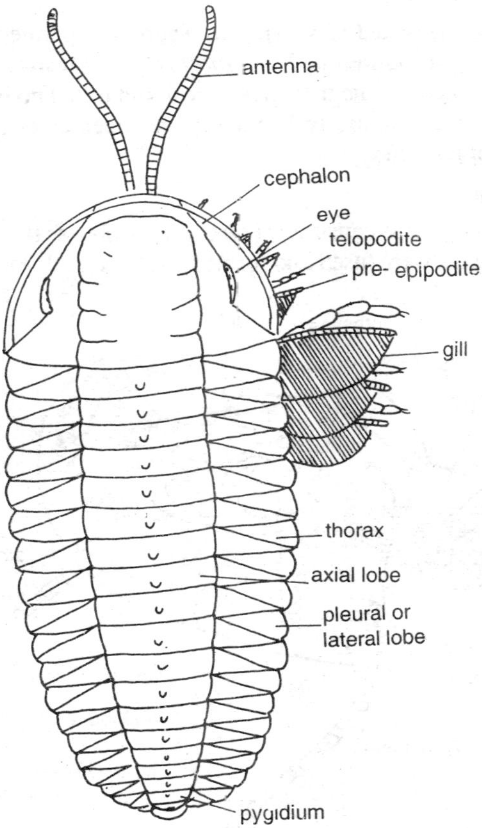

Fig. 17.2. *Triarthrus* (Dorsal view)

the trunk and pygidium. Each appendage was biramous and consisted of inner walking leg or *telopodite* and an outer gill bearing branch was *pre-epipodite.* The pre-epipodite was situated between the lateral extension of the pleural lobe and the walking leg and consisted of numerous segments, each of appendage overlapped the fringe of next appendage. These filaments have been called gills but in few species in which they have been preserved, the feather-like barbs suggesting such functions as digging, filtering or swimming depending upon species. In *Phacops* certain of the pairs of basal segment of the appendages i.e. coxae carried opposing toothed processes that must had functioned as *jaws.*

2. Trunk or Thorax

The trunk consisted of varying numbers of separately articulating segments. The number may be 2 in *Agnostus* to 29. Owing to the mobality of thoracic segments some trilobites were found rolled up into a spiral or ball. Each segment ventrally had a pair of appendages similar to the appendages of the head.

3. Pygidium

The pygidium was constructed on the same plan as the thorax except that the segments were fused and formed a solid shield. The appendages

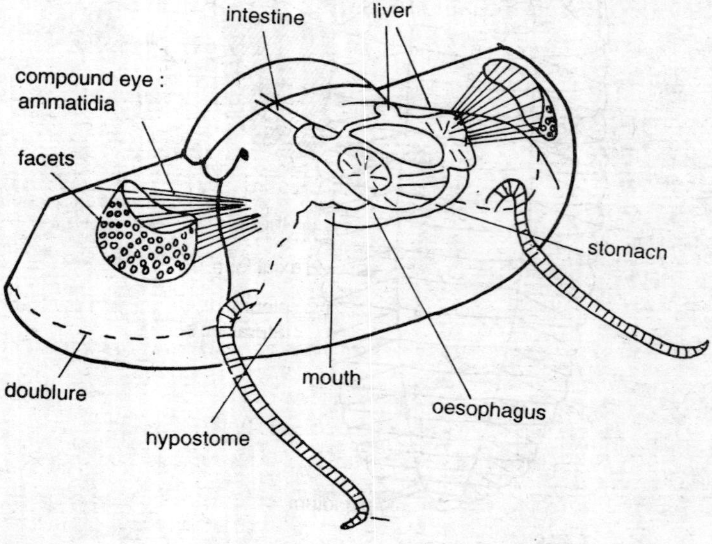

Fig. 17.3. *Trilobite.* Head seen from one side of front.

of pygidium were successively smaller towards the posterior end. The number of abdominal segments varied from 2-28 and the last segment of pygidium was called *telson*. The *anus* was situated on underneath of the telson.

Most of the trilobites were bottom dwellers and crawled over sand and mud using the walking legs. This type of existence is proved by the flattened body and dorsal eyes. Some were adapted for burrowing as is evidence by a shovel-shaped or plough-shaped head. Some groups of trilobites were apparently not confined to the bottom but took up a pelagic swimming existence. In these forms the body was narrower, and the eyes were located on the sides of the head. Although nothing is known about their appendages. The walking legs may have been flattened and adapted for swimming.

Moulds and radiographs provide some information about the digestive tract. A pear-shaped stomach surrounded by a digestive gland extended downward to the mouth. From posterior end of stomach a long narrow intestine extended to near the posterior end of the body.

Development

The larvae of several species of trilobites have been found in the fossil state. For some forms, such as species of *Sao* and *Olenus*, almost the entire larval series has been determined from fossil specimens. During their postembryological development, trilobites passed through three larval periods, each of which consisted a number of instars. The first larval stage was *protaspis larva*. It was a small planktonic form measuring from

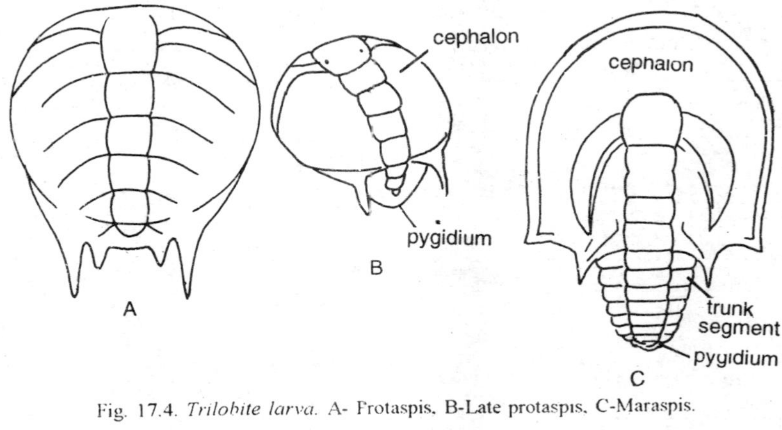

Fig. 17.4. *Trilobite larva*. A- Protaspis. B-Late protaspis. C-Maraspis.

0.5-1 mm. in length. The body had a single dorsal carapace, consisting of the acron and four postoral segements. After passing through several instars, in which additional segments were added to the carapace the protaspis larva passed into a *meraspis larva.* In this larva the pygidium was located behind the cephalon. The thoracic region appeared during successive moults, as segments were gradually freed from the anterior border of the pygidium. The final larval stage was *holaspis larva,* though still very small, contained the definite number of adult thoracic segments. Nothing is known of larval appendages and none of the stages, discovered so far, can be considered as naupli.

Position of Trilobites

The systematic position of trilobites is still uncertain but the available evidences show that they are arthropods. Theories of their relationship fluctuate between Crustacea and Xiphosura, the evidences weighing more in favour of the Crustacea.

The trilobites are, however, more primitive than any other crustacean. Their resemblance to some of the Phyllopoda is sufficient to make it probable that they have some ancestral connections. Formerly it was believed that the trilobites were the ancestors of Crustacea. In recent years, the work of *Stormer* and many zoologists began to feel that trilobites and crustaceans are not related.

18

Limulus

On the evidences available it appears quite probable that the air breathing Arachnida have been derived from marine *Limulus* ancestors. The evidences come from the resemblance between the scorpion and *Limulus*.

SYSTEMATIC POSITION

Phylum	-	Arthropoda
Class	-	Arachnida
Order	-	Xiphosura
Genus	-	*Limulus*

Distribution

King carbs or horse-shoe crabs are marine animals, but occasionally occur in fresh and brekish estuarine waters. They live in shallow waters along muddly shores. They are noctural as they become active and swim through the water at night.

EXTERNAL FEATURES

Shape and size. The body is depressed, dark brown in colour and measuring about 60 cm. in length. They are called king crabs but neither they are kings nor crabs.

Division of Body. The body is divisible into *cephalothorax, abdomen* and a *caudal spine* or *telson*. The body is covered by smooth chitin. The cephalothorax is semi-circular and consists of a pair of *median eyes* which are very close to each other. The median eyes lie on median ridges whereas the lateral eyes on lateral ridges. Seven spines are also present. On the ventral side of cephalothorax there is a *mouth* and seven pairs of appendages. Infront of mouth there is a flattened longitudinal ridge called *camerostome* and beyond the mouth is oval structure called *promesosternite*. The mouth is surrounded by seven pairs of appendages. The first pair is called *chelicerae*. Each chelicera is three segmented structure which is followed by 5 pairs of walking legs called *pedipalps*. Each pedipalp is made up of six segments. In female the fourth anterior pair of pedipalps is chelate in appearance whereas in the male the first pair has claws which are curved. The 5th pair in male differs from the female by possession of a peculiar curved spine called *spatulate process* or *flabellum*. All the appendages are postoral in position but at later stages the first pair becomes preoral in position.

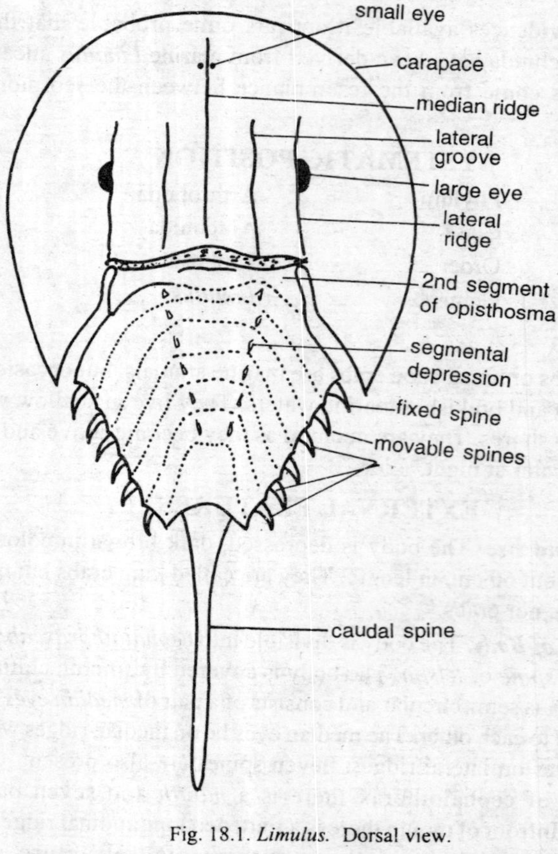

Fig. 18.1. *Limulus.* Dorsal view.

The abdomen is jointed to cephalothorax or prosoma by a transverse *hinge*. It is broad, hexagonal with hardly any trace of segmentation. It is made up by the fusion of an anterior 6 segmented *mesosoma* and a posterior 3 segmented *metasoma*. The mesosoma bears 3 spines in a single median row on dorsal side and two lateral rows of 6 small pits. Each lateral side is serrated and produced into 6 immovable spines. Abdomen consists of 6-pairs of appendages. They are lamellar in form and resemble the crustacean limbs. The anterior most abdominal appendages fuse to form the *genital operculum*. The genital operculum acts as a protective cover over the gills. Among the 5 pairs of abdominal appendages each consists of endopodite and exopodite i.e. biramous.

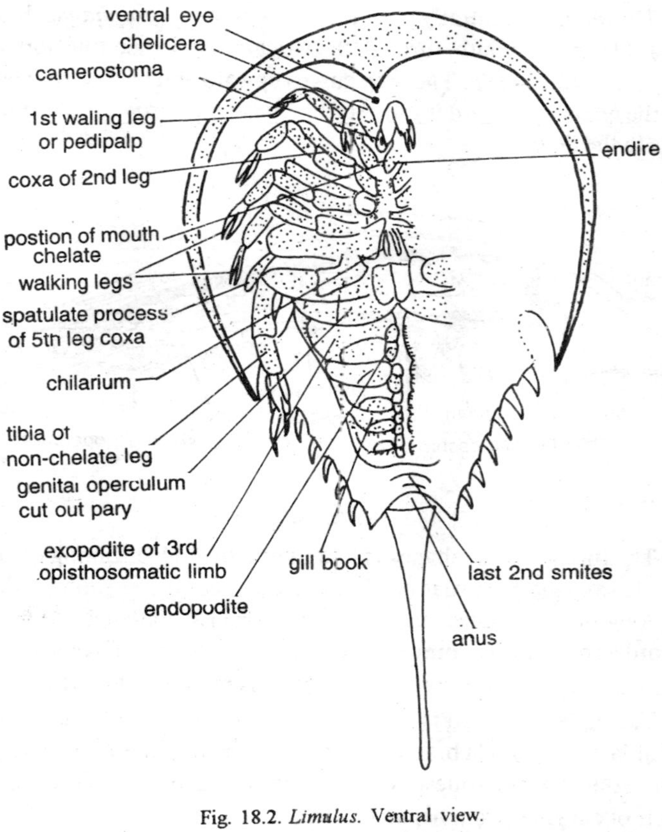

Fig. 18.2. *Limulus*. Ventral view.

Each exopodite bears a lamellated sureture known as *book-gill*. The book-gills are external in *Limulus* in contrast to the interal appearance of book-lungs of scorpion. Each book-gill has 120-200 leaf-like lamellae. The abdomen of *Limulus* is correspond to preabdomen and postabdomin of scorpion. Posteriorly the abdomen bears a much pointed, triangular, S-shaped *telson*. The telson is not true as it does not bear the anus. It consists of fused tergites or dorsal plates of the abdominal segments. The telson is not used for defence but is thrust into mud. In *Limulus* internal chitinous structure, the *endosternite* is present in between ventral nerve cord and alimentary canal. It gives attachment to muscles. A similar structure found in scorpion and spiders besides a few crustacean like *Apus.*

Digestive System

The *mouth* is ventrally placed and followed by *oesophagus.* It passes upward from the mouth, through the vascular ring, then runs forward to the more swollen *crop*. The crop curves upward and backward, merging into the *gizzard.* Gizzard is followed by *stomach-intestine.* A short *rectum* leads to the *anus*.

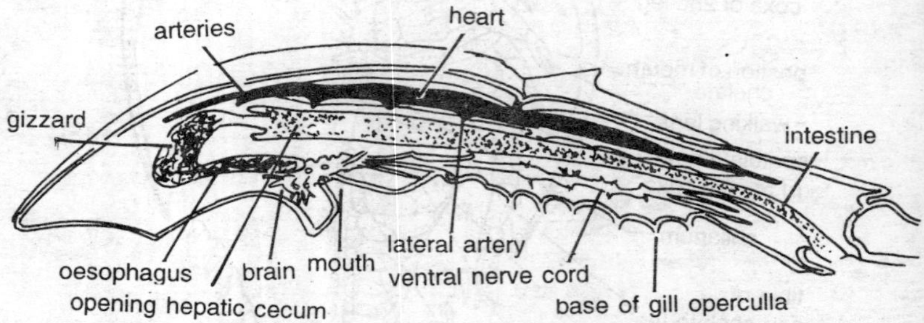

Fig. 18.3. *Limulus.* Sagittal Section.

The lining of oesophagus is longitudinally folded. Lining of crop is wrinkled and the gizzard is much less distensible, and its lining is thrown into longitudinal ridges studded with somewhat flattened teeth. From gizzard a conical valve projects into stomach-intestine. Except stomach-intestine, the internal lining of other parts is made up of cuticle.

The digestive gland is known as *hepatopancreas*. It consists of one pair of large and much branched hepatic caeca which remify. Two ducts leading into stomach-intestine. Digestion occurs in stomach-intestine by means of enzymes from hepatopancreas.

The animal is scavenger feeding on worms, soft mollusks and algae.

Circulatory System

A heart lies in prosoma and is surrounded by a thin-walled pericardial sinus. It is suspended from the walls of pericardial sinus by 9 pairs of elastic ligaments. The heart is tube-like and provided with 8 pairs of ostia.

The arteries, leave the heart, are four pairs of *lateral arteries*, one pair of *aortae* and a *median ventral artery*. Among them the first pair is known as *aortic arch* and fuse to form *vascular ring*. The ring gives rise to small vessels which supplies the walking legs and other median and internal structures. The artery arising from the vascular ring extends posteriorly and gives rise smaller vessels in gills. The blood from the median artery is collected by *supra anal vessel* which bifurcate to give rise the *collateral vessels*. Collateral vessels communicate the heart by 4-pairs of lateral vessels. The blood from the gills is collected back by *longitudinal vessel*.

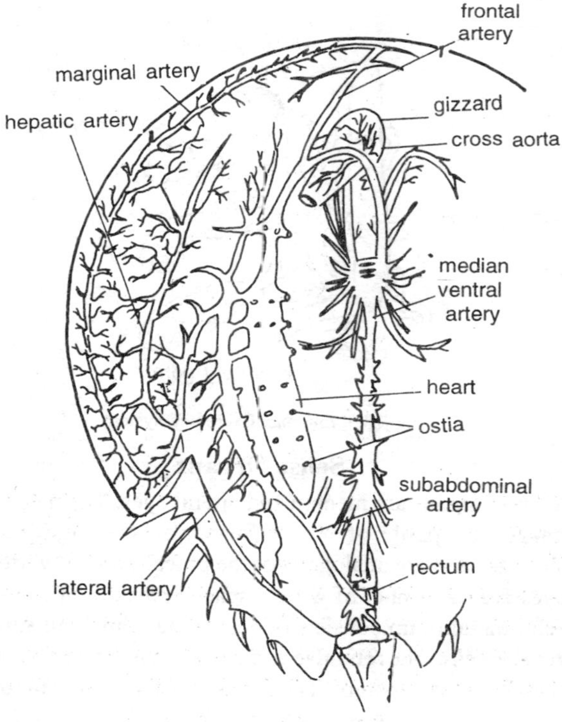

Fig. 18.4. *Limulus*. Arterial System.

The colour of blood is bluish due to the presence *of haemocyanin*. Amoeboid cells are also present in blood.

Excretory System

There is a pair of *coxal glands* which are brick-red in colour. Each gland consists of four lobes and inside each lobe there is a network of tubules. At the base of 5th appendage, the lobes of a side open by means of a duct through a bladder.

Nervous System

It is highly concentrated and consists of two pairs of ganglia or *brain* united by commissures. The commissures form a ring which lies within vascular ring. The anterior part of brain sends separate pairs of nerves to the eyes and olfactory organs. The nerve ring posteriorly gives rise double *ventral nerve cord* which extends thoughout the abdomen and bears 5-*branchial ganglia, 3- post-branchial ganglia* and *lateral nerves*. The central nervous system lies within the vascular system, so that it is dipped in blood at all the times.

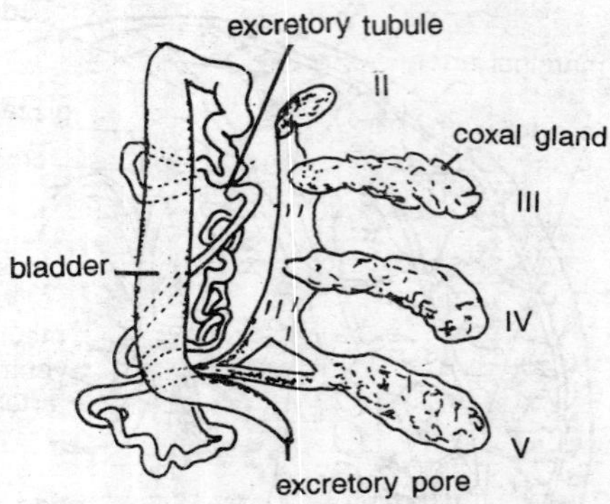

Fig. 18.5. *Limulus.* Excretory System.

Sense Organs

The sense organs are median and lateral eyes. The median ocelli are rudimentary, incapable of image formation but highly sensitive to ultraviolet light and are analogous ot the pineal body of vertebrates. The lateral eyes bear a number of widely spaced *ommatidia* beneath a single lens. Each ommatidium consists of central ganglion cell surrounded by 9-11 *retinular cells.* The retinulae and ganglion is formed by hypodermis. The lens is cuticular in nature. The image formed is of compound nature.

Reproductive System

The sexes are separate. Males are smaller than the females and their first walking legs are non-chelate but clawed, serving as claspers during pairing.

In male the *testis* consists of an intricate reticulum of tubes with small follicles on the wall of many of the follicles. A globular *sperm sac* is present in which early stages of spermatogenesis occur. *Vasa deferentia* are located in the hepatopancreas. Each *spermatozoon* is provided with a long, vibratile tail.

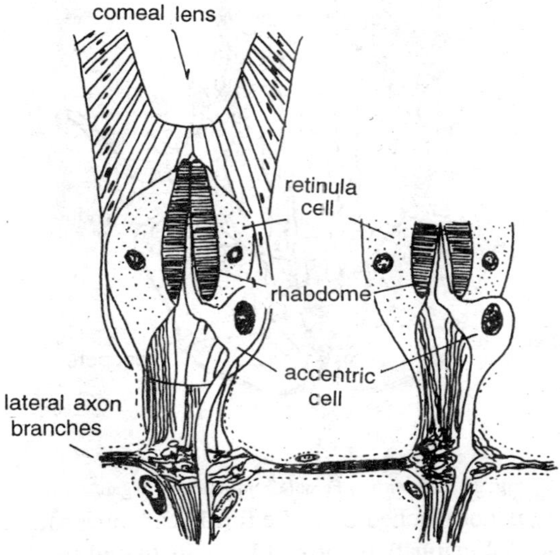

Fig. 18.6. *Limulus*. L.S. of eye through two ommatidia.

In female the *ovary* is very similar in structure and extent to the testis of the male. The *oviducts* are large, muscular tubes, each of which starts medial and posterior to the block of muscular supply of the 6th appendages and passes posterioventrally to its opening on the genital operculum.

The male genital openings on the operculum are round and each is located on a well-defined papilla. In female each opening is a transverse slit and there is no definite papilla.

The breeding season is Spring and early Summer. For mating and egg laying the males and females migrate into shallow water. Spawning occurs at night, when male climbs onto the abdominal carapace of the female. Meanwhile, the female digs out a series of holes in the sand and deposits 200-300 large eggs in each hole. The eggs in each hole are fertilized by the male during their deposition. The eggs are centrolecithal and are covered by a thick chorion. Cleavage is holoblastic. A *trilobite larva* emerges from each egg. It actively swims about and burrows in sand. The little caudal spine does not project beyond the abdomen. Only 2 pairs of book-gills are present instead of 5 pairs. With the successive moults remaining book-gills appear. The caudal spine increases in length.

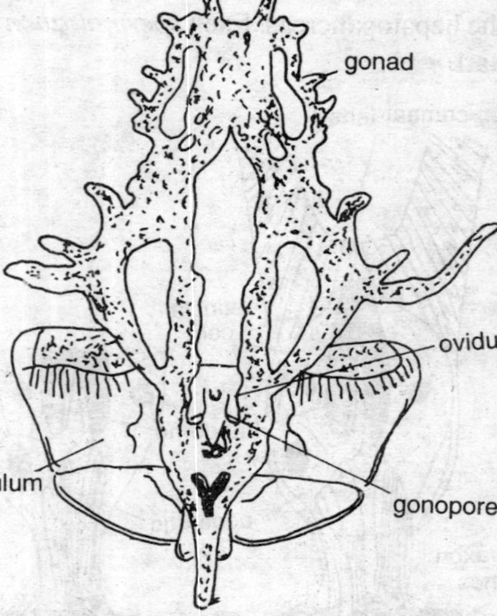

Fig. 18.7. *Limulus.* Female reproduction organ.

Sexual maturity is not reached until the third year. In Japanese species attainment of sexual maturity requires 13 moults in male and 14 moults in female.

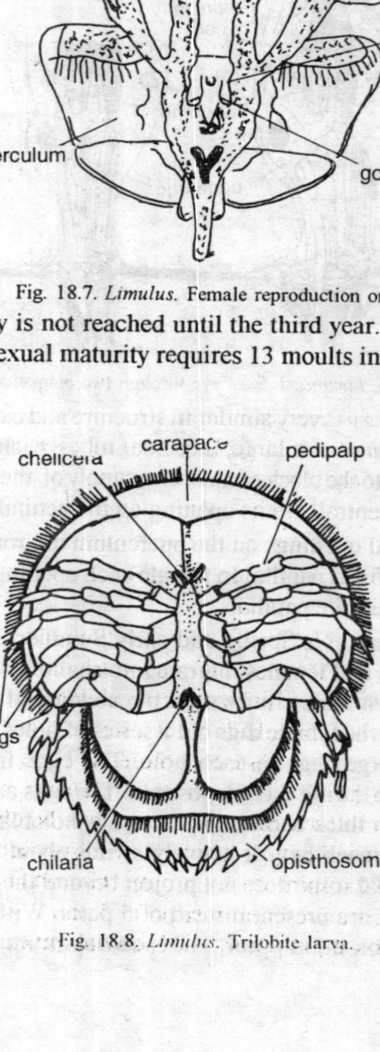

Fig. 18.8. *Limulus.* Trilobite larva.

Affinities of *Limulus*

King crabs or *Limulus* are an ancient group, now nearly extinct. Their large size and isolated taxonomic position have long excited the interest of zoologists. They are usually referred to as *living fossils*.

I. Affinities with Arachnids

Lankester and associates have compared scorpion and *Limulus* in very much details and established a close relationship between the two. The points of resemblance between the two are:

1. The cephalothorax bears 6 pairs of appendages.
2. The appendages are postoral but for the first becomes preoral in position, during development.
3. Chilaria of *Limulus* is respresented by a pentagonal plate. In scorpion it lies infront of operculum.
4. Both bear median and lateral eyes. Antennae are absent.
5. Abdomen of *Limulus* corresponds with the pre- and post-abdomen of scorpion.
6. In both the first pair of abdominal appendages form genital operculum.
7. The second pair of abdominal appendages is modified into pectines in the scorpion and rest of the appendages disappear but in *Limulus* all the appendages are present and bear book-gills.
8. Structure of book-gill comparable to the book-lung of scorpion.
9. Hepatopancreas is present in both with reproductive organs embeded.
10. Coxal glands (one pair) are present in both.
11. There is a similarities between the development of *Limulus* and scorpion

II. Affinities with Trilobita

Lankester put *Limulus* and Trilobita in the same phylum. The larva of *Limulus* shows superficial resemblance to a fossil Trilobite.

1. Body is trilobed divided by a pair of longitudinal grooves into a median and two lateral lobes.
2. Abdomen distinctly segmented.
3. Presence of lateral eyes.

It, howerver, differs from Trilobita in the absence of antennae and the appendages show marked differences. In *Limulus* the appendages show marked differences in Trilobita the appendages are similar and biramous.

III. Affinities with Eurypterida

Eurypterids are extinct relatives of *Limulus*. The chief resemblance between two are:

1. Both are aquatic.
2. Respiration by gills.
3. Carapace dorsally bears paired median, simple and lateral compound eyes.
4. Position of appendages similar.

VI. Affinities with Crustacea

Untill recently, the Xiphosura have been placed with Crustacea because both the groups are aquatic, respire by gills and possess a thick exoskeleton. Compound eyes are present on head. Biramous abdominal appendages are present.

However, they differ from each other in several characters. In Xiphosura, the prosoma is made up of 6 or 7 segments while it is made up of 5 segments in Crustacea. Antennae are absent and nauplius larva is absent in developmental stage in Xiphosura but present in Crustacea.

Conclusion

It seems that marine *Limulus* ancestors gave rise to the terrestrial air breathing animals like scorpion. During the change of aqatic to a terrestrial life changes took place in respiratory system. So that the book-gills modified into book-lungs. The structure of book-gill and book-lung is however, similar though they are differently situated. The book-gills are external in position and lie on the external abdominal appendages in *Limulus* whereas the book-lungs are situated inside the pulmonary sac or pits. These pulmonary sacs are necessary for terrestrial mode of life. Probably the tracheae of other arachnids arose as modification from the pulmonary sac. The evidence that the tracheae developed from pulmonary sac come from the fact that in certain terrestrial isopods the tracheae develop from endopodites. Therefore, it seems proper that *Limulus* like ancestors which were marine, gave rise to air breathing arachnids like scorpions and later gave rise to other tracheal air breathing arachnids probably by modification of pulmonary sac into trachea.

19

Scorpions

The scorpions are the oldest known *terrestrial arthropods* and they have been the first members of this phylum to have conquered land. The scorpions are most common in tropical and sub-tropical areas except New Zealand and oceanic islands. They are most abundant in the Gulf states and Southwest. About 800 species have been described some common genera are *Palamnaeus, Buthus* etc.

Systematic Position

Phylum	-	Arthropoda
Class	-	Arachnida
Order	-	Scorpionida
Type	-	Scorpion

Habits and Habitat

Scorpions are generally secretive and nocturnal, hiding by day under stones and wood and in burrows in the grounds. But there are some species associated with vegetation. They are often found near dwellings. Scorpions are popurlarly believed to inhibit desert regions, but although many desert species exist, they are by no means restricted to arid regions. Many scorpion species require a humid environment and live in tropical rain forests and similar jungle habitats. They are carnivorous and feed on small arthropods, especially spiders and nocturnal insects, cannibalism is also practisised. These are solitary animals and never live in groups. All the scorpions are *viviparous* and the females carry the young ones on their backs for some times, for about a week.

EXTERNAL FEATURES

Shape and Size

The body is long, narrow and dorso-ventrally flattened. The size varies from 3 to 8 cm. in length. The smallest species is the Middle Eastern *Microbuthus pusillus* which is about 13mm. long. The largest one is African *Pandinus imperator* which measures about 18cm. in length. The largest Indian species, *Palamnaeus swammerdami* grows to a length of about 15 cm. while *Buthus tamulus*, a common species, reaches a length of nearly 8 cm.

Colouration

The colour is variable, usually corresponding with the habitat of the animal which is generally blackish dorsally and slightly light coloured ventrally. The species, which are living in tropical jungles, are of shining

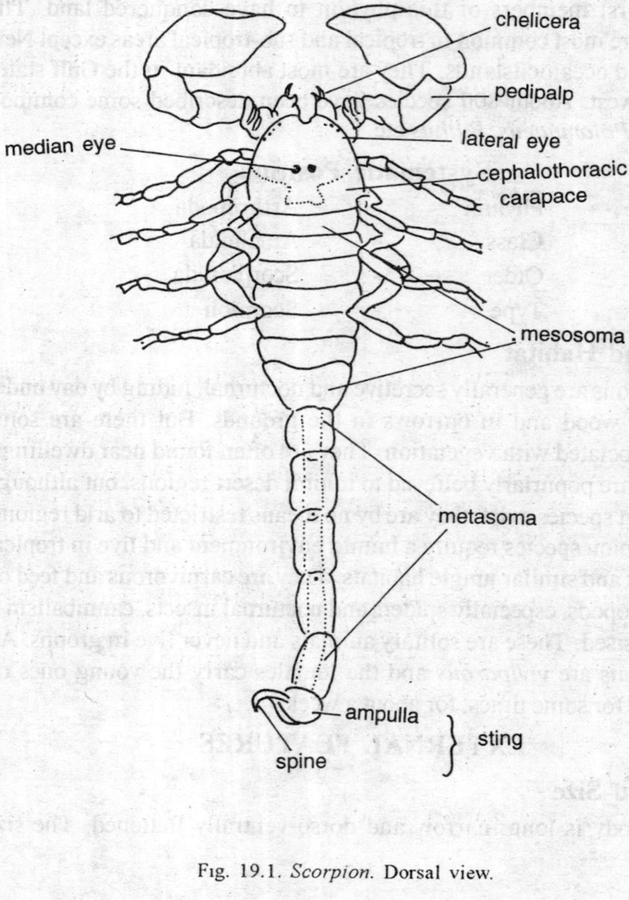

Fig. 19.1. *Scorpion.* Dorsal view.

black colour, and the species which are found in the sand are pale-yellow in colour usually the dorsal surface is much darker in colour than those of the ventral surface.

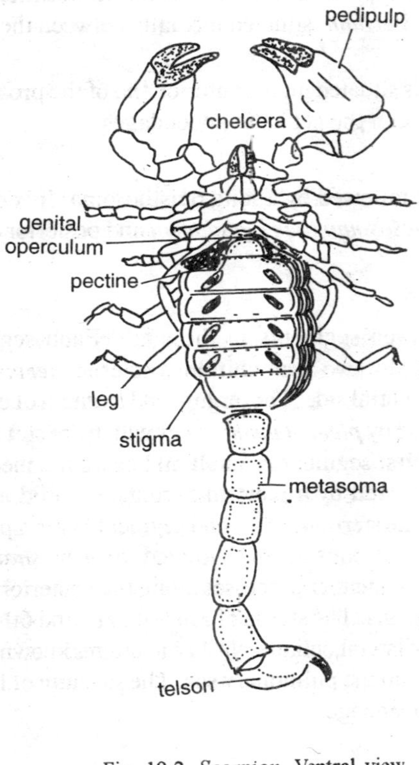

Fig. 19.2. *Scorpion.* Ventral view.

Division of body

The scorpion body consists of a *prosoma* or *cephalothorax* covered by a single carapace and a long *abdomen* or *opisthosoma.*

(i) Prosoma

It is the broad flat, anterior region composed of the *head* and *thorax* which are fused together. It is formed by the fusion of a preoral and six post-oral segments but only six are distinguished in the adult. The segments are marked by the presence of six pairs of appendages on the ventral side. Dorsally, the prosoma is covered by a large, shield like exoskeletal plate, somewhat squarish in form. It is known as *carapace*. Its anterior margin is divided into *right* and *left frontal lobes* by a median notch. Two to five pairs of *lateral eyes* and one pair of *median eyes* can be distinguished along the anterio-lateral margins and in the middle of the carapace. These are simple eyes. On the ventral surface there is a narrow triangular plate, the *sternum*, situated medially between the coxae of 3rd and 4th pairs of legs.

A small *mouth* is situated on the anterior end of the prosoma towards the ventral side between the cephalic appendages.

(ii) Opisthosoma

The prosoma is followed by a long opisthosoma. It is divisible into two parts: anterior *mesosoma* or *pre-abdomen* and posterior *metasoma* or *post-abdomen*.

(a) Mesosoma

It consists of seven segments in the adult. Each segment of the mesosoma is covered by two firm chitinous sclerites *tergum* on dorsal side and *sternum* on ventral side. The tergum and sternum of each segment are laterally connected by *pleural membranes* which are soft and flexible. The sternum of the first segment is small and bears the median *genital aperture* which is covered by a plate-like, rounded, bifid and movable *genital operculum*. The sternum of second segment bears a pair of comb-like *pectines*. A pectine consists of 3 jointed *stem* or *shaft* or *handle* bearing a row of 4 to 36 slender processes along the posterior border. The pectines are tectile organs. The sterna of 3rd, 4th, 5th and 6th mesosomal segments bear a pair of lateral,oblique, slit-like apertures known as *stigmata*. Each stigmata leads into the pulmonary sac. The sternum of last segment does not bear any appendage.

(b) Metasoma

The slender *metasoma* or *post-abdomen* is often wrongly called a *tail*. It is held upraised, arching over the back in a moving scorpion. It consists of five cylindrical segments, each enclosed within a complete chitinous ring. The last segment bears the *anus* on its post-ventral side and a stinging apparatus or *telson*. The telson consists of a swollen *base*, the *vesicle* or *ampulla*, and a curved and pointed spine, the *aculeus*. It remains attached to the last segment by means of a flexible membrane. Inside the vesicle lies a pair of *poison glands*, the ducts of which open by a pair of minute apertures at the tip of the spine.

The venom of most scorpions, although sufficiently toxic to kill many invertebrates, is not harmful to man. At most, the sting in equivalent to that of a hornet. However, certain species exist that possess a highly toxic venom that can be fatal to man. *Androctonus australis* of the Sahara Desert has venom equivalent in toxicity to cobra venom, and this venom can kill a dog in seven minutes. The neurotoxic venom of scorpions is very painful and may cause paralysis of respiratory muscles or cardiac failure in fatal cases.

Appendages

The cephalothorax bears 6 pairs of appendages, these are a pair of *chelicerae*, a pair of *pedipalpi* and four pairs of *walking legs*. Of these the chelicerae are preoral and remaining appendages are postoral.

(i) Chelicerae

The chelicerae are short but stout appendages. They lie close together at the anterior end of the cephalothorax. Each consists of three *podomeres*. The basal podomere is ring-like and concealed beneath the carapace. The middle and distal podomeres together form a grasping apparatus called the *chela* or *pincer* for holding the prey during feeding. The middle podomere is large, bears setac and is produced on the inner side into a conical toothed process that forms the *immovable finger* of the chela. The distal or interal segment is small, curved and toothed. It forms the *movable finger* of the chela.

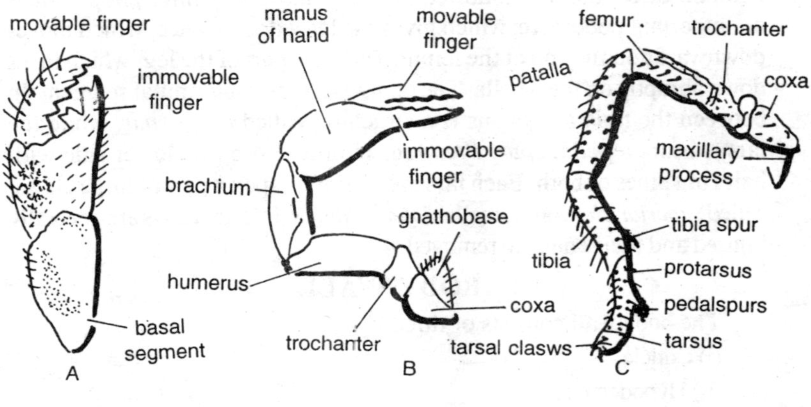

Fig. 19.3. *Scorpion*. Appendages. A- Chelicera, B-Pedipalp, C-First Walking leg.

(ii) Pedipalps

Behind the chelicerae there are a pair of powerful and clawed appendages called *pedipalps*. Each pedipalp is made up of six segments, viz., *coxa*, *trochanter*, *humerus*, *brachium*, *manus* and *movable finger*. The coxa is small and situated on the side of pre-oral cavity beneath the carapace. On each coxa there is a gnathobase towards the mouth. The gnathobases of both sides serve as jaws, protruding in the preoral cavity and help in squeezing and crushing the body of the prey.

The *trochanter* follows coxa is irregularly shaped having spinous anterior margin. At the right angle to longitudinal axis of trochanter, there is a stout and powerful *humerus*. The *brachium* lies above the humerous and is directed forward and is provided with granular crests. The *manus* is longest segment and it forms a powerful chela with the movable finger which is movably articulated and help in seizing the prey.

(iii) Walking legs

There are four pairs of walking legs, attached to the cephalothorax. All the legs are alike and are used for walking. Each leg consists of seven podomeres, *coxa, trochanter, femur, patella, tibia, prototarsus* and *tarsus.*

The coxae of lst and 2nd pairs are provided with forwardly- directed triangular maxillary processes or gnathobases or endites. Coxae are very small, movable, except those of the fourth legs, which are longer. The coxae of the second pair meet with each other in the middle line while the coxae of the 3rd and 4th pairs are immovable and joined together but they do not meet in the middle line of the body and are separated by sternum of the cephalothorax. The *trochanter* is a short and stout segment, moving in all the directions. It is followed by a long, movable femur. The *patella* is an important podomere, which gives the leg a double knee, which moves downwards on the end of the femur. *Tibia* is a part of the leg, which turns down abruptly on the patella. The tarsus is a long subterminal segment. In between the tibia and tarsus is a structure, called *tibial spur*, while the tarsus bears a *pedal spur*. *Pretarsus* is provided on its lower side with hairs or spines or both. Each tarsus bears at its apex a pair of large claws, called *superior claws* and smaller *inferior claws*. Inferior claws are generally injured and their ends are removed.

BODY WALL

The body wall consists of three layers:

(i) Cuticle,

(ii) Hypodermis,

(iii) Basement membrane.

(i) **Cuticle.** The outermost layer is a brilliantly coloured, non-living chitinous layer which is secreted by the hypodermis.The cuticle is made up of three layers. (a) a superficial cuticle, *telostracum,* (b) the middle layer, *epiostracum* and the lower lamellous layer, *hypostracum*. The cuticle is traversed by canals which open outside on the surface. In the openings are lodged the duct of secretory glands which are located in the hypodermis.

Various kinds of projections are found on the cuticle :

(a) *Dermatidia.* These are small chitinous projections on the chela.

(b) *Coelodermatidia.* They occur in the form of hollow projections on the integument. They form hairs, spines, bristles, spur and claws etc. of tarsus.

(ii) **Hypodermis.** It consists of a single layer of columnar cells filled with pigment granules.

(iii) **Basement membrane.** It is a thin and structureless layer.The hypodermal cells rest on this membrane.

(c) *Dermal glands.* These are known as alveolar glands occur in the chelae, pedipalps and telson.

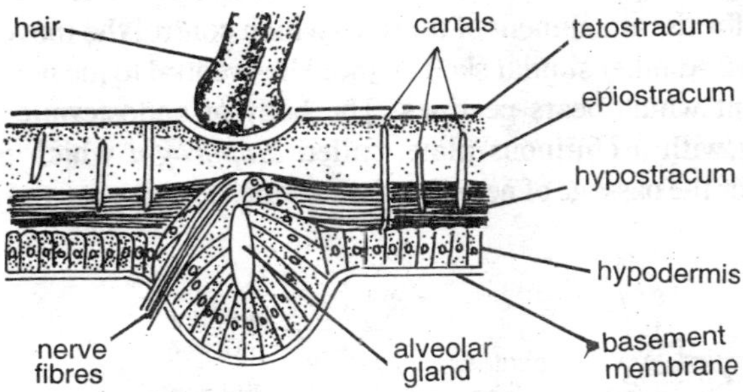

Fig. 19.4. *Scorpion.* V.S. through body-wall.

Glandula Plicata

The males of some species of genus *Bothriurus* are provided, on the dorsal surface of the poison vesicle, with a scutelliform depression. The chitin of this place is very thick and underneath, the hypodermis forms a special organ consisting of numerous longitudinal folds of cuboidal epithelium. These folds are longer in middle and low at margins and are glandular. As they are found only in males they are supposed to play some role in sexual reproduction.

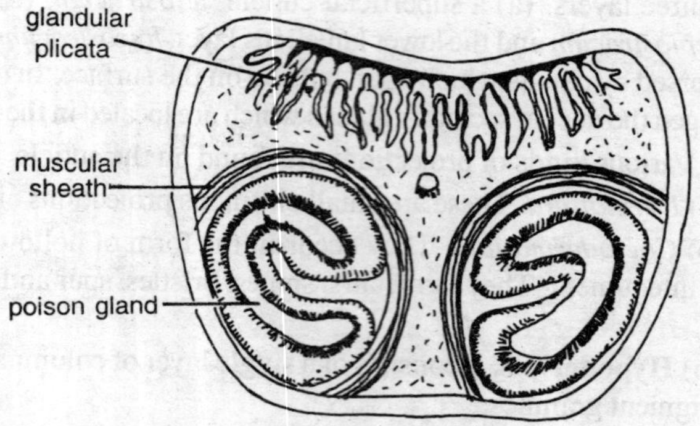

Fig. 19.5. *Scorpion.* T.S. of poison vesicles with two poison glands and glandula plicata.

ENDOSKELETON

The endoskeleton of prosoma and mesosoma is present in the form of cartilaginous plate called *endosternite*. It is placed above the nerve cord and alimentary canal. It possesses apertures for the passage of the intestine and aorta. Its edges are produced into *anterior*, *posterior* and *lateral processes* for the attachment of muscles which control the movements of appendages. Another similar skeletal piece lies ventral to the nerve cord in the segment which bears pectines. The body of endosternite is unites posteriorly with a chitinous plate, called *diaphragm* which possesses apertures for the passage of nerve cord, intestine, dorsal aorta and muscles.

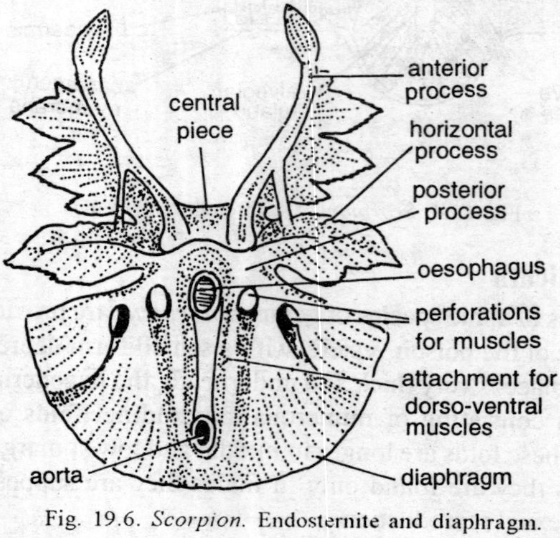

Fig. 19.6. *Scorpion.* Endosternite and diaphragm.

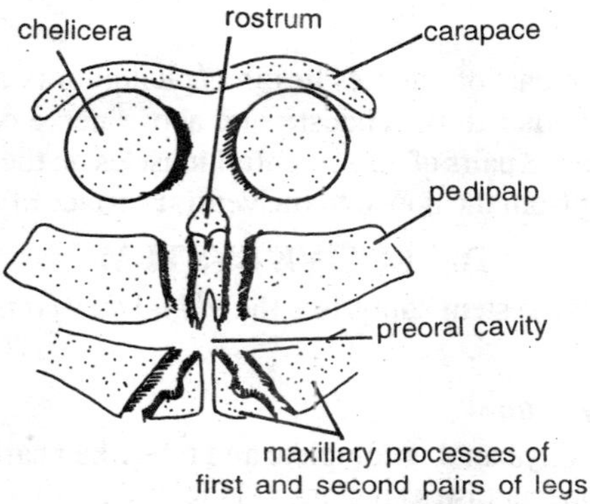

Fig. 19.7. *Scorpion.* T.S. through preoral cavity.

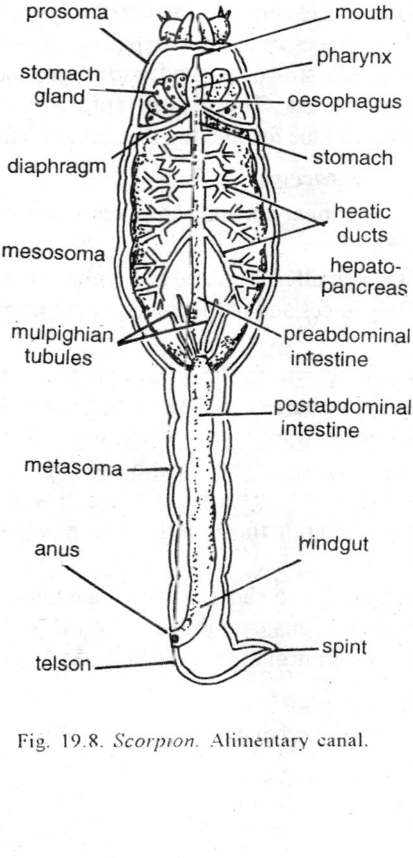

Fig. 19.8. *Scorpion.* Alimentary canal.

Musculature

There is one pair of muscles lies in cephalothorax attached to it by one end and by other to the endosternite above nerve cord. In the pre-abdomen there are 8 pairs of dorso-ventral muscles on the left and right of midgut running from the dorsal to the ventral surface of the body.

DIGESTIVE SYSTEM

The digestive system comprises the *alimentary canal* and *digestive glands*.

I. Alimentary Canal

The alimentary canal is complete and tube like nearly uniform. It is differentiated into four district regions. (1) Pre-oral cavity, (2) Stomodaeum, (3) Mesenteron, (4) Proctodaeum.

1. Pre-oral cavity

It is formed infront of mouth. It is enclosed above by the rostrum, infront by the coxae of chelicerae, on the sides by the coxae of the pedipalps and by the maxillary processes of first two pairs of walking legs. A laterally compressed cushion-like structure called *rostrum* protrudes into the preoral cavity from the posterior side. The rostrum is provides with a set of muscles which can roll it into a tube for sucking the fluid into the pharynx.

2. Stomodaeum or foregut

It includes mouth, pharynx and oesophagus and is lined internally by the cuticle.

(i) *Mouth.* It is a small narrow and transverse aperture just behind the preoral cavity. Only juices and pulps can enter the narrow mouth. Mouth leads into pharynx.

(ii) *Pharynx.* The pharynx is a large, pear-shaped and muscular structure situated obliquely below the mouth. Due to the presence of several bundles of muscles radiating outwards from pharyngeal wall to the wall of the cephalothorax, its walls are highly elastic and capable of great dilation. The muscles serve the pharynx to act as a sucking organ so that the liquid food is sucked in through the mouth. The pharynx is followed by oesophagus.

(iii) *Oesophagus.* It is a short, narrow tube that passes through the nerve ring. The oesophagus extends into the midgut to from the sieve valve which prevents regurgitation of food. It leads into stomach.

3. Mesenteron or midgut

It includes stomach and intestine. These are lined with epithelium.

(i) *Stomach.* The stomach is a short thin walled and dilated tubular structure situated in the cephalothorax. It extends upto the diaphragm. A trilobed, brownish gland, lying in the cephalothorax opens into the stomach. *Huxley* regarded it as *salivary gland* but *Blanchard* and *Pavlovsky* called it *stomach gland.*

(ii) *Intestine.* It is the longest part of the alimentary canal and extends upto the last abdominal segment where it joins the hindgut. The intestine is wide tube-like structure with glandular wall. It is divisible into two parts: *Preabdominal* and *postabdominal.* At the junction of these two is a constriction and from here one or two pairs of *Malpighian tubules* arise. These tubules are excretory in function.

4. Proctodaeum or hindgut

It is the smallest part of the alimentary canal, restricted to the last metasomatic segment. It is internally lined by chitin and opens to the exterior through the *anal aperture* situated ventrally at the base of the stinging apparatus.

II. Digestive glands

There are a pair of stomach glands and a large hepatopancreas.

(i) Stomach-gland

It is a trilobed brownish gland and opens into stomach. It secretes digestive enzymes like pepsin, lipase, erapsin etc.

(ii) Hepatopancreas

It is a large, brownish and lobulated gland occupies the entire preabdominal cavity. It has a mid-dorsal longitudinal groove in which lodges the heart. Intestine and other organs are found embedded in it. The wall of the lobules is composed of a layer of epithelium arranged on a thin membrane or *tunica properia* and a peritonial covering. Two types of cells are found in the epithelium (a) large numerous *absorptive cells* and (b) fewer smaller *ferment cells.* Hepatopancreas sends 5 pairs of ducts open into the stomach.

Food and Feeding

Scorpions are carnivorous and predaceous animals. They also practise cannibalism. They feed upon small animals like insects and spiders. The prey is seized and held by powerful chelate pedipalps or being paralysed by the sting if the prey is large. The pedipalps pass the killed prey to chelicerae which tear it to pieces. The food is then transferred to the preoral cavity which helps in squeezing the prey and oozing liquids are sucked in by the muscular pharynx. The process of feeding is very slow as the scorpion takes at least two hours to devour a cockroach. Scorpion can live without food for months together (about 6 months).

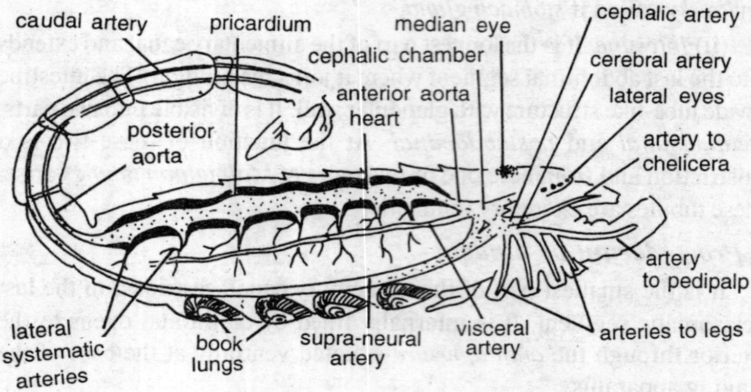

Fig. 19.9. *Scorpion.* Circulatary system.

Digestion

Inside the preoral cavity the food is reduced to a pulp. Its proteins are partially digested by the secretion of certain *alveolar glands* which are present in the maxillary processes of first two pairs of legs. The food is reached to the stomach in partly digested state. Here the food is mixed with the secretion of stomach gland. The secretion contains pepsin, erepsin and lipase enzymes. Most of the digestion completed in stomach. In stomach, hepatopancreas pours its secretion that contain anylase, lipase and proteinases enzymes. The digested food is absorbed into the absorptive cells of the hepatopancreas for storage. The food is sent to the intestine where water and remaining digested food material is absorbed. The undigested food is sent into the hindgut from where it goes out through the anus.

CIRCULATORY SYSTEM

In scorpions, the circulatory-system is open type. It consists of : Heart, Arteries, Sinuses, Pulmonary Veins and Blood.

I. Heart

The heart is an elongated and muscular tube of a faint greenish colour. It lies in the pre-abdominal region. just beneath the terga. in a deep median dorsal groove of the hepatopancreas. The heart is divided from outside into seven chambers by shallow transverse constrictions. The heart has seven pairs of dorso-lateral and valvular apertures or *ostia*, each chamber possessing one pair.

The heart lies suspended by means of ligaments in a thin-walled, membranous sheath, the *pericardium*. Ligaments divide the pericardial cavity into four compartments : one dorsal, one ventral and two lateral.

II. Arteries

The heart is continued anteriorly as *anterior aorta* and posteriorly as *posterior aorta*. A series of paired lateral *systemic arteries* are also given off from the heart.

1. *Anterior aorta*. From anterior aorta a pair of *visceral arteries* are given off to the hepatopancreas and the intestine. After piercing the diaphragm it expands to form a small chamber behind the brain. The arteries to all the cephalic and thoracic appendages are given off from this chamber. It also gives off a pair of *ventro-lateral arteries* which curve round the oesophagus and unite to form the *supra-renal artery*. This artery runs behind above the ventral nerve cord, and supplies posterior region of the body.

2. *Posterior aorta*. It is also known as *caudal artery* and arises from the posterior end of the heart. It lies on the dorsal side of intestine and supplies blood to the intestine, muscles, telson etc. It ends in poison glands.

3. *Systemic arteries*. A systemic artery is given off on either side of each chamber of the heart. They ramify to form a networks and supply the blood to different organs found in the abdomen.

III. Sinuses

The ultimate branches of arteries open into blood spaces or lacunae among the viscera. The blood from the *lacunae* is collected into five larger *sinuses*. Out of these, a *pericardial sinus* surrounds the heart, *dorsal sinus* above pericardial sinus, two *lateral sinuses* are present on lateral sides and a large *ventral sinus* is situated on the ventral side.

IV. Pulmonary veins

The oxygenated blood from each book-lung is carried by a pulmonary vein to the pericardial sinus.

V. Blood

The blood of scorpions is slightly bluish due to the respiratory pigment *haemocyanin* dissolved in plasma. It contains many nucleated corpuscles. Blood transports food, oxygen, carbon dioxide, hormones and nitrogenous wastes.

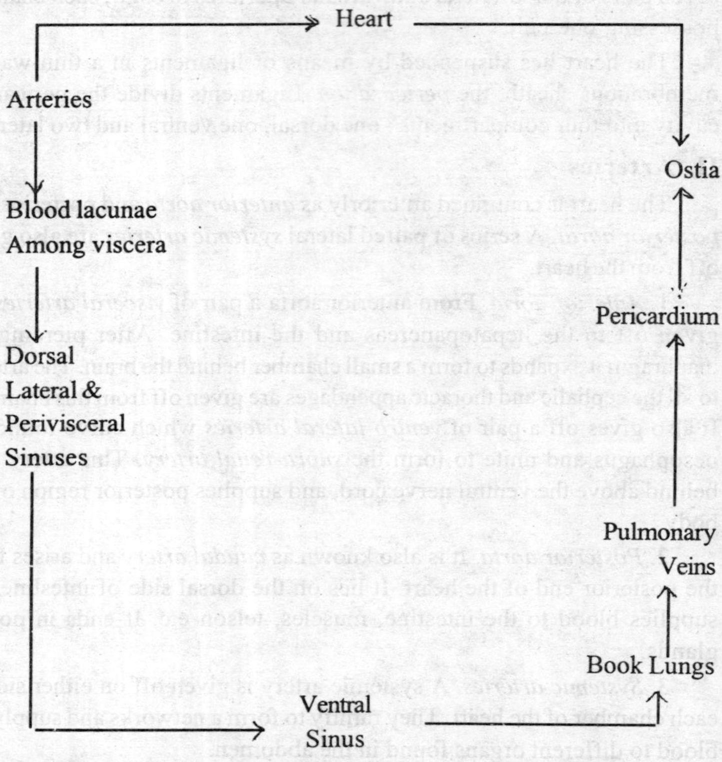

Chart Showing course of blood circulation in Scorpion

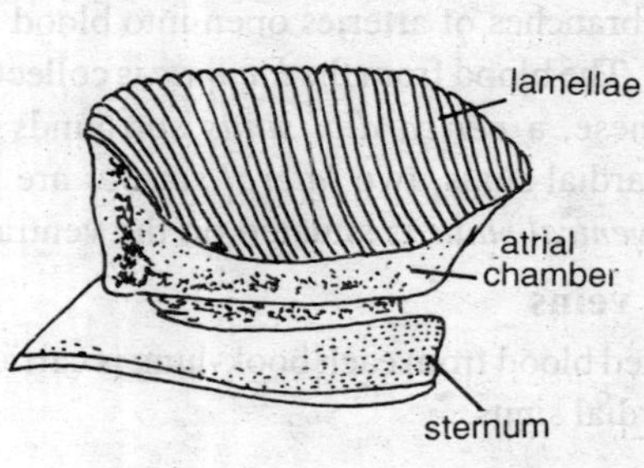

Fig. 19.10. *Scorpion.* Book-lung in dorsal view.

Course of circulation

With the contraction of the ligaments, the cavity of heart enlarges and the blood of pericardial sinus passes on to the heart through ostia. Now the heart contracts, the ostia are closed and the blood is pumped to various organs of the body through the anterior and posterior aortae and the systematic arteries. The impure blood is collected in ventral sinus from

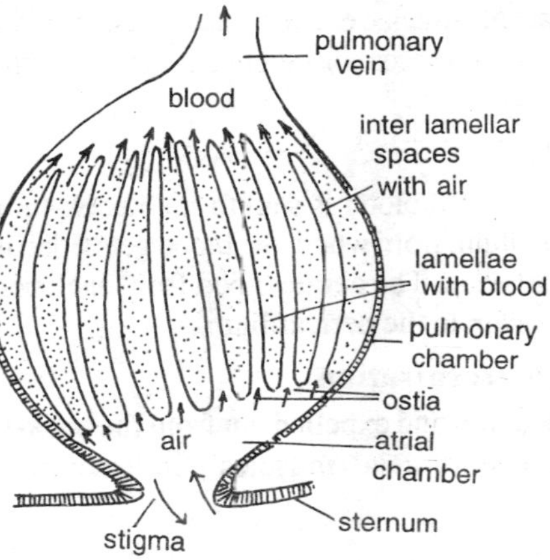

Fig. 19.11. *Scorpion.* Book-lung in ventral view.

where the blood is sent to the book-lungs and is oxygenated. From book-lungs pure blood is carried by pulmonary veins to pericardial sinus from where blood enters the heart through ostia.

RESPIRATORY SYSTEM

The respiratory organs are *book lungs* in the scorpion. The respiratory system consists of 4 pairs of book-lungs situated on the ventro-lateral sides of 3rd, 4th, 5th and 6th abdominal segments. Each book-lung opens to the outside by a narrow slit- like aperture called *stigmata.*

Each book-lung is distinguished into two parts

(i) Atrial chamber and

(ii) Pulmonary chamber.

(i) Atrial chamber

It is the ventral or proximal part of the book-lung. It is small and dorso-ventrally compressed and contains air. Its proximal and opens to outside by stigmata. There is a row of small pores called *ostia* in the roof of atrial chamber communicates with the interlamellar spaces of the pulmonary chamber through these ostia.

(ii) Pulmonary chamber

It is the dorsal spacious oval chamber filled with about 150 vertical folds or *lamellae*, attached to the posterior side of the chamber. These are arranged parallel to each other and one over the other like the leaves of a book. Each lamella is a hollow structure formed of two thin layers of cuticle unites at their edges and thus enclosing *interlamellar spaces* filled with blood. The adjacent lamellae enclose interlamellar air spaces filled with air. The interlamellar air spaces communicate with the atrial chamber through ostia.

Blood supply

The deoxygenated blood is sent to each book-lung from the ventral sinus by diverticulum, from where the blood gets filled in the intralamellar spaces of the lamellae. The aerate blood from each book-lung is drained by a *pulmonary vein* to the pericardium.

Mechanism of respiration

Air is drawn into and expelled out from the book-lungs by the action of dorso-ventral and atrial muscles. Contraction of these muscles

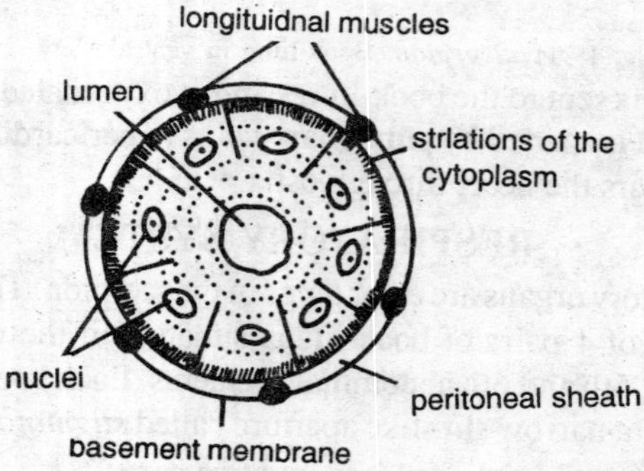

Fig. 19.12. *Scorpion.* T.S. of Malpighian tubule.

compresses the book-lungs this expells out the foul air of the interlamellar spaces into the atrial chambers and then to the outside through the stigmata when these muscles relax, the pulmonary sac regains their normal shape as a result of which the fresh air rushes the atrial chambers through the stigmata and then fills the interlamellar spaces.

The lamellae contain deoxygenated blood and are surrounded by fresh air. Thin walls of the lamellae permit the exchange of gases by diffusion.

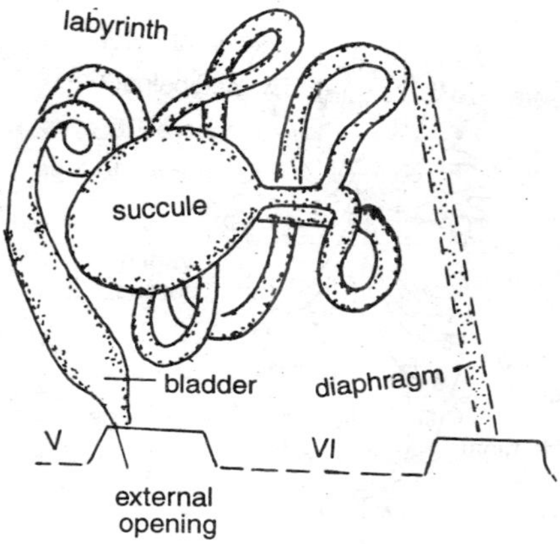

Fig. 19.13. *Scorpion.* Coxal gland.

The blood takes up oxygen from air and gives up carbon dioxide to the air. The oxygenated blood and befouled air are then replaced and process is repeated.

EXCRETORY SYSTEM

The excretory system of scorpion consists of :

(i) Malpighian tubules

These are one or two pairs attached to the junction of mesenteron and hindgut. They float in blood from which they eliminate nitrogenous waste material and discharge it into the alimentary canal for elimination with the faeces.

(ii) Coxal glands

A pair of coxal glands is situated, one on either side in prosoma, close in front of endosternite. Each gland consists of three parts: a large *saccule* or *end sac*, a long coiled duct or *labyrinth* and a small *terminal bladder* or *reservoir.* The bladder opens outside by means of a minute pore situated on the posterior surface of the coxa of the fifth walking leg.

The coxal glands of scorpion are considered homologous with the antennary glands of prawn. The saccule and tubule of each coxal gland collect excretory nitrogenous waste from blood and pass it out side through the excretory pore. Urate crystals are noticed in the saccule.

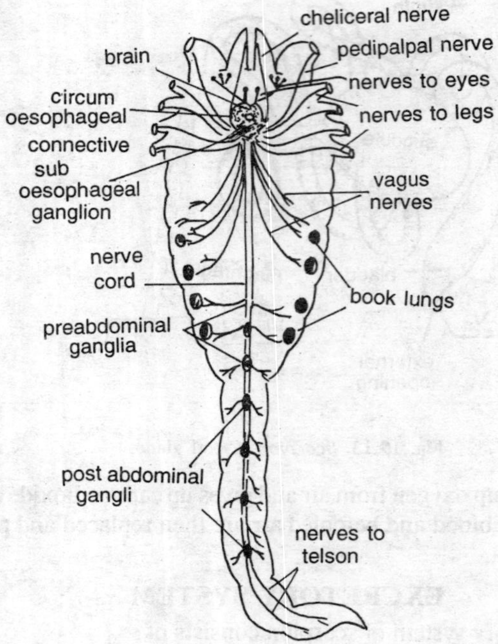

Fig. 19.14. *Scorpion.* Nervous system.

(iii) Hepatopancreas

According to *Pavlovsky* the hepatopancreas of scorpion also serves as an excretory organ. According to him if ammonium carmine is injected in the body cavity of scorpion is collected in the form of small, bright, red granules by the cells of hepatopancreas.

(iv) Nephrocytes

Large nephrocytes and lymph tissue organs, present beneath the body wall in mesosoma, are both excretory as well as phagocytic in nature.

NERVOUS SYSTEM

The nervous system consists of :

1. Brain or supra-oesophageal ganglion
2. Circum-oesophageal connectives
3. Sub-oesophageal ganglion
4. Ventral nerve cord

The brain is small bilobed mass situated in the prosoma just beneath the median eyes. From the brain arises a pair of thick, short and stout circum-oesophageal connectives which encircle the oesophagus and unite ventrally in a *sub-oesophageal ganglion*. The sub-oesophageal ganglion is continued backwards into the abdomen as double *ventral nerve cord*. It extends upto the 4th segment of the postabdomen. The nerve cord bears three segmental ganglia in the preabdomen and four in the postabdominal segments. The nerve cord is rounded and slender in the preabdomen but flattened and ribbon shaped in the postabdomen.

Brain gives off a pair of *optic nerves* to the marginal eyes. Besides, the brain gives off numerous median delicate nerves to the rostrum, pharnyx and oesophagus. From the cerebral ganglion and circum-oesophageal connectives arise six pairs of *lateral nerves*. These nerves innervate the six pairs of cephalothoracic appendages. The sub-oesophageal ganglion sends two to four pairs of *vagus nerves*. These run posteriorly into the preabdomen and supply the genital operculum, pectines and the first two pairs of book-lungs.

The segmental ganglia of the preabdomen sent out nerves to the last two pairs of book-lungs, heart and muscles of the abdomen. The segmental

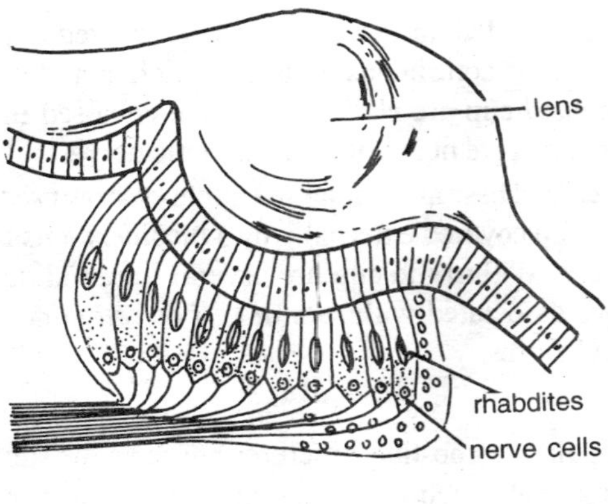

Fig. 19.15. *Scorpion*. V.S. of median eye.

ganglia of the postabdomen supply nerves to the segmental muscles. The last postabdominal ganglion also supplies the sting.

SENSE ORGANS

The sense organs of scorpion include the eyes and pectines.

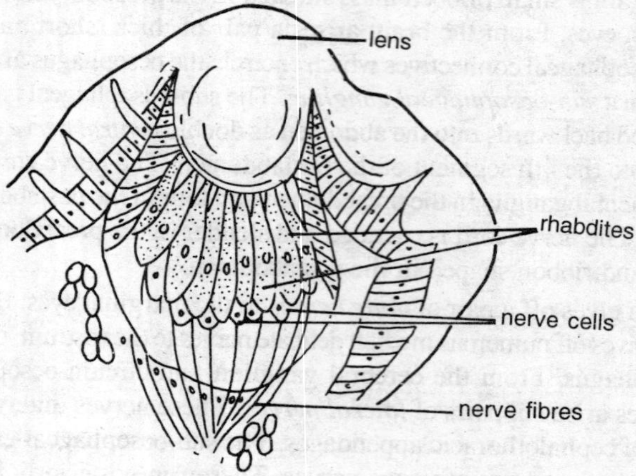

Fig. 19.16. *Scorpion.* V.S. of lateral eye.

1. Eyes

There are four pairs of eyes in scorpion - one pair of median eyes and three pairs of lateral eyes.

(i) *Median eyes.* Median eye is like a cup covered externally by a cuticular lens which is continuous with the cuticle but is much thicker. Inside the pigmented cup are *rhabdomes*, each enclosed inside several retinal cells which receive nerve fibres of an optic nerve.

(ii) *Lateral eyes.* These are similar to simple eyes of insects. These are like a pigmented cup covered externally by a biconvex lens formed from transparent cuticle. Within the cup are several longitudinal optic rods called *rhabdomes* associated with *retinal cell* or *retinulae*. The retinal cells receive nerve fibres.

2. Pectines

These are a pair of comb-like structures borne on the sternum of 2nd preabdominal segment. Their processes have many sensory cells. They are tactile organs. They are also considered to be sensitive to the ground vibrations and changes in humidity.

It is also regarded that scorpions have some preception of sound. In some species of scorpions *stridulating organs* are found on the coxae of pedipalps in the form of ridges, across which file-like surface can be drawn to produce some sort of sound.

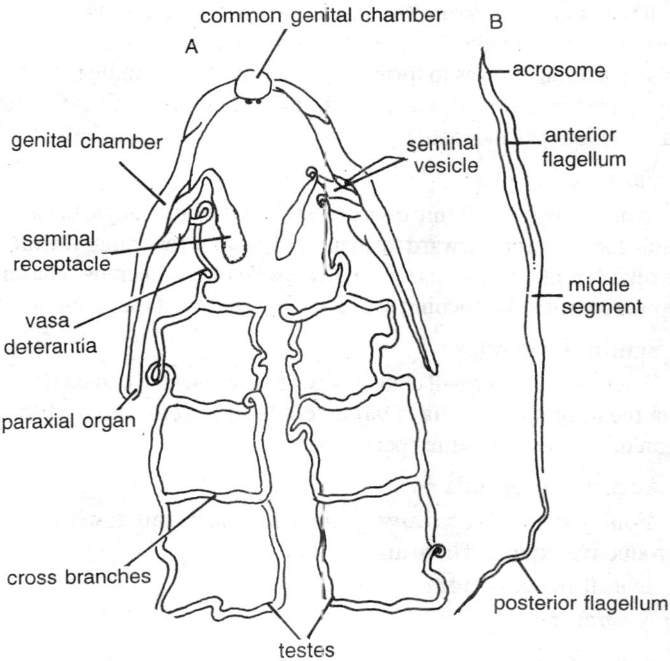

Fig. 19.17. *Scorpion.* A- Male reproductive organs, B-A sperm.

REPRODUCTIVE SYSTEM

The sexes are separate. Externally male and female are alike but they can be distinguished on the basis of following characters :

(i) The female is generally larger than the male.

(ii) The abdomen of female is broader than that of males.

(iii) The pectines of female are smaller than of male.

(iv) The pedipalps of female are smaller than those of male.

Male reproductive system

The male reproductive system comprises. (1) Testes, (2) Vasa deferentia, (3) Seminal vesicles, (4) Accessory glands, (5) Genital chamber, (6) Paraxial organs.

1. Testes

A pair of testes are long mesh-like structures which lie embedded in the hepatopancreas from 3rd to 6th segments of mesosoma. A testis consists of two slender longitudinal tubules connected together by four transverse tubules so as to form three squares. Each tubule is lines with a layer of germinal epithelium. The cells of epithelium divide to form sperms which are filiform and motile.

2. Vasa deferentia

Anteriorly, a small thin duct arises from the outer angle of each testis. It runs forward and outward to atrium. Prior to its opening, it widens and this dilated part is known as *terminal ampulla* of vas deferens. The terminal ampulla receives the opening of accessory glands and the seminal vesicles.

3. Seminal vesicles

Each vesicle is a small club-shaped structure which extends backwards from the terminal ampulla of vas deferens and opens into it. It acts as an organ of storage of mature spermatozoa.

4. Accessory glands

Four pairs of accessory glands are associated with the male reproductive organs. These are:

(a) Dorsal annex glands
(b) Ventral annex glands
(c) Oval glands
(d) Cylindrical glands

The fluid secreted by all these glands is collected in the terminal ampulla and helps is reproduction.

5. Genital chambers

The genital chambers are confined to the first and second mesosomatic segments. They extend obliquely towards the middle-line and open into the *common genital chamber* which opens out by means of male genital aperture. The aperture is situated on the ventral surface of the first mesosomatic segment and covered over by the *genital operculum*.

6. Paraxial organs

Each genital chamber is produced behind into a *paraxial organ* which contains a tightly-fitting chitinous rod, the *flagellum*. The structure of the flagellum varies in different scorpions. It is provided with many spines and a longitudinal groove on the inner side. The two flagella together form the so called double *penis* of the scorpions and they are said to evert out

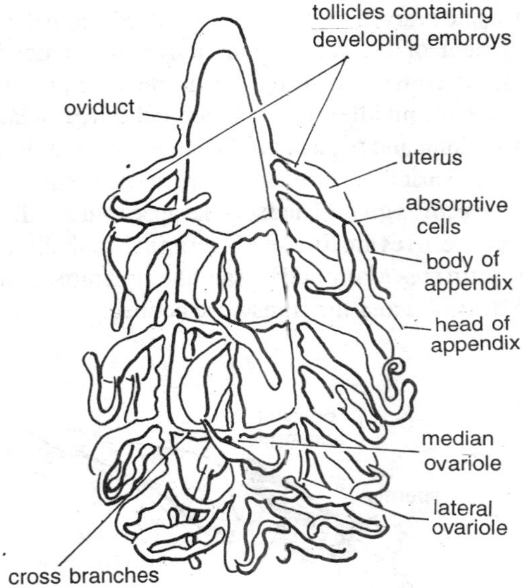

Fig. 19.18. *Scorpion.* Female reproductive organs.

through the male genital aperture to serve as claspers during copulation.

Female Reprocutive System

The female reproductive organs comprise:

1. Ovary,
2. Oviducts,
3. Common genital chamber.

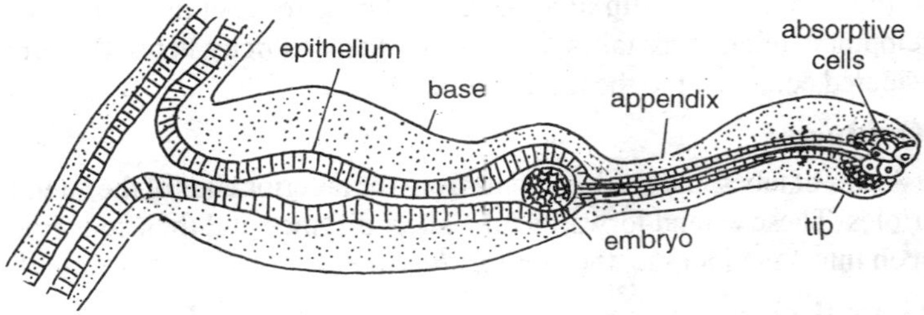

Fig. 19.19. *Scorpion.* Diverticulum of ovarian tubule.

1. Ovary

There is a single ovary in scorpion which extends from the 3rd segment to the 6th segment of the mesosoma. It remains embedded in the hepatopancreas. It consists of three long and narrow tubules called *ovarioles*. Of them the middle one is smaller and called *median ovariole*. The other two are long and present on the lateral sides and called *lateral ovarioles*. All the ovarioles are connected with each other at four places by *transverse connectives* giving a ladder-like appearance. All the ovarioles and transverse connectives are lined with germinal epithelium, the cells of which divide to form *ova*. Many hollow sac-like structures, called *follicles*, arise from the ovarioles and the transverse connectives. Each follicle is

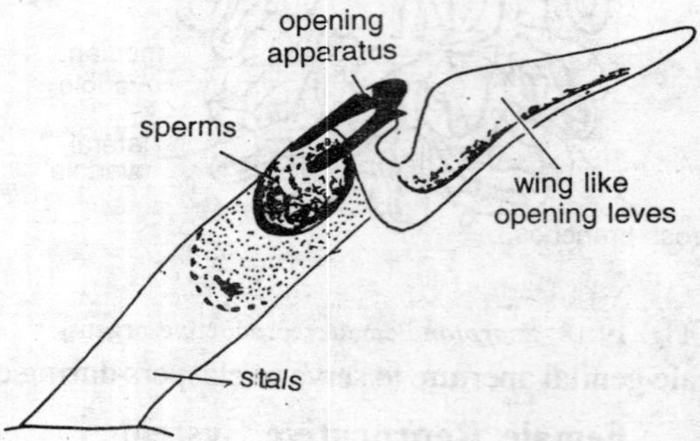

Fig. 19.20. *Scorpion.* A spermatophore.

made up of two parts a proximal swollen part, the *base* and a narrow distal part, the *appendix*. The tip of appendix enlarges to form the *head*. The development of embryos takes place within these follicles hence they are considered equivalent to the uteri of other animals.

2. Oviducts

A pair of tubes, the *oviducts*, start from the anterior ends of the lateral ovarioles. These extend forwards for a short distance and then converge to open into a median sac, the *genital chamber*.

3. Genital chamber

The genital chamber is situated in the first mesomatic segment. It opens out side by the *female genital aperture* situated on the ventral side of first mesosomatic segment being covered by the bifid *genital operculum*.

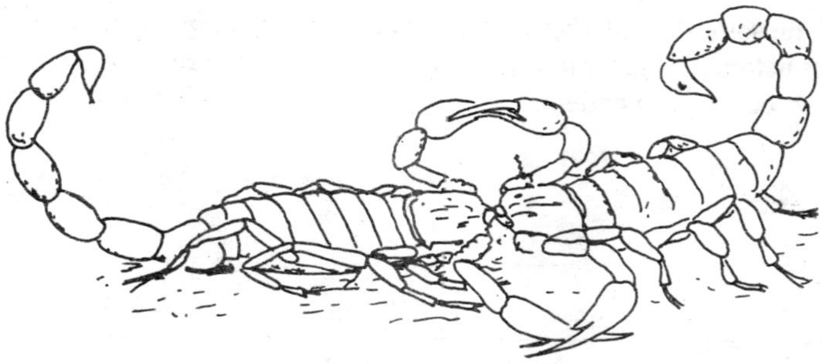

Fig. 19.21. *Scorpion*. Copulation.

Spermatophore

The sperms are transferred to the female by means of a *spermatophore* that is attached to the ground by a stalk. The spermatophore bears a wing-like lever at its free end. Connected to the lever is a pocket carrying sperms and also an opening aparatus. As the lever is pressed the opening apparatus is released and the sperms are triggered into the female genital aperture.

Copulation and fertilization

Fabre has described the mating habits of scorpion. Male and female face each other, raise their postabdomens high into the air, and move about in circles. The male then holds the pedipalps of the female with his own pedipalps and both interwine their postabdomens above them. They now walk together forward and backward and in rings. While moving in a ring, the male retreats, dragging the female after it. Courtship may last for hours or even days. Finally, the male digs a burrow and both enter there and mate. The male is often eaten up by the female after mating. Eventually the male deposits a spermatophore, that is attached to the ground. The male manoeuvres the female so that the genital area is over the spermatophore. The sperms enter into the female orifice and fertilization takes place. The Indian scorpions breed during mansoons.

Development

Development takes several months or even a year or more. The scorpions are *viviparous* and the mother gives birth to two or three dozens of young ones at a time. The egg is oval or spherical which is rich in yolk and is surrounded by a thin membrane lies in a follicle that arises as outgrowth of the walls of the ovariole or transverse connective.

After birth the young ones are carried by the mother scorpion on her back for some days till they can care themselves. Young ones moult several times before they grow into adults. Young ones than leave the mother and then become independent.

Revision Questions

1. Give an account of external structure of scorpion.
2. Describe various appendages of scorpion.
3. Give an account of digestive system and mode of feeding in scorpion.
4. Describe respiratory system of scorpion.
5. Describe the reproductive system of scorpion and its mating behaviour.
6. Describe the male and female reproductive organs of scorpion.
7. Write short notes on (a) Book-lungs, (b) Pectines, (c) Pedipalp, (d) coxal gland (e) Trichobothrium.

20

Peripatus

Peripatus or the "*walking worm*" belongs to the small phylum *Onychophora*. Its position has been controversial in some texts. *Onychophora* has been ranked as class where as in others as sub-phylum. In modern texts it is treated as separate phylum that includes only about 70 living species. It has characteristics of both the *Annelida* and *Arthropoda* and is referred to as the "*connecting link*".

Historical account

Phylum *Onychophora* is represented by a single genus, *Peripatus* which was wrongly placed in phylum *Mollusca* by *Guilding* (1826) because of its slug-like appearance. But soon after the discovery of tracheae in the animal by *Mosely* (1874) it was classified as a class of phylum *Arthropoda*. *Sedgwick* has given a monograph on *Peripatus*.

Geographical Distribution

Peripatus is practically having its wide range of distribution both in tropical and subtropical areas. As shown in the table below, it is one of the best example of *discontinuous distribution*.

No.	*Geographical distribution*	*Nomenclature given by Sedgwick*	*Name*	*No. of species*
1.	Mexico and W. Indies	*Neoperipatus*	*Peripatus*	29
2.	W. Africa and Congo	*Congoperipatus*	*Mesoperipatus*	1
3.	Malaya		*Eoperipatus*	4
4.	S. Africa, Netal to Cape cod	*Capoperipatus*	*Paraperipatus*	7
5.	Malaya Penninsula	*Malayan Peripatus*	*Paraperipatus*	1
6.	Australia, Tusmania, New Zealand	*Austroperipatus*	*Ooperipatus*	1
7.	Chili Island	*Chiloperipatus*	*Opisthopatus*	1

Habits and Habitat

Peripatus is exclusively a terrestrial and walking form and living in a

Fig. 20.1. *Peripatus.*

damp leaf moulds, barks of trees, under the stones, logs and in the crevices of rocks, but not in salt water medium.

They are nocturnal in habit, usually they come out at night for feeding. They are carnivorous and capture prey by shooting out a glue-like substance from openings on either side of the mouth. Small animals such as crickets and spiders quickly become entangled in the sticky glue.

EXTERNAL FEATURES

These individuals possess a reddish colour on the ventral side, while darker hues on the dorsal side. The body is cylindrically elongated,

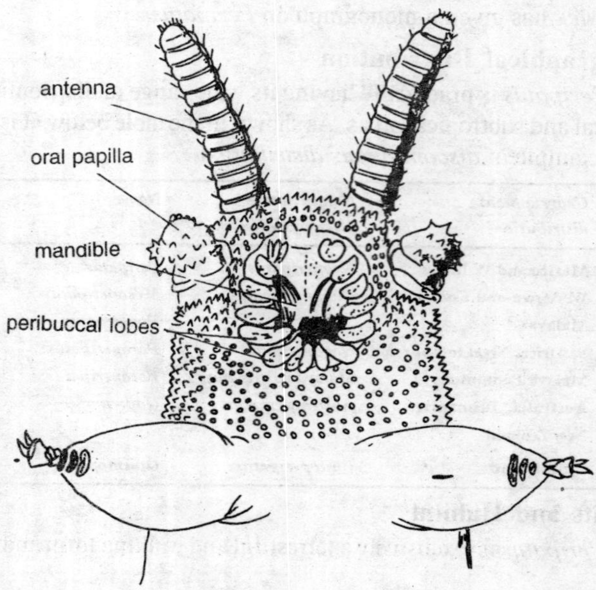

Fig. 20.2. *Peripatus.* Anterior region.

bilaterally symmetrical and small. It measures 1.5- 15 cm. in length, according to species. The skin is velvety, and thrown into transverse ridges bearing number of wart-like papillae furnished or armed into chitinous spines. Externally the body is not much marked segemented but each possess a pair of appendages.

First three segments form head, which is not distinct. It bears a pair of segmented *antennae*, a pair of *eyes* situated dorsally. *Mouth* is mid-ventrally situated bearing a pair of small horny jaws surrounded by a *circular lip* or *circumoral fold* and a pair of blunt *oral papillae* on either sides. Each oral papilla bears an aperture for *slime gland* which produces a sticky glue-like secretion for catching prey.

A series of short, stumpy, conical *legs* arise ventro-laterally from the trunk. All the appendages are alike but for the last 2 or 3 are reduced. Their number varies from 14-43 pairs from one species to another. Each leg is short, unsegmented having tubercle and spiny pad on the ventral side. Each leg is hollow with terminal foot bearing a pair of curved claw.

Genital opening always lies infront of the anus which is situated at the posterior end of the body.

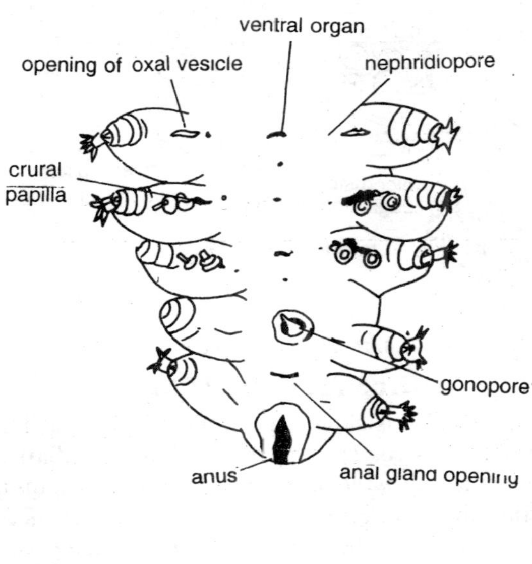

Fig. 20.3. *Peripatus*. Posterior region.

BODY WALL

The body wall is of annelidan type so the skin is formed of three layers, *cuticle*, *epidermis* and *dermis*. The exoskeleton is absent. The cuticle is thin flexible, chitinous and waterproof covering. It bears numerous velvety processes. It is secreted by cuticle. Beneath the cuticle is a single layer of epidermis, resting on a thin basement membrane. The dermis is made up of connective tissue and there is a complex musculature composed of circular, longitudinal and transverse muscle layers. Beneath dermis is epithelial layer which lines the coelom or body cavity and invests the contained organs. By means of incomplete muscular partition the body cavity is divided into a median and two lateral compartments.

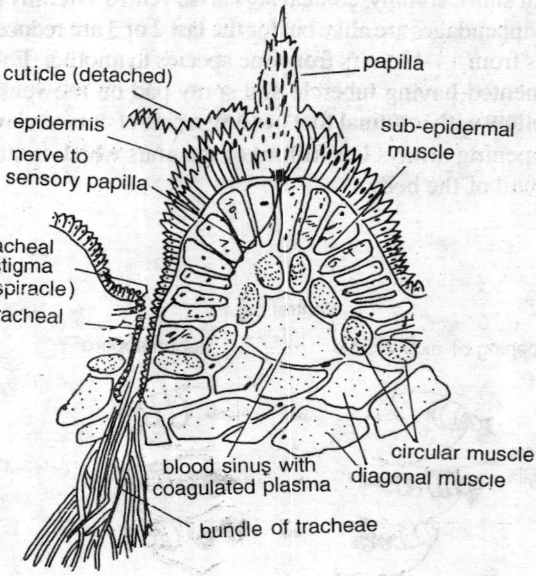

Fig. 20.4. *Peripatus.* Section of skin.

DIGESTIVE SYSTEM

The alimentary canal is simple and straight as in annelids but the buccal cavity is formed secondarily by the union of a papillary ring and folds surrounding the true *mouth* into a circular lip. It contains a *tongue* and the *jaws*. Into the posterior end of buccal cavity there is a median opening of salivary glands. The *salivary glands* are elongated, narrow and tubular, representing modified nephridia. Buccal cavity follows a thick.

muscular *pharynx*. Pharynx is followed by *oesophagus* a wide *midgut* and a short *hind gut* or *rectum*. The rectum opens out through the *anus*.

A pair of elongated and branched *slime glands* are present in the coelom. These open on the tips of oral papillae and secrete a glue-like adhesive substance which hardens on contact with air. This secretion helps in catching the prey. Digestion is inter-cellular and takes place in midgut.

CIRCULATORY SYSTEM

The heart is an elongated muscular, mid-dorsal, contractile tube lying in the *pericardial sinus*. It is open at both the ends and bears several pairs of lateral ostia, blood flows forward into various organs, there are no blood vessels to return the blood. The blood in collected into various sinuses and then passes to the heart.

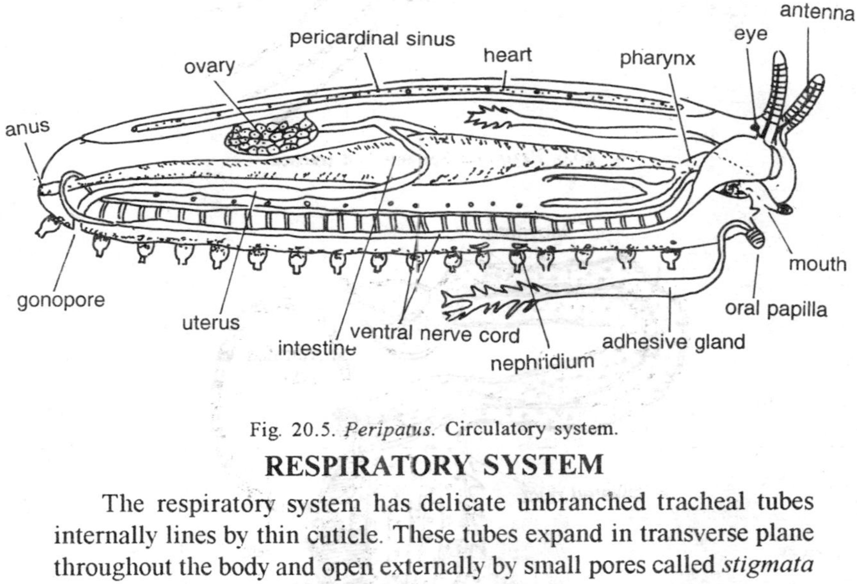

Fig. 20.5. *Peripatus*. Circulatory system.

RESPIRATORY SYSTEM

The respiratory system has delicate unbranched tracheal tubes internally lines by thin cuticle. These tubes expand in transverse plane throughout the body and open externally by small pores called *stigmata* or *spiracles*. The stigmata lie in little depressions of the integument, the *tracheal pits*. The number of spiracles varies from species to species, through these spiracles the air is drawn into the body.

EXCRETORY SYSTEM

The excretory system is annelidan type which is effected by the segmentally arranged *nephridia*. A nephridium consists of distally a

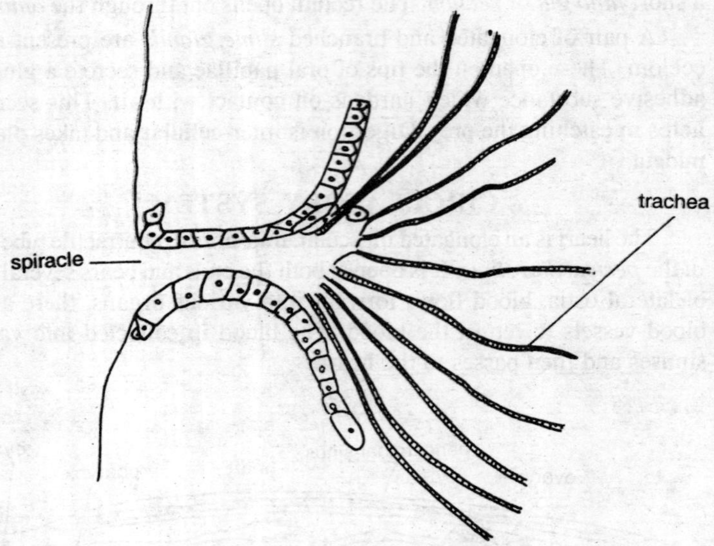

Fig. 20.6. *Peripatus*. Spiracle and trachea.

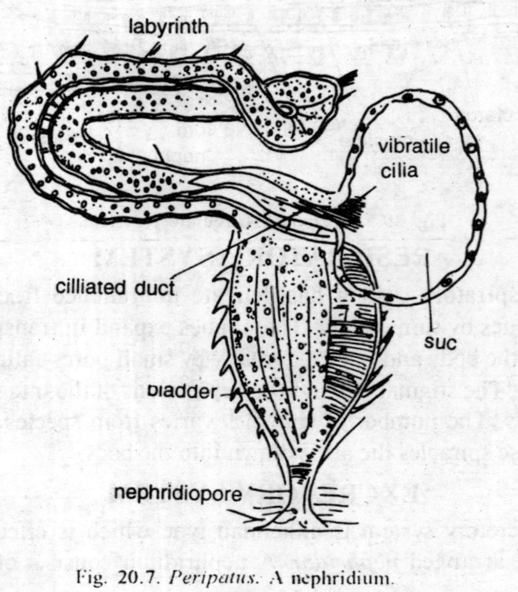

Fig. 20.7. *Peripatus*. A nephridium.

bladder extending into a *coiled tube*. It opens externally by *nephridiopore* situated at the base of leg. Internally it opens into coelomic sac by a *ciliated funnel*. The bladder is originated from the ectoderm and the rest of the part from mesoderm. Other than nephridia there are *coxal* or *crual glands* lying in the lateral component of the body cavity whose ducts open on the lower surface of the legs just outside the nephridiopore. These glands are absent in females. The *Peripatus* is *uricotelic*, excreting chiefly uric acid as nitrogenous waste product.

NERVOUS SYSTEM

The nervous system is primitive and is similar to that of annelids. It consists of two oval ganglia, the *brain* in the anterior part of the body lying over the pharynx. Two parallel longitudinal *ventral nerve cords* joined by many commissures assuming ladder-like in appearence.

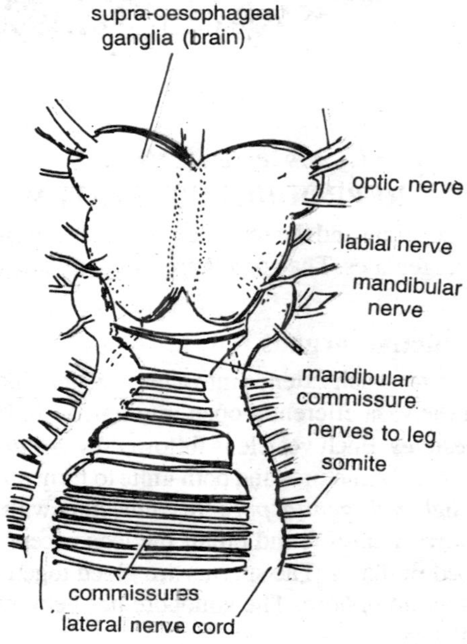

Fig. 20.8. *Peripatus*. Nervous system.

EYES

The eyes are typical, resembling those of annelids in all respects. It is obvious when the animal is irritated their secretion is discharged in the form of fine viscid thread which might be the measure of defence.

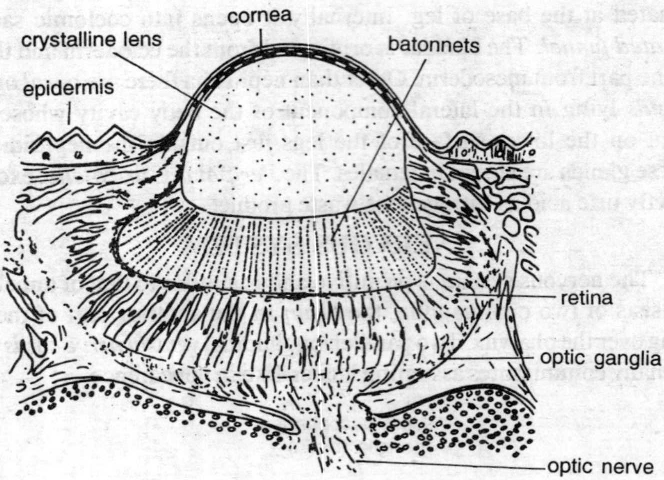

Fig. 20.9. *Peripatus.* V.S. of eye.

REPRODUCTIVE SYSTEM

Sexes are separate and the reproductive ducts are ciliated. The males are smaller than females. The gonads are paired and lie posteriorly above the intestine.

Male reproductive organs

A pair of *testes,* each extends into a narrow tube-like *vas eferens,* are present. Both the vasa efferentia open into narrow bulb-like structures, the *seminal vesicles*. Each vesicle is followed by *vas deferens*. The vasa deferentia are coiled structures and both unite to form an *ejaculatory duct* opens out through *male genital pore*. In connection with the reproductive system are *crural glands* found on all the legs except the first and are better developed in males. The sperms are glued together in ejaculatory duct to form *spermatophore*. The gonopore lies between the last pair of legs on ventral side.

Female reproductive organs

It comprises a pair of tubular *ovaries,* a pair of *oviducts* and two long *uteri*. Uteri unite behind into a median *vagina* that opens out through genital pore. Some species may have a *seminal receptacle* connected with the uterus and communicating with a funnel shaped *recepticulum ovorum* embracing the ovaries. In oviparous forms a long ovipositor is formed from the vagina.

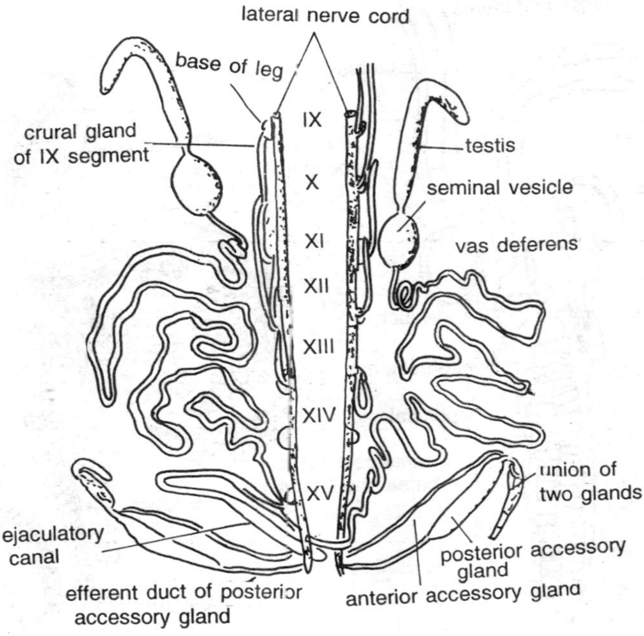

Fig. 20.10. *Peripatus*. Male reproductive organs.

Copulation

The real process of copulation is not yet observed and nothing is known about sperm enterance so far. The fertlization is internal sperms are not found in the uterus and oviducts so it appears probably that they reach directly to the ovary by boring through the skin or traversing the body cavity. Mostly they are *viviparous* but a few are *oviparous*. In viviparous forms embryos receive nourishment from the uterine wall through a placenta or they develop at the expanse of yolk. Oviparous forms lay eggs with advanced embryos. The young ones are alike to their parents except in size and colour. So that the mother does not appear to pay any special attention to her young ones, which wander away and get their own food.

AFFINITIES

The *Peripatus* is very important from the zoological point of view

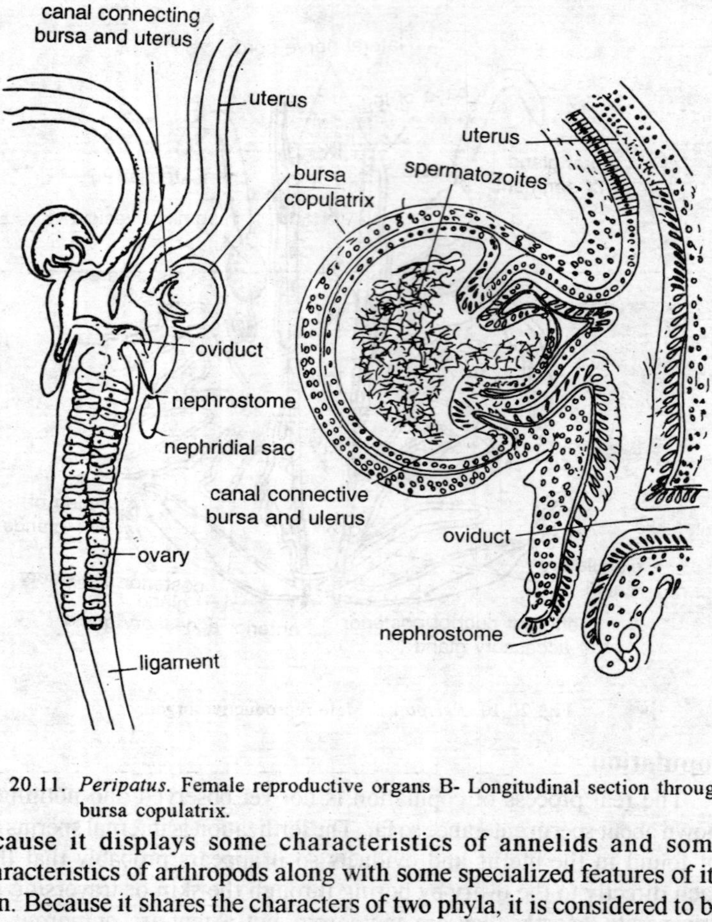

Fig. 20.11. *Peripatus*. Female reproductive organs B- Longitudinal section through bursa copulatrix.

because it displays some characteristics of annelids and some characteristics of arthropods along with some specialized features of its own. Because it shares the characters of two phyla, it is considered to be a *connecting link* between the two phyla showing that arthropods have arisen from annelids.

I. Resemblances with Annelids

1. Body is vemiform, with truncated extremities.
2. True head is absent.
3. Body wall consists of a thin cuticle beneath which presence of circular and longitudinal muscles.
4. Eyes alike and simple.

5. Like the parapodia of annelids, the hollow, stumpy legs which are the extensions of the body wall.
6. Straight and simple alimentary canal.
7. Nervous system is primitive, originated from ectoderm.
8. Paired nephridia which are segmentally arranged.
9. Cilia are present in excretory and reproduction ducts.

II. Resemblance with Arthropods

1. Antennae are present.
2. Jaws are nothing but modified appendages.
3. Locomotion by definit legs, having definit musculature and provided with claws.
4. Haemocoelomic condition is present similar to arthropods.
5. True coelom is reduced to the cavities of excretory and reproductive organs.
6. Respiration by tracheae.
7. Presence of peculiar salivary glands which are modified nephridia.
8. Cuticle has thin chitin deposition.
9. Structure of reproductive organs and development mainly arthropodan type.
10. The dorsal tubular heart communicating with pericardium by lateral ostia.
11. Absence of perivisceral section of coelom.

III. Its own characters

Peripatus exhibits certain characteristics of its own to recognize it as neither an annelid nor an arthropod.

1. Absence of external segmentation of the body.
2. Velvety and soft nature of skin.
3. To some extent the antennae of *Peripatus* can not be homologous to that of other arthropods.
4. Nature of 3-segmented head seems to be mid-way of arthropods (head in made up of 6 segments).
5. Restriction of jaws to a single pair.
6. Quite unevenly distributed spiracles and tracheae.
7. Absence of true ganglia, but separate two ventral nerve cords.
8. Difference in the distribution of reproductive organs.
9. Structure of eyes is less complicated.

Taxonomic Position

With the peculiar nature and some annelidan characters it appears that probably annelids is one of the source and that it is originated from the marine Polychaeta, which given up the marine habit and has become modified for locomotion on the land without jointed appendages. It seems to be entirely against to the rest of the arthropods, yet it has been kept under phylum *Arthropoda*, giving it an separate class rank.

According to *Lankester* (1904), of aquatic region, the Hyparthropoda (a hypothetical undefined group presumably without preoral segments) were the first to appear and from these arose the protarthropoda with one pre-oral segment, a terrestrial discendent having survived as *Peripatus*.

According to *Menton* (1950-52) linkage of this group Onychophora with two phyla will not be probably evidence of simplification and moreover, the habitats of Onychophora provides ample reasons for the persistance of flexible body wall, leg like jaws together with the elaboration of a subcutaneous connective tissue instead of cuticular scutes.

According to *Snodgrass* (1938) the Onychophora and marine *Aysheaia* are possible discendantly from an extinct group of marine lopopod annelids and they may have no intimate relationship with trilobite chelicerate of precambrain arthropods. Onychophore bears chitinous mandibles which only might be preserved as fossil nothing of other parts known so far due to its soft nature.

No doubt the *Periptus* is a living example and strictly terrestrial, and the only fossil referred to the class have been questioned. Of these *Aysheaia* from middle Cambrain is the best known fossil. It was a marine form, probably devoid of tracheae. Moreover, 10 pairs appendages but these bear with 6 claws, while the recent onychophores bear only two claws on their legs. It seems obvious that this *Aysheaia* may represent a very primitive onychophore which is more closer to the annelids than the arthropods.

Apart from that it must have been derived from archiannelidian stock because of the ventral nerve cords were primitive, widely separated conditions or it may more be correct to say that the present day annelids and *Peripatus* have got their origin from a common ancestor from which they independently evolved.

It is due to that common features appeared between the onychophores and present day annelids are similar, being evolved from common ancestral stock in which such characters were probably present.

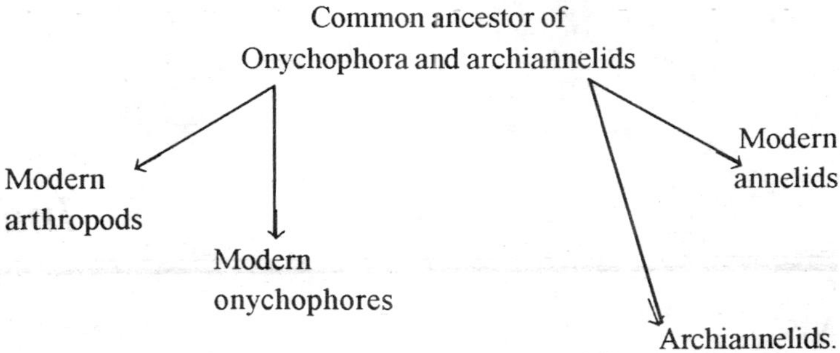

However, according to modern concept (based on *Barnes, Marshall* and *William*) this group was more wide spread and diversified. The view is supported by its *discontinuous distribution* and discovery of several fossils from middle Cambrian. It is, therefore, believed that Onychophora is an ancient independent group of segmented animals cantemporary to arthropods and as such has been placed under separate phylum *Onychophora*.

Revision Questions

1. Describe the habits, habitat and external morphology of *Peripatus*.
2. Discuss the affinities of onychophora suggest a suitable place for this group in the Animal Kingdom.
3. Give an account of geographical distribution, general organisation and affinities of Onychophora.
4. Describe the important features of *Peripatus* and mention its taxonomic importance.

Index

A

Abdominal muscles, 91
Absconding, 250
Absorptive cells, 327
Accord barnacle, 36
Alary muscles, 166, 198, 217
Alcepas, 6
Alima, 41
Alveolar glands, 328
Amphipneustic, 223
Anal circus, 192
Anteenal spine, 77
Anteenary gland, 109
Apical lobe, 184
Appendicular muscles, 91
Apodemes, 91
Apophysis, 81
Appendix interna, 88
Appendix masculina, 88, 120
Apposition vision, 119
Apus, 3
Argulus, 30, 147, 148
Arista, 212
Artemia, 3
Arthrobranch, 101
Arthrodial membranes, 77, 156
Articular membrane, 153
Asperate lobe, 181
Atrium, 199
Autotomy, 56, 125
Auxillary hearts, 217

B

Balanus, 35
Bee dance, 252
Blastopore, 124
Biramus appendage, 79
Blastopore, 124
Blatta, 152
Book gills, 309
Book lungs, 331
Branchial basket, 291
Branchiostegite, 77
Branchipus, 3, 24
Brine shrimp, 24
Bursa copulatrix, 248
Buthus, 317, 318

C

Camerostome, 307
Camouflage, 54
Cannibal, 152
Cannibalism, 260
Carapace, 76
Cardiopyloric strand, 104, 121, 123
Carp louse, 30
Carpora adiposa, 161
Caterpillars, 20
Centipede, 47
Cephalic apodeme, 92
Cephalic gland, 255
Chela, 84
Chironomus, 287

Chrysalis, 275
Coarctate, 223
Coccon, 275
Collateral glands, 182, 206, 148
Collophore, 49
Commensalism, 43
Conchoecia, 4
Corbicula, 245
Corpora cardiaca, 176, 300
Coxal gland, 349
Cricket, 52
Crystalline cone, 117
Cursorial, 152
Cuvier, 1
Cyclops, 28
Cypridina, 4
Cypris, 35, 36, 37, 149
Cythereis, 4
Cytherella, 4

D

Danaus, 65
Daphnia, 3, 26
Derocheilocarus, 4
Deuterocerebrum, 172
Diapause, 207
Dioptrical region, 117
Diphyllobothrium, 29
Discontinuous distribution, 353
Dorsal shield, 76
Dracunculus, 29

E

Ecdysis, 172, 186
Ecdysone, 178, 300
Elytra, 18, 157
Endophragma, 91
Endopodite, 84, 155
Endosternite anterior, 92
Endosternite posterior, 92
Entocythera, 4
Epimeron, 77
Epipodite, 82, 84
Epiproct, 192
Estheria, 3
Exopodite, 84, 155
Expiratory spiracles, 199

F

Fenastrae, 154
Feelers, 79
Filter plate, 96
Fish lice, 145
Flagellum, 212, 228, 338
Flexor muscles, 91
Foragers bees, 253
Frons, 154
Furcula, 49

G

Gambusia, 239
Germarium, 182
Giant water bug, 62
Gnathobases, 81, 82
Gnathochilarium, 46
Gonapophyses, 181, 184
Goose barnacle, 34
Green glands, 1, 109
Gregareous, 51
Guppy, 239
Gynatrium, 182
Gynovalvular chamber, 158

H

Habronema, 225
Haemocoel, 1, 90, 193, 214
Haemocyanin, 104, 311
Haemoglobin, 287
Haemolymph, 193
Halters, 19, 214, 231
Haustellum, 213
Hemelytra, 62
Hepatic cells, 98
Hepatic spine, 77
Hepatopancreas, 94
Hermit crab, 42
Heteromorphosis, 125
Hippa, 43

Histoblasts, 234, 299
Histogenesis, 224, 236, 299
Histolysis, 224, 234
Holaspis larva, 306
Holoptic eyes, 246
Human louse, 57
Hydrofuge hairs, 292
Hypermetamorphosis, 298

I

Imaginal buds, 224
Imago, 236
Incisor process, 81, 93
Instar, 222, 294
Inspiratory spiracles, 199
Intestinal bulb, 97
Intima, 169
Iris pigmented cells, 118

J

Johnson's organ, 228
Juvenile hormone, 177

K

Kentrogen, 37

L

Labella, 213
Labyrinth, 333
Labyrinth plexus, 110
Lacinia, 153
Lacunar type, 197
Lady bird beetles, 67
Leaf insect, 54
Lepidurus, 3
Leptodora, 3
Lepisma, 84, 294
Lingua, 156
Living fossil, 316
Lucifer, 138
Lynceus, 3

M

Maggot, 222
Malpighian tubules, 1, 163, 171, 196, 200, 217, 327
Mandibular palp, 81
Mayfly, 58
Megalopa, 45
Meraspis larva, 306
Mesophragma, 92
Mesopleura, 191
Mesosternum, 191
Metanauplius, 4
Metapneustic, 223, 232
Micropyle, 206
Millipede, 46
Mimicry, 54
Mosaic vision, 118
Mole crab, 43
Mole cricket, 53
Mycetocytes, 172
Mysis, 31

N

Naiads, 295
Nauplius, 3, 5, 31, 35, 36, 37
Nephridia, 347
Nephridiopore, 347
Nephroperitoneal sac, 111
Nuchal sense organ, 24
Nupital chamber, 269
Nymph, 13, 186, 207, 297

O

Oenocytes, 172
Olfactory setae, 79
Ommatidia, 117, 312
Omnivorous, 152
Ootheca, 184
Oothecal chamber, 158
Opercular valve, 36
Opisthosoma, 21, 73, 301, 319
Orbital notch, 77, 116
Organogenesis, 299
Organ of Berlese, 262
Osmoregulation, 111
Oviparous, 351
Ovipositor, 159, 184

P

Palmate bristles, 233
Palpifer, 155
Paraeopods, 82
Parasitic castration, 38
Parthenogenesis, 26, 30
Patella, 322
Paurometabolic, 186
Paurometabolous, 295
Pecten, 233
Pectines, 320
Pericardium, 214, 218, 329
Peritreme, 198, 288
Peritrophic membrane, 165
Petal spur, 322
Phallic organ, 181
Phyllobranch, 102
Plant lice, 64
Pleurobranch, 101
Pleuron, 77
Pollen basket, 245
Pollen packer, 245
Polycopa, 3
Polyvoltine, 273
Prementum, 156
Preying mentis, 53
Proboscis, 31
Propolis, 255
Prosoma, 21, 319
Protaspis larva, 305
Prothoracic gland, 300
Protopodite, 79, 155
Pseudocephalon, 223
Pseudopenis, 181
Pseudoscorpion, 73
Pseudotracheae, 38, 213
Pulvillus, 162, 191, 195
Puparium, 223
Pyloric plates, 96

Q

Queen substance, 252

R

Ram's horns, 73
Raptorial, 62
Receptor region, 117
Rectal gills, 291
Red cotton bug, 62
Relapsing fever, 58
Renal aperture, 77
Renal gland, 79
Renal sac, 169
Reservoir, 164, 333
Retinal cells, 117
Rhabdome, 117
Rock barnacle, 36
Root headed barnacle, 37
Rostrum, 77, 326

S

Sacculina, 137, 149
Sand flea, 32
Scaphognathite, 81
Scout bees, 253
Scutellum, 231
Scutum, 231
Sensillae, 175
Sericin, 275
Silver fish, 48
Slime gland, 345
Solitary, 51
Spatulate process, 308
Spermatophore, 182, 341, 350
Spinneret, 275
Spiracles, 159, 214, 347
Spring tail, 49
Stadium, 186, 222, 294
Statocyst, 79
Stenopodium, 79
Sternal plate, 76
Sternum, 77
Stick insect, 55
Stink gland, 159
Stridulating organ, 68
Stylet, 31

Stylocerite, 79
Subgenital plate, 192
Swimmerets, 88
Sympod, 89
Squama, 80

T

Tadpole shrimp, 26
Taenidia, 169
Tegmina, 157
Telopodite, 304
Telson, 98, 305
Tenant hairs, 214
Tentaculum, 49
Tentorium, 160
Tergosternal muscles, 169
Tergum, 77
Termitarium, 265
Titillator, 181
Tormogen cells, 176
Trachein, 170
Trench fever, 58
Trichogen cells, 176
Trilobite, 21
Trilobite larva, 313
Tunica properia, 98, 327
Typhus, 58

U

Univoltine, 273
Urate cells, 172
Uricose glands, 172
Uropods, 89
Utricularia, 239
Utriculi breviores, 180
Utriculi majores, 180

V

Van Siebold, 1
Ventriculus stomach, 215
Vertex, 153, 188
Vitelline membrane, 206
Viviparous, 317, 342, 351

W

Water flea, 27
Water scorpion, 61
Wiggles worth, 167, 291
Wire worm, 46
Wood lice, 38

Z

Zoea, 31, 45
Zymase, 165
Zymogenic cells, 165